Infrastructural
Attachments

Infrastructural Attachments

Austerity, Sovereignty, and Expertise in Kenya

EMMA PARK

Duke University Press   *Durham and London*   2024

Designed by Courtney Leigh Richardson
Typeset in Portrait by Westchester Publishing Services

Library of Congress Cataloging-in-Publication Data
Names: Park, Emma, [date] author.
Title: Infrastructural attachments : austerity, sovereignty, and expertise in
Kenya / Emma Park.
Description: Durham : Duke University Press, 2024. | Includes bibliographical
references and index.
Identifiers: LCCN 2024010849 (print)
LCCN 2024010850 (ebook)
ISBN 9781478031109 (paperback)
ISBN 9781478026846 (hardcover)
ISBN 9781478060093 (ebook)
Subjects: LCSH: Infrastructure (Economics)—Political aspects. | Infrastructure
(Economics)—Social aspects. | Economic development—Kenya. | Technology and
state—Kenya. | Telecommunication—Kenya. | Roads—Kenya—Design and construction. |
Kenya—Politics and government. | BISAC: SOCIAL SCIENCE / Ethnic Studies /
African Studies | TECHNOLOGY & ENGINEERING / Telecommunications
Classification: LCC HC865.Z9 C3 2024 (print) | LCC HC865.Z9 (ebook) |
DDC 338.96762—dc23/eng/20240716
LC record available at https://lccn.loc.gov/2024010849
LC ebook record available at https://lccn.loc.gov/2024010850

Cover art: Jackie Karuti, *Etching XXII, Variation 2*, 2018. From a set of
copperplate etchings produced while thinking about fossils. Aquatint print
on watercolor paper, 10 × 10 cm. © Jackie Karuti. Courtesy of the artist and
Circle Art Gallery, Nairobi.

Contents

Preface

The sources for this project, and the modes of reading them, were heterodox. In reading between and across these sources, I have worked to generate a thick description of the politics of infrastructures in colonial and postcolonial Kenya. National archives were consulted in the United Kingdom, Kenya, and Uganda. Corporate archives—namely those of the Imperial British East Africa Company, the British Broadcasting Corporation, the Marconi Archives, and the Cable and Wireless archives—were consulted in the United Kingdom. I also had the opportunity to work with the small remaining collection of audio recordings of early radio broadcasts housed at the Kenya Broadcasting Corporation. Mission archives also offered invaluable insights. In Kenya, I worked with the archives of the Anglican Church, as well as the papers of Leonard Beecher and Louis Leakey, which are in Nairobi.

Archival materials were read alongside oral historical research undertaken in Kenya. For me, conducting oral histories with radio men was critical to understanding both the cultural politics of infrastructures and their complex poetics. This is important in its own right, but more specifically because it enabled an understanding of the consistency with which African expertise had been sidelined in the archives from which we construct our histories. The importance of telling these stories of sidelined expertise, insofar as I was able, was heightened by the ethnographic component of my research, which mainly involved working and chatting with a friend, Peninah, at her small M-Pesa kiosk, and working with others in the so-called "informal" sector (M-Pesa agents, "hawkers," and *matatu*-industry workers) some of whose stories are surfaced here, others I hope to tell elsewhere. For the final portion of the book, I also conducted interviews with digital technologists and industry experts. Their perspectives and insights were coupled with the anonymous voices that make themselves heard on Kenya's lively blogosphere, as well as communications captured and preserved by WikiLeaks. These insights, in tandem with everyday talk, formed the foundations of much of the third section of the book.

This method has, I think, been a boon rather than a liability. Moving between these source bases alerted me to the importance of being as concerned with locating continuities and repetitions as I was with understanding ruptures and change over time. While this is certainly a story about transformation, claims to change do a good deal of work in the world, especially as regards infrastructures and technologies.[1] Keeping an eye on these claims to rupture guided my reading of the archive, forcing me to ask whether and when narratives of rupture obscure more than they reveal. In this instance, the ways in which infrastructures' long history as "public goods" in the region has always been crosscut by the aspirations of capital firms; or the ways that infrastructural work has repeatedly been rendered merely banal and unremarkable labor; or the fought-over processes by which infrastructures have been enrolled to enact competing social and political goals.

While I work across sites in the pages that follow, readers will note that Central Kenya and the Rift Valley, as well as the cultural and technical work of Kikuyu-speakers feature heavily. The reasons for this are historical. Those from the center of the country often accessed infrastructures before, and sometimes to the exclusion of, other regions and communities. In the present, this is popularly associated with the dispensations that are held to be the result of historical connections of Kikuyu-speakers to the center of administrative power. In the colonial period, by contrast, infrastructural densities in these regions reflected not only the presence of white settlers in the area but also the intensity of commodity production and these areas' proximity to Nairobi. What is today framed as an unfair advantage, in other words, was in the colonial period experienced by many as the state's overzealous incursions into the lifeworld of the region's African communities. By bringing these stories and all their attendant unevenness together, I hope to have done justice to the characters and processes that animate these pages, generating a series of tales of infrastructures, and their attachments, that are three-dimensional in scope.

Map of Kenya and the East Africa region

Acknowledgments

It is as hard to mark beginnings as it is to mark ends. As a book, this manuscript came together following my move to NYC, when I started at The New School in 2017. As an idea, this project was born at the University of Michigan (UM). As a cluster of ideas and nebulous political commitments this project germinated following my first trip to Kenya and my disillusion with the limitations in what was considered the domain of the political. I began working through this problematic at Concordia University where I completed an MA. As a maximalist who tends to think in terms of continuities, and who considers knowledge work to be a fundamentally collaborative and collectivist endeavor, I choose the longest timeline. Life is but a series of (ideally) socially reproductive debts, and mine are many.

I am deeply indebted to the many friends and interlocutors in Kenya who made this work possible, particularly the family of Judith Maina. Judy was my first friend in Kenya and on our second meeting was working with an NGO collecting the life histories of the oldest living residents of Mathare. While initially skeptical of me, something she expressed with a keen wit, she became a fast friend, soon welcoming me into her family and teaching me about Kenya along the way. These dynamics became more intimate and complicated as Judy suffered a series of health crises. By the time I arrived to do the longest stint of fieldwork in 2014, she had lost her vision. I then lived with Judy's sister-in-law, Peninah Mugure Njenga, her brother, Vincent Maina Kariuki, and their son, Kamaish. We stayed on the same compound where Judy and her mother lived, with Judy's father, Humphry Maina Githiro, joining some weekends. Judy's partner, her champion and stalwart caregiver, Claudio Torres, joined her family in trying to make her life normal. In spite of the unfolding tragedy through which they were all living, they welcomed me as family. I traveled with them to their shamba up-country, and spent many evenings chatting with Judy and her mother. Other afternoons, I worked with Peninah at her M-Pesa kiosk. Judy died in 2020. This book is, in large part, for her and the network that sustained her.

Outside of the succor and warmth provided by this family, I extend my gratitude to Gitau Kariuki who chattered with me as my Kiswahili improved, patiently answered my myriad questions about the most minute of dynamics, and worked with me as a research assistant. I am also very grateful to the tireless work of the late Richard Ambani, who made all sorts of knowledge production possible for all sorts of people in his capacity as a freelance archivist at the Kenya National Archives (KNA). There was not a person with deeper institutional knowledge of the KNA and I grieve his passing. Huge thanks are also owed to Vivian Wanjiku who, aside from being a steadfast friend, also acted as a sometimes research assistant, and to Peninah for explaining to me the intricacies of her work. Thanks to Stanley Macharia, Fuadi Mbigi, and their comrades who taught me about the politics of life, space, and economy in Nairobi from their vantage as "hawkers" at the Figtree Market and workers in the *matatu* industry, respectively.

The earliest seeds of this project grew under the generous and generative advising of Andrew Ivaska, who I now happily count as a friend. But this project really grew and took shape at the University of Michigan. I cannot imagine having more incisive, kind, and supportive advisors than Gabrielle Hecht and Derek Peterson. The creativity and political clarity with which Gabrielle approaches knowledge work is unparalleled—Gabrielle impressed on me that anger in the face of durable historical inequalities can be an engine of creative output. My thinking has been indelibly shaped by her influence. She is at once an intellectual and professional role model to whom I am forever grateful. Derek's deep commitment to eastern Africa, his ethics as a researcher and teacher, and the rigor and focus of his thinking and research remain a well of inspiration to me. I could not have asked for more generous intellectual parents, who variously nurtured my person and pushed me as a thinker. Nancy Rose Hunt is among the most capacious historians I know, and she impressed on me the need to be attentive to the peculiar, the out of the ordinary, and the idiosyncratic. In the spaces between, there are connections to be drawn and meanings to be surfaced. This project also benefited from the astute insights of Jatin Dua, whose work and thinking exemplifies historical anthropology at its very best. Big thanks are also owed to Will Glover, whose insights regarding power and the built environment shaped this project.

My time at the University of Michigan was contoured by the insights of its faculty, including Omolade Adunbi, Adam Ashforth, Kelly Askew, Howard

Brick, John Carson, Paul Edwards, Geoff Eley, Webb Keane, Mike McGovern, Martin Murray, Anne Pitcher, Howard Stein, and Rudolph Ware III. I was surrounded by a generous and kind group of colleagues throughout my time at UM, including Adam Fulton Johnson, Brady G'Sell, Dan Hirschman, Sara Katz, Doreen Kembabazi, Benedito Machava, George Njung, Davide Orsini, Nana Quarshie, Tasha Rijke-Epstein, Ashley Rockenbach, Jonathan Shaw, Edgar Taylor, Maria Taylor, Christopher Tounsel, and Daniel Williford. Paul Love fed me more times than I can count. Andres Pletch became one of the best friends I've had in this life. Though our paths diverged following graduate school, his care, laughter, and brilliance sheltered, nourished, and challenged me throughout.

The New School was precisely the place I was supposed to land. Aaron Jakes made clear from the jump that he was my new family. He brought me into home and hearth, and there was always a spot for me at the table he shared with Tania Abbas. This table has grown with the addition of Isaac and Gabriel. I couldn't be more grateful for their enduring friendship, warmth, and intellectual stimulation. Within the Department of Historical Studies, I have wonderful colleagues who have supported me: Elaine Abelson, Federico Finchelstein, Natalia Mehlman Petrzela, Claire Potter, and Eli Zaretsky. Special thanks are owed to Oz Frankel and Jeremy Varon, each of whom helped me navigate through some tricky periods, and Julia Ott who was Chair of the Department when I was hired and who shepherded me through both the third year review and tenure process. I have grown tremendously since arriving at The New School for Social Research owing in large part to the learning I've done with an incredible group of graduate students, including Adhip Amin, Ella Coon, Julian Gomez Delgado, Aditi Dey, Baiyi Du, Ethan Dunn, Santiago Mandirola, and Miri Powell. None of this would have been possible without the tireless work of Salima Koshy who was instrumental in getting the first draft of this manuscript in order, offering incisive feedback in the process.

I am also tremendously grateful for my friends and colleagues outside of the Department of Historical Studies. Aside from being a dear friend whose frankness has saved me on more than one occasion, Cinzia Arruzza has been an ethical and political center to me. Benoit Challand's humor and kindness have sustained me. Frequent beers with Quentin Bruneau have brought much laughter. Far too infrequent hangs with David Bering Porter and Clara Mattei have enlivened my thinking. Roller-skating and laughter with Daniel Rodriguez-Navas have brought joy. The generosity of Caroline Dionne has nourished me, particularly during the wrenching days of the pandemic.

I am lucky to have long been surrounded by friends within or adjacent to the academy who are as brilliant as they are good. My deepest gratitude to Robyn d'Avignon, Amiel Bize, Meghna Chaudhuri, Kevin Donovan, Basil Ibrahim, and Matt Shutzer, who joined Aaron Jakes in reading the manuscript in full. It goes without saying that my thinking has been sharpened by each of these humans, all of whom I am lucky to count as dear friends. Many thanks also to Abi Celis, Zebulon Dingley, Jen Johnson, Sara Katz, Marissa Mika, and Edgar Taylor, with whom I have been in ongoing conversations for many years. My deepest debt, though, is to my best friend, most stalwart cheerleader, research buddy, and partner in crime, Kevin Donovan. He has countless times swooped with good humor, incisive thoughts, and care. I couldn't ask for a better brother and thinking partner. To many, many more conversations, dear friend. Thanks are also owed to Emily Brownell for always welcoming me to the table she shares with Kevin and Jo with such generosity, kindness, and brilliance.

This research was made possible by the Social Sciences and Humanities Research Council, the Eisenberg Institute for Historical Studies (UM), the African Studies Center (UM), the Department of Afroamerican and African Studies (UM), the Rackham Graduate School (UM), and the Sweetland Writing Center (UM). At The New School, this research was supported by GIDEST and the Heilbroner Center for Capitalism Studies.

Portions of this book were presented at the Heilbroner Center for Capitalism Studies, the International History Workshop at Columbia, the Global Histories of Capital workshop, the American Anthropological Association Annual Meeting, the African Studies Association Annual Meeting, Johns Hopkins, the University of Sussex, the Institut des mondes africains, Concordia University, the University of Chicago, Princeton University, the University of Wisconsin, the Platform Economies Research Network, GIDEST, and the Mellon-funded Michigan-Wits collaborative workshops. There are too many to name them all, but for those conversations, I would like to thank all those who participated. Big thanks are also owed to the team of the Mellon-funded Sawyer Seminar "Currency and Empire: Monetary Policy, Race, and Power" Aaron Jakes, Gustav Peebles, Sanjay Reddy, and Paulo dos Santos, as well as our brilliant research assistants, Josephine Baker and Erin Simmons, and all those who participated, from whom I learned a great deal.

Many dear friends have made life possible over these past number of years. To all of those who offered me intimacies and reminded me that laughter and joy were still possible in the midst of a detachment that was as painful as it was necessary, thank you. Stephanie Beamer, Claire Bennett, Maria Cutrona, Larissa Diakiw, Natasha Grace, Lyndie Greenwood, Rachel James, Catie Lamer, D'Arcy Leitch, Andrea Lipsky-Karasz, Alan Lucy, Tess Owen, Juliana Rubbins-Breen, Amy Seigel, Kelsey Tyler, and Alice Waese. Particularly big thanks are owed to another brother, Aidan Cowling, whose care, brilliance, wit, and play-lists have sustained me.

I was lucky to work with the dedicated team at Duke University Press to bring this book to print. Elizabeth Ault expressed enthusiasm regarding this project from the outset, and patiently guided me through the process. Thank you also to Benjamin Kossak who was immensely helpful in walking me through the logistics of getting the book out. Big thanks to Lisa Lawley, Erin Davis, and Jamie Thompson for their considered help (and patience!) through the editorial process. Tim Stallmann worked with me tirelessly as he made maps for this book. He taught me the joy of translating argument into visualization. I could not have asked for more generative and thorough feedback than that provided by the two anonymous reviewers who helped me make this a sharper and more focused book.

Aside from my chosen family, my family by birth has been an unwavering source of support to me over the course of this journey. My sisters, Jessica Hay and Sari Park, are fierce in their loyalty and acceptance. Jess, your generosity and perseverance inspire me. Sari, your dogged commitment to justice is exceptional. I am grateful for the admittedly too infrequent trips to northern Ontario and time spent in Toronto and Mexico. Kris Bronstad began as a sister, and I now count her as a dear friend. To Millie and Birdie, thank you for bringing the joy of possibility. My stepfather, Gary Kraemer, has been a consistent intellectual interlocutor and friend. But this book owes its greatest debts to my parents, Norman Park and Alison Fleming. Dad, your radical acceptance of me at every turn and your intellectual curiosity have at once stabilized and inspired me. Mum, I have no words. The many, many late-night conversations made life possible both in joy and in grief. My love for mentoring was no doubt shaped by many years of watching you grow alongside your own students. Your commitment to community, generosity of spirit, brilliance, and evenness are qualities to which I aspire. This book is for you.

Introduction

One afternoon in August of 2014, I traveled from Pangani to Zimmerman, a lower-middle-class neighborhood some thirty minutes from Nairobi's Central Business District, to visit Peninah. As on most days, the neighborhood was alive with activity—"hawkers" were selling clothing from makeshift structures, *mama mboga* (mama vegetable, the Kenyan shorthand for the largely female workforce of vegetable sellers) were serving customers, as *matatu* (minibus) conductors were soliciting waiting commuters to board their brightly painted vehicles, blasting music from the radio as they stopped along the one paved road that cuts through the neighborhood. When I arrived, Peninah was standing outside of her M-Pesa kiosk chatting with neighbors as her son, Kamaish, played at her feet. At the time, Peninah worked as an M-Pesa agent for Safaricom.

Once an unremarkable mobile phone company, Safaricom is now the largest corporation in East Africa.[1] M-Pesa, a mobile-to-mobile service that allows users to store, transfer, and withdraw money, was a harbinger for this growth and enabled Safaricom's rapid rise to dominance. As an M-Pesa agent, Peninah's work officially involves signing up new users, collecting their ID and mobile numbers, and recording user transactions as she moves money in and out of the system. For this work, Peninah receives a small commission based on the total number of transactions conducted over the course of a month. But Peninah's work, as I came to learn while spending many afternoons with her in her shop over the course of 2014 and 2015, entails much more than this: It requires building and maintaining social relationships that comprise an unacknowledged but crucial component of the infrastructure upon which Safaricom's profitability depends.

Greeting Peninah and Kamaish, the three of us entered her small "Safaricom green" kiosk. Sitting next to Peninah behind the small wooden counter and iron grating that separates her from Safaricom's users that sunny afternoon, an elderly man walked into the shop. Peninah greeted him warmly. He handed Peninah KSH 1,000, asking to deposit KSH 800. She completed the transaction as they chatted, handing him the change. "He's my customer," she told me as he left.[2] "He comes here every evening and makes a deposit." His status as "her customer" was not simply a matter of routine, she explained. "He can't see very well, so he asks me to read transaction texts to him." For Peninah, these acts of care are crucial to retaining "her customers." But according to Safaricom, M-Pesa agents simply act as undifferentiated interfaces between the corporation, its customers, and their shillings. Indeed, the importance of these interpersonal attachments notwithstanding, developmentalist market makers are explicit in devaluing this work, referring to the over one hundred thousand people who take charge of Safaricom's M-Pesa kiosks, using a peculiar but telling nomenclature: "Human ATMs." On this reading, M-Pesa agents merely act infrastructurally. Half-human, half-machine, it is the job of these people to take the place of a largely absent banking infrastructure, thereby enabling the digital money-transfer system to work. This framing of human action, while troubling, is revealing of a more fundamental reality that Safaricom's labor regime works to conceal.

Safaricom's infrastructure, as Peninah well knows, is by no means a purely technological network. It relies on people like Peninah to act as what I refer to as "infrastructural prosthetics," a concept that highlights the forms of largely unremunerated but transformative work required to make Kenya's infrastructural landscape hang together, and in many instances to expand. Peninah acts as an infrastructural prosthetic in two interrelated ways. First, she gathers data and generates knowledge about people who visit her shop, data which Safaricom mobilizes as it generates new services that enable it to seize value from those at the "bottom of the pyramid."[3] Second, it is Peninah's work that enables M-Pesa and Safaricom to themselves function as infrastructural prosthetics, filling the gap in a context where brick-and-mortar banks are the exception rather than the rule. She is both a systems builder and maintainer. And the work of Peninah, and the over 160,000 women and men who work as M-Pesa agents, has been essential to Safaricom's profound financial and technological success. Over the past fifteen years, Safaricom has emerged as the largest corporation in the region, having established a near monopoly over both mobile telephony and digital financial services. In 2022, Safaricom's annual profits totaled more than $417 million.[4]

Not only is Safaricom not merely a technological network, but it is also not merely a market actor. The firm began as a state-held corporation, incrementally privatizing under pressure from the International Monetary Fund (IMF) and the World Bank. Today, the corporation is jointly held by the United Kingdom based firm Vodafone (through its South African subsidiary, Vodacom), the government of Kenya, and largely Kenyan shareholders. This distributed ownership, notes Keith Breckenridge, gives the "state a double-dipping interest in the company's enormous profits: first as shareholder and second as tax collector."[5] Safaricom itself then operates as an infrastructural prosthetic that shores up the reproduction of the Kenyan state. Neither wholly public, nor wholly private, Safaricom is emblematic of what I call the "corporate-state."

I begin with Peninah (the putative Human ATM) and Safaricom (the corporate-state), because together they offer a contemporary window into a set of historical dynamics that sit at the heart of this book. While the informalization of labor and the outsourcing of infrastructural provisioning to a corporation might seem unique to the "neoliberal" present, these are dynamics that have deep and specifically colonial roots that are disclosed by a study of Kenya's infrastructural past.

Infrastructural Attachments works to reconstruct a history that takes seriously a longer-term view of the Kenyan state as, from its founding inception, premised on austere forms of statecraft that have depended on the intimate interweaving of the political and the economic, the state and the market, the public and the private. This historical account of the region's infrastructures—roads in the late nineteenth and early twentieth centuries, radio broadcasting from the 1920s through the 1950s, and digital financial services in the present—demonstrates that an ambiguous blending of corporate and state power founded the institutions of governance in the region, and this has had implications for the status of infrastructural work. The story of Safaricom and its labor regime are *not*, then, aberrations unique to the neoliberal present. They are historically constituted forms of statecraft that are better understood as inheritances of colonial capitalism.

Scholars of the postcolonial state have long recognized the enduring role of colonial institutions in shaping the modalities of governance in contemporary Africa. The focus has largely been on decentralized governance (first undertaken in the colonial period under the mantle of "indirect rule"), the domination of cultural categories (namely "tribe") in structuring political affinity, the outsourcing of social provisioning to the family, and the primacy of "informal" labor.[6] These accounts of the African state concerned with the colonial origins of the postcolonial detected a set of logics that, today, are commonly associated with neoliberalism and austerity—devolution, speculative planning, identitarian politics, and the delegation of risk and responsibility from the state to the family and the worker.[7] Scholars concerned with neoliberalization in

Euro-America have argued that these processes required redrawing the lines between the market and the state, between the economic and the political.[8] However, as the literature on the colonial origins of the postcolonial state demonstrates, the boundaries between the private and the public, between the economic and the political, have never been neatly drawn in Africa. The problem with accounts of neoliberalism when read through this literature, then, is both the temporal claim ("neo"), and the political and economic claim ("liberalism").[9] Put differently, what we often think of as a story of neoliberalization has a different and longer genealogy in this part of the world.

One of the reasons that the genealogy laid out in *Infrastructural Attachments* has not always been visible is that studies of the state have focused on colonial modalities of governance and discursive formations, occluding the principle importance of infrastructures in mediating the relations between economy and polity.[10] Infrastructures bring these abstractions into being in concrete ways; they are the site where the demarcation between the state and the market have unfolded.[11]

As I show, since the earliest days of the British presence in the region, striking a balance among capital accumulation, state formation, and the "public good" has been difficult for a state whose origins were corporate in character.[12] In 1888, facing limitations in the British fiscus and public skepticism of overseas adventures, the Crown granted a Royal Charter to the Imperial British East Africa Company (IBEA) authorizing the firm's operations as the *de jure* sovereign in the region. The Crown's charter was matched by a concession from the territorial sovereign, the Sultan of Zanzibar, who was similarly facing fiscal constraints. In entering into these agreements with the IBEA, the twin sovereigns outsourced the cost and risk of administering the region, enrolling private capital to enact the dual goals of "commerce and civilisation."

There was nothing particularly unusual about this distribution of sovereign authority in the nineteenth century. As British jurist Henry Sumner Maine summarily wrote in 1892: "Sovereignty is a term which, in international law, indicates a well-ascertained assemblage of separate powers or privileges . . . there is not, nor has there ever been, anything in international law to prevent some of those rights being lodged with one possessor and some with another. Sovereignty has always been regarded as divisible."[13] In the case of the early colonial state, this divisible vision of sovereignty saw an assemblage of sovereign prerogatives, privileges, and rights transferred from the Crown and the Sultan of Zanzibar to the IBEA. The IBEA was, then, a complex entity; neither self-evidently "private" nor "public," it was a hybrid legal form.[14]

While corporate sovereignty may have formally passed with the collapse of the IBEA, the outsourcing of large-scale infrastructures to private firms did not. Indeed, though short-lived, the patterns of divisible sovereignty inaugurated with the IBEA were remarkably durable, subjecting Kenya and Kenyans to varying forms of company rule.[15] The expansion of radio broadcasting from the 1920s through the 1950s was shaped by early agreements that bound the colonial state to the networks of the British monopoly Cable and Wireless Ltd. (C&W), contracting the firm to provide services for remunerative white listeners. By the 1940s, it was clear that C&W's monopoly status was a liability, frustrating the colonial state's aspirations to expand broadcasting to the African majority. As for Safaricom in the present century, the firm's partial privatization has not entailed the retreat of the state, but its reconfiguration—today, Safaricom is jointly held by British multinational Vodafone, the Government of Kenya, and a shareholding "public." Across the long twentieth century, the government's ambitions to use infrastructures as a tactic of statecraft have been shaped by agreements that attached its fate to corporate firms whose interests were driven by profit motives. We err, in other words, in seeing state planning as opposed to market formation.[16]

If the separation of the "political" from the "economic" is one of capitalism's founding abstractions, in the colonies these boundaries were evidently never so neatly drawn. In Kenya, a former settler colony, the devolution of state sovereignty to corporate firms has been the norm. This book aims to put the world back together by disclosing how the infrastructural state in Kenya has consistently tied its fate to private firms. It is not simply that states produce markets, but that corporations shape the fate of states.[17]

If enacting the infrastructural state has relied on state-corporate collusions, it has also crucially depended on the exploitation and expropriation of Africans' infrastructural work. While this work has critically enabled infrastructural expansion, maintenance, and repair, the centrality of these contributions has been systematically effaced by metropolitan observers and recognized experts.[18] Building on a growing literature that explores Africans' technological action, I center the crucial and transformative work of African knowledge-workers and technological experts.[19] In foregrounding these workers and their work over the long twentieth century, I interrogate the various ways that the corporate-state has subsumed Kenyans' knowledge and expertise to underwrite the development of (in some instances global) infrastructural expertise, the circuits of finance on which (post)colonial infrastructural expansion has been premised, and the forms of profit-making it has enabled.

Over the long twentieth century, these infrastructural dynamics, I show, have been contoured by the highly stratified racial, ethnic, and economic orders that guided (post)colonial administration and shaped everyday life in Kenya. In the colonial period, both access to infrastructures and the conditions of infrastructural work were critically shaped by the racial hierarchies and ethnic topographies that structured the settler colony. Creating lines of connection under these conditions produced a landscape characterized by what Manu Goswami has described as "internal differentiation and fragmentation," leading to the development of Kenya's uneven infrastructural topography.[20] Since independence, "ethno-regional patterns of stratification" have continued to ensure that people's experiences of infrastructures as "public goods" reflect a striated vision of the public; however, ethnicity and class have largely (though not completely) supplanted race as the primary frameworks through which many Kenyans interpret infrastructural exclusion.[21] Infrastructures bring these dynamics into stark relief; they are the material substrate intermediating attachments between the state, markets, and "the public."[22] As a result, as people have made and continue to make claims on the infrastructural state, they have engaged in forensic work, comparing how their infrastructural access measures up against that of various others. Put differently, Kenya's infrastructural politics have been shaped by administrative practices and corporate profit-driven strategies that produced a divisible public, at times impeding the formation of a nation.

These compounding legacies notwithstanding, the corporate-state has routinely attempted to distance the work of governance from the work of profit generation by positioning infrastructures as "public goods" crucial to the region's "development." Indeed, this long-angle view reveals that developmentalist thinking—whether "civilizational," "social welfarist," or "neoliberal"—has been as much concerned with shifting strategies of marketization as it has been with poverty alleviation. Prospective profit generation has been at the center of these projects of social "uplift." But state administrators and designers did not hold a monopoly over the developmentalist aspirations embedded in new infrastructures. Across the chronology laid out here, Kenyan communities have read globally circulating theories of social change through alternative epistemologies of individual and collective transformation.[23] Infrastructures, then, are not merely materializations of abstract state or corporate power. They are also intimate objects—the everyday networks of circulation and blockage that structure people's quotidian lives, sometimes becoming the terrain where social and political subjectivities are formed.[24] Unraveling these histories helps us understand why, today, access to infrastructures is a key metric by which people evaluate and debate the meaning of social and political belonging.

Austerity, Infrastructures, and the State: A View from Kenya

Scholars working in Europe and North America have argued that infrastructures are one means by which states materialize their power over space and people, forming the material substrate of "modernity."[25] The nineteenth century, which saw a frenzy of large-scale infrastructure building, was by many accounts a harbinger for these transformations.[26] In the United Kingdom, between 1841 and 1850, Parliament authorized the construction of over 10,000 miles of new track.[27] Stocks and bonds floated in the centers of finance were gobbled up by investors eager to sink idle capital into infrastructures, banking on future returns that the new forms of circulation promised.[28] Despite the private provenance of the capital used to fund infrastructural expansion, these lines of communication are largely credited with producing "state space"; the networks themselves emerging as key sites where relations between states and citizens materialized.[29]

Kenya's infrastructural history has been shaped by similar dynamics, but with distinctive colonial characteristics and imperial material and genealogical inheritances.

A key claim pursued in these pages is that the infrastructural state in Kenya has, from its founding moment, been marked by conditions of austerity, which have critically shaped the contours of Kenya's infrastructural landscape over the course of the long twentieth century. My use of the term "austerity" is intentionally anachronistic, for reasons both historical and political. For scholars working in Europe and America, austerity, while sometimes associated with the deprivations of the Second World War, more typically names a bundle of policies enacted following the fiscal crisis of the 1970s. As many states and international financial institutions (IFIs) advocated for the supposed need to balance budgets and reign in state spending, government services and public infrastructures were gutted, with private firms left to profitably take up the task of governance.[30] By contrast, scholars of Africa have accorded little attention to austerity as a concept or an analytic.[31] In part, this is because it arrived in its official and postcolonial form under the mantle of "structural adjustment." Irrespective of geographical orientation, conditions of imposed fiscal constraint—whether at the hands of IFIs or fiscally conservative governments—are firmly tethered to the period beginning in the late 1970s. In those years, state indebtedness was reframed not as an investment, but as a problem. The solution: A reduction in public spending and the informalization of labor, with the market left to fill in the gaps opened up by the state's withdrawal.

If implicitly, this literature takes what Thomas Piketty referred to as "the Great Compression"—from 1945 to the early 1970s—as the norm against which to contrast the deprivations of the austere, neoliberal present.[32] In these accounts, those twenty-five years—which saw the expansion of social welfare and state provisioning of infrastructures and services, and a reduction of inequality—are framed as the normative ideal against which later austerity is critiqued.[33] For these authors, austerity is irruptive, applying to "situations where societies and individuals that formerly enjoyed a higher standard of consumption must now make do with less."[34] As a "theory of history," writes Nathan Connolly, these state-centric accounts posit a "supposedly clean chronology moving from laissez-faire capitalism to New Deal and Keynesian liberalism . . . [with] 'neoliberalism' [appearing] to close out capitalism's biography." Beginning with the supposed "arrival" of financialization in the 1970s and 1980s, policies of austerity inaugurated the "start of an epoch when [national] governments that once protected citizens took to defending corporations, at times even mimicking their structure."[35]

But the historical geographies from which we narrate our stories matter to the types of stories we are able to tell. Not only is this a partial story it is also largely "a white story," as Connolly argues; its narrative arc is structured by the blinders of a white, worker, and male subject position that never applied to most of the world.[36] As Piketty himself argued, the Great Compression was an aberration in what was otherwise a longer, and decidedly more austere, trajectory. On this framing, what is irruptive is not austerity, but its inverse—robust state provisioning for infrastructures and services and the leveling out of inequalities. The long history of the infrastructural state in Kenya offers a useful corrective, allowing us to tell new stories that challenge the typical chronology and geographies that frame histories of austerity.

As *Infrastructural Attachments* argues, the delegatory technopolitics of austerity are not unique to the neoliberal period, but were baked into the logic of statecraft from the earliest days of colonial occupation, calling into question the novelty of austerity as a mode of governance and lived experience. Structural adjustment was an important shift to be sure, but in seeing it as an epochal break, we lose sight of important lineaments that connect the colonial to the postcolonial. This reorientation affords important insights. For too long, poverty has been a characteristic feature of representations of Africa. This view risks naturalizing what is better understood as manufactured deprivation, as scholars of underdevelopment- and world systems-theory have long recognized.[37] Austerity, then, is not synonymous with scarcity, which signals an absolute or objective lack.[38] Instead, my use of the term austerity refers to a discretionary mode of fiscal (and infrastructural) governance that has enabled the enrichment and provisioning of some people and some places at some times at the expense of others. And these others are often, though not always, racialized. As this suggests, imperial austerity depended on producing disjunctures across geographies.[39] In the colonial period, these dynamics structured relations between the colony and metropolitan firms, generating the uneven geography of historical capitalism that underwrote "imperial state space."[40] Within Kenya, infrastructural provisioning was shaped by explicit policies of discretionary (under)investment in the lives of the African majority, ensuring the development of lumpy infrastructural networks that mapped onto and reproduced the racialized topography of accumulation that structured the settler colony more generally.[41] Today, policies of austerity imposed on Kenya by IFIs have enabled the thickening of asymmetric linkages between the independent state, Kenya's elite, and private British infrastructure firms, generating profits that are largely borne on the backs of the poor.

Infrastructural Attachments: Expertise, Prosthetics, and Work

In charting the long history of the austere infrastructural state in Kenya, a key concept I develop is "infrastructural attachments." I conceive of infrastructural attachments as the material and conceptual tethers that bind institutions and people to new networks, ideologies, markets, and forms of circulation and blockage that they engender. As I show, the historical actors who animate these pages were engaged in projects geared toward generating infrastructural attachments. While they did not invoke this concept, of course, the analogs that they used are suggestive. Corporate officials and colonial administrators forecasted that new road networks would integrate the region and its people into global markets, enabling the inroads of "commerce" and simultaneously "civilising" people and their lands. Radio enthusiasts in the 1950s hoped to use the broadcasting network to shake people out of their supposed "parochialism" by enacting "national units." Safaricom engaged in attachment building of a particular kind, as it leveraged "local" knowledge in a bid to embed its networks and services in Kenya, while distancing the work of the corporation from the work of the state. At its broadest, then, I use this analytic to explore the legal mechanisms, ideological frameworks, and economic logics that enabled the emergence of various iterations of the corporate-state.

As a methodological stance, though, the utility of infrastructural attachments lies in its capacity to operate as an "interscalar vehicle," directing attention away from ambitious plans to attend to the prosaic technical, material, and ideological mechanisms mobilized to make infrastructural attachments stick.[42] These attachments have often taken residence in "less visible locations," "unfamiliar technical forms," and seemingly mundane sites—contracts, treaties, legal precedents, shareholder agreements, currencies, taxation regimes, technological features, kilowatts, wavelengths, algorithms, and, crucially, forms of infrastructural work are all material at this scale.[43] As this suggests, attachments are not just generated out of relations of affection or fondness. While infrastructural attachments are sometimes sought out as a means to make claims or to forge more desirable futures, these attachments are often coercive—severing relationships with some people, some places, some things, some ways of being in the world in order to secure, extend, and preserve others. This makes infrastructures and their attachments—real or prospective—something worth fighting over.

If the concept of infrastructural attachments allows us to move from grand ambitions to prosaic material forms, it also enables us to track the distinct, historically contingent mechanisms and forms of work that have enabled attachment building in the three periods addressed in this book. These differences reflect the particularities of the infrastructures and their respective affordances, the shifts in ideologies of statecraft and development that underwrote their expansion, the budgetary decisions out of which they emerged, the possibilities for profit-making they portended, the publics that assembled around them, and the labor regimes on which they relied.

Just as administrators were forced to confront perennial limits in state financing, drawing on private capital to enact infrastructural rule "on the cheap," technologists hoping to forge attachments had to contend with eastern Africa as a unique material zone.[44] As elsewhere in the colonial world, administrators posited their knowledge as universal—claiming to have generated "principles true in every country."[45] "Expertise," put simply, was part of the armature of empire. Colonial subjects within this discursive and material field were expected to live under colonial rule for a "necessary period of pupilage," to shake them out of "backwardness" and to "steward" them into "modernity."[46] Expertise, like all master categories, then, is a relational claim, foregrounding some forms of knowledge, some forms of work, and some forms of being in the world, while rendering others inconsequential or quaint at best, and invisible and subject to expropriation at worst.[47]

But these infrastructural architects came armed with visions of how these networks should work based on experiences, models, and designs developed abroad. These conditions simply could not be counted on in eastern Africa. In both material and ideological terms, infrastructural plans and infrastructures themselves had to be modified—"tropicalised," to use the language *en vogue* in the 1940s—if they were going to become embedded in this particularly Kenyan milieu.[48] Technological diffusionist pretentions notwithstanding, under these material conditions expertise and technologies imported from abroad, while claiming global purchase, reached definite limits in practice. In the seams of this lumpy infrastructural state, alternative types of experts and forms of expertise asserted their autonomy.

Indeed, as *Infrastructural Attachments* makes clear, in negotiating conditions of austerity, designers and engineers have relied on the contributions of African infrastructural workers—topographical experts in the case of roads, technologists and knowledge-workers in the case of radio, and lay ethnographers and data gatherers in the case of digital financial services—to enact and shore up this ideologically dense but materially thin infrastructural state. This is typical of capitalism operating under conditions of austerity, which "proceeds through the devaluation of labor; decentralized speculative planning; and improvised low-tech investments."[49] Over the course of the histories narrated in the chapters to follow, these "low-tech investments" have required the dynamic infrastructural work of African knowledge-workers and experts. These men and women have routinely been called to act as infrastructural prosthetics. That is, they have been forced to fill the gap between the developmentalist aspirations of the corporate-state and realities that have been characterized by the interruption of unforeseen materialities, partial knowledge, arterial networks, and limitations in financing. This work, while constituting forms of expertise, has routinely been devalued—both materially and conceptually—as merely rote labor. This devaluation has enabled the development of shifting regimes of exploitation and expropriation, which have been critical to the operations of the austere infrastructural state.

A word on prosthetics. As noted, today, developmentalist market makers are explicit in naming Kenyans' work with digital infrastructures prosthetic labor, referring to people who take charge of Safaricom's M-Pesa kiosks as Human ATMs. Conventionally, a prosthetic is defined as an addition, application, or attachment. On this framing, a prosthetic executes the task of a missing limb, its labors leaving the whole more or less unchanged. Drawing on science studies, in the chapters that follow, I argue that we need to think of prosthetics not simply as replacements that execute the functions and daily tasks of missing parts but as a "fundamental category for understanding."[50] Infrastructural prosthetics and prosthetic work are extensions that fundamentally transform the systems to which they are attached, sometimes irrevocably changing them, at other times putting them to uses unforeseen by designers. As it pertains to infrastructures in now postcolonial Kenya, a focus on infrastructural prosthetics and prosthetic work enables us to reframe our understanding of large-scale technological networks by placing front and center the work of making them hang together, and its transformative effects.[51] The work of these men and women, I show, was (and is) critical to generating infrastructural attachments. These histories, in turn, allow us to chart how infrastructures—ostensibly generic and universally applicable networks—became "tropicalised" and "peculiarly" Kenyan.[52]

Accessing this expertise is difficult. While I can approach contemporary knowledge-workers such as M-Pesa agents ethnographically, the presence of African infrastructural experts in the archives are as subjects—invoked out of frustration, curiosity, or flourish by Euro-Americans and recognized experts—not as authors of their own documentary trails. The partial visibility of these experts is a result of pernicious representational practices that frame Africa as a place without technologies, as a place without technological experts.[53] Over more than a century, technologists and market makers have presented this expertise as banal, making it appear to be *merely* prosthetic.[54] It goes without saying that the occlusion of African infrastructural expertise in these archives shaped the histories I have been able to narrate in these pages. Given these limitations, readers will note that the ethnographic density of the stories of these African men and women increases as we move across the chronology narrated here. This expository lumpiness is to be lamented. But to not tell the stories of these people simply because they were deemed unexceptional (or were actively devalued) by chroniclers would be to leave untroubled the racist underpinnings that guided the composition of the archives with which I worked. Focusing on the difficulty of making infrastructures hang together, and looking to the breach that divided aspiration from reality, offered one opening, allowing me to surface new histories of work and new histories of expertise. In this regard, while partial, the stories offered in these pages treat (post)colonial "middle figures": infrastructural knowledge-workers and experts.[55] Their work, far from being generic or banal, was (and is) transformative work that constituted (and constitutes) an unrecognized and under-remunerated—when paid at all—form of expertise.

This, in turn, is essential to understanding the repetitions that guide the workings of capitalism in the present—enthusiasts' claims to rupture notwithstanding—and the types of inequalities that it shores up. Proponents of contemporary developmentalist thinking claim that capitalism in an "altruistic" register can "transform" the lives of the poor by locating value at the "bottom of the pyramid."[56] I show that appropriating this expertise, and the social infrastructures on which this value has been based, has a long history in the region. In understanding these processes, I move away from a vision of "people as infrastructure" to explore how it has come to pass that people are compelled to act infrastructurally.[57] This distinction is important. The first risks naturalizing the material conditions under which people as infrastructures emerge as a phenomenon. The second, and the one pursued here, works to parse out those conditions as a mode of critique of the particular articulations of capitalism—both past and present—in Kenya and beyond.[58]

Arc of the Book

The term "austerity" was not used by the historical or contemporary actors that people these pages. In each of the three sections that structure the book, I use the actor categories that I understand to be austerity's analogs—"sound finance" and colonial "self-sufficiency" from the nineteenth century through the interwar years (chapters 1 and 2), "community development" following the Second World War through the late colonial period (chapters 3 and 4), *harambee* (Kiswahili for "all pull together") in the independence period, and "the digital" in the 2010s (chapters 5 and 6). Across the chapters, I signal these important historical ideological and material transformations, while training attention on the shaping capacity of the deep "cross-historical processes" that have characterized the operations of the austere state.[59] Such a long-angle view reveals that the "intransigence of infrastructure[s]" lies not simply in their presence, but crucially in their durable absence.[60] As Joshua Grace has shown, material constraints shape infrastructures in "underappreciated ways." As underinvestment was naturalized as scarcity, the absence of "permanent infrastructure[s]" also gained momentum.[61] In Kenya, this has had implications for the shape of infrastructures of the future.[62]

Chapter 1 charts the expansion and decline of the Imperial British East Africa company (IBEA) in the late nineteenth century. The IBEA's rise as a corporate-state emerged at a peculiar conjuncture, which saw the Crown outsource the costs and risks of undertaking colonial occupation to the firm, this in keeping with the logic of "sound finance." The Crown's charter was matched by a concession from the Sultan of Zanzibar. With the delegation of sovereignty into corporate hands, the IBEA promised to bring "commerce and civilisation" to the region. The construction of a word network was at the heart of this project. A company road network would enable the firm to establish a revenue regime in pursuit of corporate profits through inaugurating novel regimes of taxation. The firm quickly realized, however, that asserting sovereignty on the ground required daily material and conceptual work. These labors turned on efforts to assert a monopoly over the meaning of the emergent distributions of sovereign authority, of debt and deference, critical to consolidating power in the hands of the corporate-state.

As chapter 2 explores, while gaining the right to seize taxes was at the heart of the IBEA's mission, getting east African communities to recognize the IBEA as the dominant administrative and fiscal authority would require a revolution not simply in regimes of sovereignty, but in regimes of work as well. As I show, the IBEA was dependent on Africans' knowledge of historical networks of mobility as it constructed a new road network, but this knowledge it hoped to subsume all the while transforming expert work into rote labor. African topographical experts resisted complying with this new labor regime. At this conjecture, company administrators tried to mobilize two new fiscal technologies—a company currency and new regimes of taxation. Working in tandem, they forecasted that these fiscal and administrative forms would operate as infrastructural attachments, binding people as labor to the road and the wage as they struggled to "find their tax." These contradictions persisted when the Foreign Office took over the administration of the region, becoming the foundational logic of the colonial state as it pursued sound finance's imperial corollary, colonial "self-sufficiency."

The middle section of the book (chapters 3 and 4) turns to the history of radio broadcasting, focusing mainly on the 1940s and 1950s, which seemed to mark the decline of colonial austerity as social welfarist models of governance made inroads in the form of the Colonial Development and Welfare Acts of 1940 and 1945. However, the durable material consequences of policies of colonial "self-sufficiency" could not easily be undone, and grand plans for centralized social welfare quickly devolved into "community development." This pivot was largely driven by parsimony, "community development" having as its core appeal being "inexpensive" because it often "relied on unpaid voluntary labour."[63] It was against this backdrop that administrators came to see in broadcasting a plausible solution to the long history of underinvestment in the lives of the African majority. Indeed, it was precisely in the absence of—let alone the funding for—"harder" infrastructures of development that many in the 1940s came to see radio as an infrastructural prosthetic; the government's messages for social and material transformation suffusing people's consciousnesses over the airwaves. The medium, they claimed, could broadcast development.

This aspiration was stymied from the outset. Absent funding from the Treasury, in the 1920s the government had entered into agreements with Cable and Wireless Ltd. (C&W), to service white settler and South Asian listeners. These early decisions—driven by the prudent logic of colonial "self-sufficiency" and a racial politics that gave primacy of place to the needs of white settlers— attached the state to infrastructural networks funded and established by a corporation. Given these material constraints, the government deployed information vans boasting receiving sets across the colony. These infrastructural prosthetics, it hoped, could mitigate the problems thrown up by the lumpy reach of the state. Such material conditions ensured that European information officers were utterly dependent on the practiced labor of African knowledge-workers and experts, something which became abundantly clear during the Mau Mau uprising of the 1950s. At one level, then, efforts to enact radio both indexed the change in direction of colonial policies that characterized the 1940s and 1950s and exemplified colonialism on the cheap. This was colonial austerity in the age of social welfare.

Chapter 4 turns to the postwar period which saw the Treasury, for the first time, make monies available to assemble a state-run infrastructure of broadcast. This was a response to labor mobilizations across the British empire in the 1930s and 1940s, which convinced the administration of the urgency of developing its own broadcasting network. In this uncertain moment, the colonial government hoped to use broadcasting to mold the social and political worlds available to Kenyans, by generating subjects of an "intermediate" scale—beyond "tribe" but before "all-embracing citizenship." This centered on efforts to produce a robust parochialism through Kiswahili-language broadcasting. These goals were complicated in practice. Not only was Kenya a unique atmospheric and topographic zone, but problems of financing remained. Material issues could not be easily divorced from the question of politics, which intensified over the course of the 1950s. As people started tuning in to broadcasts from Egypt's Radio Cairo on their shortwave receivers, they were invited to generate novel forms of affinity. Listeners seized on this opportunity, demanding that the state provision them with vernacular-language broadcasts. While this led to the consolidation of a technopolitical network geared toward forwarding "parochial" attachments, this process was driven from below.

The final section of the book (chapters 5 and 6) addresses the period following political independence and charts the consequences of the incremental privatization of Safaricom, which began its life as Kenya Post and Telecommunications Corporation (KPTC). Independence marked a rupture as "Africanization" saw white managers and state officials replaced by Kenya's black elite. While in rhetorical terms, the history of austere infrastructural governance was one that the state under Jomo Kenyatta was eager to shake off, in practice the developmentalist ideology of the early postcolonial state was largely an extension of that of the late colonial state. "Community development" was rebranded *harambee*, or "all pull together." Occupying a central place in Kenya's coat of arms, the message was clear: Austerity would be the normal state of affairs in the near and mid-term future.

Nevertheless, the promises of independence were buoyed by the boom in commodity prices in the 1950s and 60s—leading some to dub the period from 1964–1980 the "Kenyan Miracle." But this "miracle" was uneven in its reach. The discretionary logic of indirect rule remained largely intact, with the government mobilizing parastatals, such as KPTC, to reward loyal constituencies. By the end of the 1970s, price inflation and high interest rates on international loans resulted in massive deficits in government budgets.

By the 1990s, under pressure from IFIs—which cited the state's inability to fund infrastructural expansion, maintenance, and repair—KPTC was dissolved and reconstituted as Telkom Kenya. In the early 2000s, the state shuttled 40 percent of its shares to the UK-based firm Vodafone, and so Safaricom was born. In 2008, the corporation launched M-Pesa. This mobile-to-mobile money transfer system emerged as a solution to long histories of underinvestment that had seen the consolidation of a banking network that was both incomplete and that had long excluded the African majority. Safaricom's financial services, like radio, then, were devised to "leapfrog" over the absences generated by longer histories of austerity. In 2008, Safaricom went "public," giving birth to "digital Kenya."[64] These transformations inaugurated a new era of corporate statehood, while ushering in a novel political form: shareholder citizenship.

Chapter 6 explores Safaricom's austere labor regime through a textured ethnographic engagement with the daily life and work of M-Pesa agents, those women and men who industry insiders refer to as Human-ATMs. Far from this work being simply infrastructural, the daily care and practiced expertise of M-Pesa agents has been essential to Safaricom's success, enabling the firm to emerge as the largest corporation in the region. It is not simply Safaricom workers who are implicated in this regime, but Kenyans writ large whose everyday tactics and social relations have been appropriated and translated into commercially useful data on which Safaricom assembles new markets.

Put simply, if our definitions of austerity are rooted in perceptions of a normalized Keynesian developmentalist model, the era of the well-financed state that could be said to resemble a welfare state was short-lived in the region. Austerity in Kenya has largely not been characterized by a reversal of fortunes; it has been the norm. This has opened up possibilities for private accumulation, with private investments and capital being enrolled to make up for limited state budgets. In Kenya, "the market" has *always* been the critical third term mediating relations between subjects and the state through infrastructures.

Taken together, a central claim of *Infrastructural Attachments* is that histories of infrastructure, capitalism, and state formation that take Kenya as their point of departure force us to confront a different genealogy of the history of both the state and infrastructures. This genealogy is one that must highlight the role played by austerity, track the peculiar institutional forms to which it has given rise, and place infrastructural prosthetics at the center of the story. In the process, occluded forms of statecraft, expertise, and regimes of expropriation become visible.

As this suggests, the histories narrated in these pages are not primarily stories about the material networks of infrastructures themselves. Instead, I use an exploration of infrastructures always "in the making," and the discussions, disputations, and practices that they elicited and enabled—to tell new stories about the changing relations among capital, state formation, and notions of belonging; to locate new histories of expertise and skill; and to explore the long-term continuities in developmentalist thinking, and how people have reframed development through vernacular notions of individual and collective well-being.

By firmly rooting people's engagements with infrastructures and infrastructural work in the social lives, political languages, and cultural practices of the historical subjects and contemporary interlocutors who people these pages, this book works to trace out the contradictions of (post)colonial rule "on the ground." Conjoining an ethnographically informed analysis of state-building and market-making with a close reading of the cultural politics of Kenyan communities, I track transformations in political economy and developmentalist thinking as manifest in infrastructures through the micropolitics that they engendered. Holding these histories together is essential to understanding why infrastructures in Kenya, far from being the invisible and unremarked background of social life, have emerged as fraught cultural and material objects.[65] They are not neutral, nor are they banal. Both discursively and as material networks, infrastructures in Kenya have been politically charged and "multiply authored" cultural and material objects.[66]

1

A Divisible Sovereignty

*The Imperial British East Africa
Company, the Crown, and the
Sultanate in the Competitive World of
Nineteenth-Century Eastern Africa*

In 1895, P. L. McDermott wrote in glowing terms of what administrators called "Mackinnon Road." It spanned, he claimed, "200 miles of the most difficult part of the country with a wide, smooth track, suitable for any description of wheeled traffic." This route, running from Mombasa to Kibwezi, was part of a "general scheme of road-making." And "Mackinnon Road," he assured readers, was "the highest testimony to the need which existed for . . . [roads] and the excellent way in which that want [had been] supplied."[1]

Road infrastructures were central to the developmentalist ethos that guided nineteenth-century imperial expansion in sub-Saharan Africa. They were the hinge uniting the supposedly synergistic project of what David Livingstone famously referred to as "commerce and civilisation."[2] According to this wisdom, new infrastructures would usher in a revolution in both trade and the social structures of societies living in the region. While the dyad of "commerce and civilisation" would become an aphorism in the "colonial lexicon," the region's early road networks were constructed not by the colonial state but at the behest of a corporation, the Imperial British East Africa Company (IBEA).[3]

Nor were poetic renderings of infrastructures unusual for company officials. William Mackinnon, a self-proclaimed philanthropist and the IBEA's founder, pinned grand hopes on transport infrastructures. As Chairman of the British India Steam Navigation Company (BI), he had successfully established routes linking the ports that connected the Indian Ocean world—including British India, Zanzibar, and London.[4] As this network expanded, British imperial power in the Indian Ocean was increasingly dependent on the infrastructural firm to secure logistics and trade. As J. Forbes Munro writes, Mackinnon transformed the firm into "a major 'public contractor' to the Government of India, a prop to its administrative structures and a servant to their policy goals."[5]

Such arrangements of divisible sovereignty had long been the basis on which power was predicated across the Indian Ocean. As Mackinnon turned his attention to eastern Africa in the 1870s, the shipping magnate worked to territorialize these dynamics. In these years, Mackinnon cultivated a fraught relationship with the territorial sovereign, Sultan Barghash bin Said of Zanzibar, who regularly consulted Mackinnon on matters of trade and policy.[6] As Mackinnon labored to extend his maritime infrastructural monopoly to the regions controlled by the Sultan, roads emerged as extensions of shipping lanes that crisscrossed the Indian Ocean in his "technopolitical" imaginary.[7]

In pursuing his lofty ambitions, Mackinnon established the IBEA's forerunner, the British East African Association, which quickly began the work of securing treaties with local sovereigns and laying down new routes of commerce.[8] Company roads, the firm promised, would aid in fighting the slave trade, all the while inaugurating "legitimate commerce."[9] Yet the IBEA's interest in constructing company roads was, in truth, more complicated—it would be along these vectors of commerce that the firm would be permitted to levy taxes.[10] From these beginnings, a discourse linking commerce and civilization to material transformations of the region's topography became a characteristic feature of the aspiring corporate-state, the firm regularly mobilizing its road building to justify the value of its presence in the region.

In the early days, the Sultan viewed himself as patron and Mackinnon as client, a dynamic of dependence and deference Mackinnon ostensibly accepted, regularly indicating that he acted as both "friend and servant" to the Sultanate.[11] The Sultan, like Mackinnon, saw inroads the possibility of extending and consolidating his sovereign prerogatives on the mainland. These dynamics crucially laid the groundwork for the IBEA's eventual operation as the *de jure* sovereign in the region, a relationship formalized with the signing of a concession with the Sultan of Zanzibar in 1887.[12]

The IBEA's sovereign rights were deepened in 1888 when the British granted the corporation a "Royal Charter," with Mackinnon as president of the firm's court of directors.[13] The charter mandated the "private enterprise" to take up the task of "acquiring and administering [the] new territory in the name of the Crown."[14] For the British government, this decision was guided by the logic of "sound finance," an austere fiscal orthodoxy shared by Tories and Liberals alike. As policy, "sound finance" was straightforward and prudent: "balanced budgets . . . and responsible Parliamentary control of expenditure."[15] With the charter, the company was granted the right to mobilize the capital of its shareholders to execute the "national task" of extending the Crown's "sphere of influence," all the while protecting the Treasury and the "purses" of British taxpayers.[16]

Both in practice and in law, then, sovereignty was enacted as a divisible entity whose constituent elements could be strategically unbundled and distributed in myriad configurations across a variety of bodies. With the permission of sovereigns, corporations could act as states. This shape-shifting has led political theorist David Ciepley to argue that historically corporations were neither wholly private nor wholly public but were "amphibian" legal forms.[17]

This chapter excavates the hybrid origins of the colonial state—which brought together the corporation and multiple sovereigns, private capital and a series of publics—through an exploration of the protracted expansion and precipitous decline of the IBEA. Legal engineering, guided by the political and economic concerns of the Crown and the Sultanate, bound the sovereigns to the company. Taken together, the charter and the concession authorized the firm to take up the task of administering the territory, an occupation that was geared toward the generation of both a revenue regime and corporate profits. These political conditions, then, were the requisite "backstory" in this period of capitalist expansion overseas, sanctioning the operations of a corporation whose responsibilities were widely distributed—the Sultanate, the Crown, Britons, East Africans, and shareholders were all relevant constituents.[18] With the delegation of administrative and fiscal authority into corporate hands, the IBEA promised to satisfy its mandate of bringing "commerce and civilisation" to this "unknown region." Central to this project: the construction of a road network.[19]

The amphibious character of the IBEA notwithstanding, both critics and supporters of corporate sovereignty in the late nineteenth century built arguments premised on the separation of the political from the economic, a bifurcation the terms of the charter and concession belied.[20] Over the course of its tenure, the company was forced to modulate between these prerogatives, at times emphasizing its status as a delegated sovereign, at other times as an independent political authority, and at other times still as a company of merchants loyal to corporate shareholders.[21] In practice, then, the twin mandate of commerce and civilization turned on an intimate interweaving of the economic and the political, a balancing act the corporate-state proved to be unable to stabilize.

Indeed, while not peculiar in the period, the mandates underwriting this marriage of the corporation and the state(s) quickly produced tensions.[22] While the concession and the charter legally sanctioned the operations of the firm, the company quickly realized that sovereignty was not an enduring quality but a precarious achievement which required daily material and conceptual work to enact. As it struggled to convince East Africans of this new dispensation, it was forced to mimic patterns of "layered sovereignty" operative in the region, entering into complex dynamics with local leaders.[23] In pursuing corporate sovereignty, the IBEA deployed myriad infrastructural attachments—treaties, taxation regimes, gifts, and violence—that, it hoped, could tie people and their futures to the road. While it often foundered in this work, these labors turned on efforts to assert a monopoly over the meaning of the emergent distributions of sovereign authority, of debt and deference, critical to consolidating power in the hands of the corporate-state.

Within but a few years of its operations, it became clear that the IBEA had failed in these efforts, a failure that was the result of myriad sovereign entanglements. Central to these conflicts were disputes over the relationships between deference and debt, sovereignty and profit. These conflicts pitted sovereign against sovereign and ultimately heralded the IBEA's unraveling. When the IBEA was forced to withdraw from the region in 1894, the British government mobilized revenue from Zanzibar's treasury to offset the costs of purchasing the IBEA's assets and sovereign rights. This transfer of wealth and legal authority was the prerequisite for formal colonial occupation, setting the stage for the emergence of "sound finance's" imperial analog: colonial "self-sufficiency."[24]

The "Amphibious" Origins of the Colonial State

The amphibious origins of the colonial state emerged at a peculiar historical conjuncture. The terms of the Berlin Conference of 1884–85 laid out that no annexation on the African continent would be recognized unless the European powers undertook effective occupation while protecting the region as a free trade area.[25] While the Conference authorized partition, Britain had long-standing imperial connections with the region.[26] By the nineteenth century, building on networks that had long structured the Indian Ocean world, South Asian merchants had consolidated their dominance in business, banking, finance, and trade in Zanzibar.[27] But Britain had direct commercial interests of its own in the region. Trade between Britain and Zanzibar increased in the 1830s and 1840s, being estimated at $214,000 in 1846–47.[28] Particularly vexing for the British, then, was the presence of "agents" of the Deutsch Ostafrikanische Gesellschaft (DOAG, or the German East Africa Company), who began signing treaties with local "chiefs."[29] These concerns were augmented in 1885 when bin Said ceded trading rights in Dar es Salaam to the DOAG. Two years later, the German company gained the right to collect customs duties in the area.[30] From the vantage point of the British, the Sultan's "vague claims of sovereignty" appeared more ephemeral by the day.[31]

These transformations dovetailed with a period of economic turmoil in Britain, which both complicated its future in the region and made retaining a foothold in eastern Africa appear more pressing. With the onset of the Great Depression in 1873, Britain's manufacturing declined in comparison to the United States and Germany.[32] Britain was in desperate need for new markets and spaces to export capital.[33] Africa seemed to offer such fertile grounds. As one time company employee Fredrick Lugard proclaimed in a speech to the Royal Colonial Institute in 1895: "The hostile tariff[s] imposed by other nations upon our industries, the competition of foreign-made goods, and the depression of trade, have driven us to seek new markets and new fields for our surplus energy."[34] After years of negotiation, and unwilling to spend funds directly, the Crown granted the IBEA its charter in September 1888. This legal instrument enabled the Crown to outsource the governance of eastern Africa to the company, all the while managing economic and political turbulence in the metropole and fulfilling the obligations of the Berlin Conference. The firm, the Crown hoped, could function as an infrastructural prosthetic. Devolving its sovereign prerogatives to the IBEA was, then, a particularly parsimonious means of extending British commercial and political dominance in the region.

As for the IBEA, the state-like functions apportioned to the corporation were laid out in the terms of the Royal Charter, which were as broad as they were vague. The IBEA was granted the right to "raise taxes, impose customs dues, administer justice, make treaties, and generally assume the powers of government within a specified area."[35] Crucial here was the right to enact sovereign seizure—the company being delegated the authority to farm taxes on the coast according to the terms of the concession, and the right to inaugurate taxation regimes inland under the terms of the charter.[36] These provisions were essential, for taxation sat both at the heart of the company's prospective revenue regime and would form the basis of corporate profits.

The state-like status of the IBEA was further mirrored in its legal protections, which ensured that the constituent components of the corporation were not subject to prosecution in British courts for acts "which only governments may perform."[37] In exchange for these privileges the company agreed to undertake the work of "development and administration."[38] And the IBEA performed its state-like status with aplomb. As Munro notes, the company radiated "an aura of late Victorian respectability, with a Court rather than a Board, a President rather than a Chairman, its own specifically-commissioned flag [and its own currency] . . . it created for itself an image . . . that was closer to government than to . . . ordinary business. . . . [This] contributed to the general perception that the company's purposes were more 'governmental' than commercial in character."[39]

The IBEA's amphibious character was further mirrored in the composition of its founders and court of directors, whose biographies reflected the porous boundaries between commercial and political interests in the nineteenth century. It was Mackinnon's work as Chairman of BI that led him to pursue corporate sovereignty in eastern Africa in the 1870s.[40] John Kirk, for his part, was a member of Livingstone's Zambesi Expedition (1858–64) before becoming the British Consul General in Zanzibar in the early 1870s.[41] Kirk was an early enthusiast of the corporate-state, routinely "communicating his zeal for commerce as a vehicle for civilization," and joined the company's court of directors following his retirement in 1887.[42] Also on the court was Sir Thomas Fowell Buxton, the son of famed abolitionist Sir Fowell Buxton, who was appointed Governor of Southern Australia following the collapse of the IBEA, and Lord Brassey, who joined Buxton in Australia in 1895, becoming the Governor of Victoria. Belgium's King Leopold, for his part, was granted the right to appoint one director. Evidently, the political work of the empire was deeply intertwined with the profit motives of the corporation.[43] For these men, roads were the infrastructure capable of territorializing their commercial interests, leading the firm's forerunner, the British East African Association, to enter the competitive realm of eastern Africa in the mid-nineteenth century, joining German, French, and Italian firms in vying to assert commercial dominance.[44]

According to the IBEA's pundits, the wisdom of outsourcing imperial expansion to the firm was well-established by "the principles of Political Economy," which made it "manifest" that it was only by virtue of "large capital [outlays] that the lever of the powerful engine necessary for so tremendous an enterprise . . . [could] be moved."[45] The charter was the legal mechanism that made this distribution of sovereign prerogatives possible. As Lugard wrote, charters enabled "merchant adventurers" to mobilize finance capital, which they put to the task of "development," accepting "the responsibility which the government shunned." These legal arrangements "would not only afford revenue for administration and defence," however, "but also a prospect of profit."[46]

Proponents of "sound finance" among the contributors to *The Economist* tended to agree, arguing that in mobilizing the IBEA to administer overseas "plantations," the Crown could expand "national territories without . . . expending, national resources." With the blessing of the Crown, the commercial enterprise would extend British imperial interests without burdening either the Treasury or British taxpayer, both of which would be protected from "rash encroachment in the name of Empire."[47] The British public's private hoards could be put to the task of imperial expansion, not through the mechanism of taxation but in their capacity as private investors and profit-seekers.[48]

The company insisted that the charter functioned simply as a letter of no objection sanctioning its work.[49] In reality, the charter ensured prospective shareholders of the security of their investments, the Crown's political backing enabling the company to raise subscriptions valued at £240,000.[50] The IBEA was confident in the soundness of this arrangement, candidly proclaiming that it could "safely be asserted . . . that the capital so employed will . . . yield, as an investment, a dividend not only good . . . but also perfectly safe."[51] Locating new fertile fields for finance was as much a political as an economic project, the divestment of sovereignty paving the way for speculative investment.

Despite the IBEA's own confidence in these arrangements, not all were so certain. The scandals that swirled around the East India Company had called into question the wisdom of allowing commercial enterprises to act as the primary agents of empire. As one skeptic candidly wrote: "If we are to have an India in Africa, let the Crown govern it from the first, and acknowledge the responsibilities which no device of intermediate companies will enable it to shake off."[52] For other critics, the problem was one of uncertainty, with the Under-Secretary for Foreign Affairs being "forced to admit that he hardly knew what the powers of the companies were," nor whether the "frontier" of the "company's possessions" had "been definitely limited." The world of contract that European merchant mercenaries came armed with repeatedly proved to be of limited concern to the Sultan who signed concessions in a bid to shore up his regime.[53] Claims and counterclaims by competitive European powers abounded.

Other pressing issues concerned the company's limited knowledge of the repertoires of sovereignty operatives both on the coast and in the interior. While it seemed apparent that the Sultan had "all manner of nominal rights . . . no one . . . [could] exactly ascertain [their limits] . . . which," one writer for *The Economist* wrote in typically demeaning language, the Sultan "hardly . . . [understood] himself." Central to these thorny issues was the relationship between territorial control and the company's right to engage in sovereign seizure, which, to be enacted, would require a "clear understanding" as to the Sultan's powers over both "territory and taxation."[54] The company harbored these concerns as well, as expanding the Sultan's revenue regime would form the basis of its prospective corporate profits.

The men of the IBEA worked to convince critics that the company's goal was not to establish "a monopoly for any single company," but to open up the territory for the benefit of the "large commercial enterprise of the country at large." The advantages accruing to the company, officials assured skeptics, were "merely prospective." Its operations were not simply those of a private capital firm, however. It was a deputy of the Crown that had sanctioned its "aims and ends," which were "imperial and immediate."[55] The company itself, then, tried to leverage a middle position, contending that its role was both commercial and political, its concerns were both with expanding new markets and enacting governance—its goal was to extend "commerce and civilisation" to this "unknown region."[56] But the sovereign prerogatives vested in the company came from multiple vectors, emanating not just from the Crown but also from the Sultan of Zanzibar, his supposed followers along the coast, and local "sultans" inland.[57] As the company struggled to exert its influence on the ground, it deployed more prosaic attachments.

Enacting Sovereign Distributions

Enacting corporate sovereignty was an incremental and protracted process that emerged piecemeal over the course of a decade out of an entangled series of overlapping contracts. The company hoped these legal instruments would enable the firm to emulate, while also reconfiguring, the "layered sovereignties" that governed the region's spatial order.[58] And this appeared an auspicious moment for the company to extend its maritime monopoly to inland trade. As Prita Meier writes, before the rise to power of Sultan bin Said, Zanzibar "was not the capital of a state or empire," but operated as a "node in a web of interconnected port cities." According to this vision of sovereignty, political power was "measured in terms of how many subordinate clients and partners one was able to amass, not in terms of territorial holdings." Under bin Said, things began to change as the Sultan worked to consolidate his authority, "delimiting the Sultanate in spatial terms and forcing those living in his 'state' to accept his authority in all matters."[59] In attaching the Sultanate to the IBEA, the Sultan hoped he could transform his commercial empire into a political one.[60] And so, in 1877, the Sultan entered into negotiations with the company because it had "capital sufficient" to undertake, "for a certain agreed term, the 'General Administration and Government' of his whole domain."[61]

The Sultan's decision to attach his government to the firm was also a pragmatic response to the threat posed by Egypt's ambitious Khedive Ismail.[62] In 1875, Egypt's forces occupied towns on the coast as outposts from which to access the Great Lakes region.[63] Following their withdrawal, the Sultan wrote to Lord Derby requesting British "capitalists" to aid in "the development and civilization of Africa and the opening up of trade on the coast and in the interior."[64]

This inaugurated a period of protracted negotiations as the company worked to convince the hesitant Sultan of the wisdom of allowing the firm to take over his sovereign prerogatives.[65] At the center of the proposed concession was the right to enact sovereign seizure, which would grant the IBEA the power to administer the customs house, as well as collect "all duties or taxes

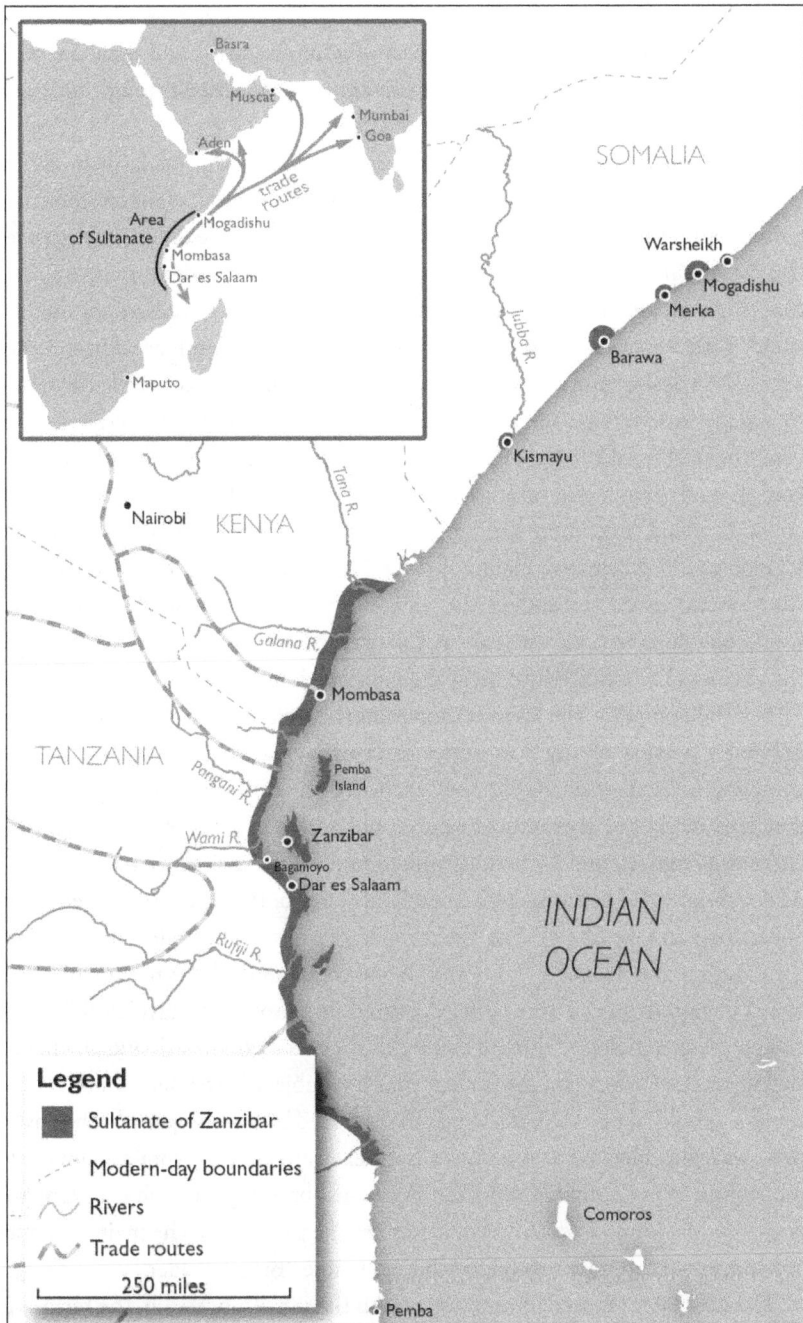

Map of the Sultanate of Zanzibar circa late nineteenth century, with present-day borders and place names for reference

either direct or indirect which may be existing *or may be established by the company*."[66] In exchange for this acquisition of administrative and fiscal authority, the company would pay the Sultan an annual rent raised through customs dues and take over a range of the Sultan's debts.

The Sultan's concerns were not unfounded. Behind his back, men of the corporate-state questioned the Sultan's soundness of mind, claiming that the Sultan had "[no] power of continuous thought." Against the Sultan's insistence that he retain his influence on the coast, company officials demurred, arguing that this was "against the essence of the bargain."[67] Did the Sultan not understand, Kirk wrote exasperatedly, that he was "the seller [and] Mackinnon the buyer"?[68] These conflicts revolved around the Sultan's refusal to cede the right to engage in sovereign seizure to the company. From the perspective of the firm, the deal would only be viable if the Sultan agreed to "give up all interference in government and taxation on the Coast," for without full access to revenues, it would "take more capital than Mackinnon dream[ed] of to make the scheme pay."[69] Normative claims of the bifurcation between the political and the economic notwithstanding, the company hoped that the world of contract would enable it to parlay its role as Customs Master—itself secured through the payment of "rents"—into fiscal *and* political dominance.

For the company, the issue centered on the Sultan's insistence that he be granted a portion of any new duties or taxes levied by the IBEA, which the company insisted must be for the "exclusive benefits of themselves."[70] The company could not accept any deal that did not include the exclusive right to enact sovereign seizure. Without this guarantee, there was little to woo investors. As one disgruntled company official explained, one of the core "inducements for Subscribers to take Shares" was the company's right "to fix collect and receive the Customs and dues of all Ports and all taxes dues and tolls." A portion of taxes seized in eastern Africa were to be translated into profits and dividends for the company's shareholders.[71] Absent this right, the IBEA would be financially hamstrung, as the Crown had made clear that its "subjects"—namely South Asian traders—could not be subject to taxes inaugurated by the company.[72] The company was thus blocked at the outset from extracting value from a significant proportion of its potential tax base. Without obtaining the Sultan's right to taxation, Mackinnon was advised, it was best he proceed on the mainland not as a bearer of corporate sovereignty but as a mere "British trader."[73]

The company realized its ambitions with the 1887 Concession, the terms of which stipulated that "His Highness" lease to "the British East African Association all the power which he possesses on the mainland in the Mrima, and in all his territories and dependencies from Wanga to Kipini," as well as the "lease of the customs of all the ports throughout" these domains.[74] Soon thereafter, the company established its headquarters in Mombasa.[75] In the first year of the company's operations, customs amounted to $56,000, or Rs. 119,000, but it was confident that these revenues were "capable of considerable increase."[76] In exchange, the company paid the Sultan an annual rent, which totaled $80,000, or Rs. 170,000, by 1891.[77] Any sum in excess of the requisite rent—which amounted to £2,595 in 1890, £4,618 in 1891, and an estimated £8,000 in 1892— would be relocated to the pockets of shareholders in the form of dividends "as their remuneration."[78] The Sultan stood to gain personally from this arrangement, having negotiated that he be granted "Original Founders' Share No. 1."[79] The concession, though, did not mark the total cessation of the Sultan's authority, with the Sultan "reserving to himself such honors of Sovereignty as he may be entitled to according to the customs of the country."[80]

In taking over the customs house, the corporation built on existing repertoires of distributed sovereignty that structured the Indian Ocean world. In 1875, Indian financier Thariya Topan had been awarded a contract to operate the customs house at Zanzibar, paying 350,000 Maria Theresa thalers per annum.[81] In exchange, Topan supplied "the Sultan with all of his public revenues—a projected figure . . . based on a combination of receipts from years past, estimates for the future, and the needs of the Sultan himself."[82] This arrangement placed the Sultan in a state of perpetual indebtedness to the customs master from whom he took out annual loans.[83] In 1880, Jairam Shivji gained control over the customs house. With Shivji, too, the Sultan regularly overdrew his account. These were not merely fiscal relations. The commercial activities of the financier and his family were not subject to taxation, leading Chayya Goswami to argue the loans were given as a "favour."[84]

Indian financiers' historic control over the customs house crucially brought together the twin mandates of administrative control and profit-making, for it was here that "the spoils of commerce merged with the imperatives of empire."[85] These relations of credit and debt were central to the Indian Ocean world. Backed by dynamics of sociality, the mutual benefits accruing to both the Sultan and his customs masters were paramount, with issues of revenue and law often taking a back seat—the world of contract perpetually being deferred by the demands of the day. Though the company labored to mimic existing distributions of sovereign authority, it rejected the more flexible dynamics of debt and deference that had grounded historical arrangements. These incongruent visions of divisible sovereignty would ultimately herald the IBEA's unraveling.[86]

Culturing Space in Two Registers

While state-like powers conferred some state-like responsibilities, the Janus-faced nature of this mandate of civilization and commerce, of "moral and material" progress—which brought together the private and public(s), the corporation and state(s)—rarely operated in lockstep.[87] In material terms, public works projects were critical to maintaining this dual vision. Both the corporation and the twin sovereigns stood to benefit from these infrastructural attachments: the corporation mainly in economic terms, the states mainly in administrative terms.

As early as the 1870s, the directors of the British East African Association tasked engineers with constructing road networks to facilitate the movement of the "vast natural wealth"—namely ivory—from inland areas to the Swahili Coast.[88] The company thereafter bound itself "to the opening up of such roads as may be necessary for commerce . . . as soon as the . . . interests of the country may require it."[89] But company administrators did not hold the goals of commerce to be distinct from the goals of civilization. They imagined roads as the infrastructural basis for expanding territorial control through the creation of new markets, *and* as core materializations of "civilisation," for without "better facilities of communication and transport . . . the development of the country could not be carried far."[90]

In bin Said, administrators seemingly found a willing ally in the twin project of commerce and civilization and materialized in the form of road networks. The Sultan was himself something of an infrastructural fetishist, launching vast public works projects in Stone Town, Zanzibar's commercial center, during his tenure.[91] Bin Said expressed particular excitement at the company's commitment to undertaking "works of public utility" which formed the object of the concession.[92]

Roads were at the center of bin Said's vision of "works of public utility." In 1879, the Sultan wrote an emphatic response to a request from Mackinnon: "As regards your asking our assistance in . . . joining you in your program of making roads in our country, we shall be glad to offer you any assistance in our power . . . we are desirous of helping anyone, who takes the trouble and feels interest in promoting the cause of civilisation, and commerce in these countries. We fully appreciate . . . [the] advantages resulting to a country from good roads."[93]

New road networks snaking inland would enable the Sultan to consolidate his tenuous authority on the mainland, while increasing the revenues farmed

Sketch of Stone Town, Zanzibar, 1886

through the customs house. Bin Said's interests in the firm's infrastructural projects, then, were complicated, located as he was at the nexus of the twin logics of civilization and commerce, the state and the corporation, sovereignty and marketization.

The goals of the company were somewhat different from the goals of the Sultan, who hoped to leverage the firm as a means of extending *his* sovereign prerogatives—particularly the right to tax—further inland. Bin Said was not being naïve in maintaining these aspirations. Kirk had promised that, in entering into agreements with the company, the Sultan would increase "his own authority . . . by opening up the country to civilisation and commerce."[94] But bin Said was standing on a knife's edge. Rather than being the vector of sovereignty and patron of the company, he risked becoming a dependent client of the corporate-state and its European backer.

Multiple modes of valuation and readings of the relationship between political authority and space were also at play outside of the circumscribed realm of high politics and beyond the 10-mile radius on the mainland that constituted the Sultan's domains. By the time the firm began its work in the 1870s, Kamba traders had asserted themselves as central commercial actors in the region. Well-organized caravans of Kamba porters transported ivory to the coast, returning inland with commodities they traded with neighboring communities.[95] As J. R. Cummings notes, "Out of this relationship evolved an international

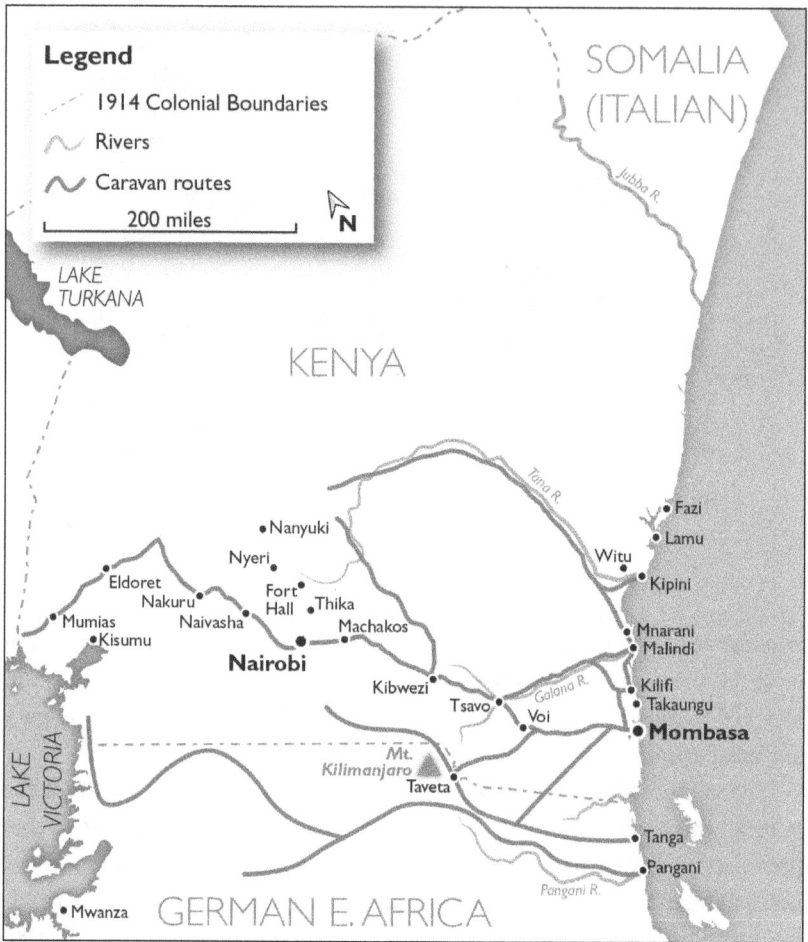

Map of nineteenth-century caravan routes through present-day Kenya

export trade which grafted itself onto the pre-existing interregional trade and transport system."[96]

Historically, participation in this trade had conferred considerable political authority. Councils of elders—composed of men, women, and "seers"—sanctioned the caravans, offering guidance on routes to take and providing ritual protection to porters.[97] Village communities were governed by the trading authority of chiefs who had "a mobile force of hunters and warriors" at their command, and followers "willing to undertake long journeys for economic rewards."[98] Wealth gained through trade was widely distributed, with professional porters ensuring their position by sharing coastal goods with followers and sending "baskets filled with coastal products to powerful local leaders elsewhere."[99] For these communities, as for the Sultan, commerce and politics were deeply interwoven and were sustained by durable relations of debt and deference. But this was a moment of flux. Before the mid-nineteenth century, Kamba traders moved coastal influence to Ukambani. However, by mid-century, as coastal traders, European merchants, and the Sultan began expressing increasing interest in inland commerce they began eroding Kamba dominance of inland trade.[100]

As they moved through the countryside, administrators confronted this lively and changing social topography. And people brought their own ideas regarding the relationship between the production of cultured space and sovereign authority to bear as they interpreted the IBEA's presence. To get at these other grammars of sovereignty, we must follow these company men as they tried to assert their sovereign authority, a protracted process revealed daily as they moved across the territory. The reflections of John Ainsworth, a British employee of the IBEA, who became the colony's first Native Commissioner in 1918, offer insights into the layered sovereignties operative in the region.[101] Reflecting on the grammars of authority that governed the territories that Kamba, Maasai, and Kikuyu communities called home, Ainsworth wrote that "wherever possible . . . communities left considerable strips of land unoccupied between themselves up to 30 miles in width, divided the A'Kamba from the A'Kikuyu."[102] Ainsworth presented these as unplanned spaces, writing that "these neutral zones were never the result of any mutual arrangement, they came about through the natural hostility of one people to another."[103] This representation was premised on a misrecognition.

Unlike the supposedly empty and homogenous space of capital, not all spaces were equally open to human intervention.[104] These uninhabited zones were popularly held to be spaces that fell outside of proprietary claims, but this was not tantamount to their being asocial or without utility. As Ainsworth noted at the end of the report: "[T]hese areas were in no way looked upon as lands belonging to any of the tribes concerned."[105] If company administrators imagined roads as lines of communication, these empty spaces, too, were zones of communication; spaces that were set out and protected in a bid to ensure cordial relations *among* people—to ensure that lines of communication were not crossed.

And these spaces were not always empty. Shifting alliances, the ebb and flow of seasons, and periodic drought, famine, and disease meant that sometimes these spaces were peopled, at other times left empty.[106] Such was the case of the zone dividing what would come to be called Ukambani and Maasailand:

The road from the shangi to the Wambikati passes [through] open and fairly easy country. The natives say that it was once densely populated, and the whole had been cleared and cultivated with good success until the time of the great famine which devastated and depopulated Ukambani some seven years ago. After this it was again peopled by a large influx of Masai during the period of the cattle disease . . . an appreciable number have settled down to cultivation with the Kamba. I have met them, and [find] them to be quite reconciled to the change . . . They are in communication with their own tribe and act as middle-men in procuring supplies for it.[107]

Administrative representational practices notwithstanding, self-identification was historically defined by mode of production, kinship, experience, expertise, and the virtues ascribed to them, not by some abstract notion of "tribe."[108] And these zones were routinely crisscrossed materially and ideationally by robust networks of trade and durable practices of kinshipping. "It should be borne in mind," Ainsworth wrote, "that native law and custom usually provides for natives of any other tribe being received into and adopted by another tribe on the stranger complying with certain formalities and paying his footing."[109] These zones were deliberative but permeable gaps mediated by dynamics of exchange and deference that reflected the shifting and layered sovereignties governing the region's spatial order. These epistemologies of space and social life—both shifting and durable—shaped popular responses to the presence of IBEA agents roving through the region.

Company administrators had to conform to these boundaries of social life and sovereignty. Company employees' movement inland was authorized by the writ of the Sultan. One such letter read: "Be it known . . . Englishmen are going into your country. They wish to make roads and to trade and to do you good. . . . these men are going with our good wishes."[110] In mimicking existing repertoires of authority, company officials tried to impress on local leaders that they "came to administer . . . posts as a representative[s] of the Sultan," placing themselves "in the position of a son to the old and a brother to the young."[111] The use of kinship metaphors was not incidental. The IBEA sought to establish relations of intimacy that appeared to mimic existing dynamics of deference and authority, while incrementally centralizing political and fiscal authority in corporate hands.

Company officials quickly learned, though, that the authority of the Sultan was thin on the ground. As one moved away from Zanzibar, his claims to sovereignty waned, giving way to other visions of distributed authority. Regarding Kikuyuland, Count Teleki reported that each "ridge" was "virtually an independent sovereignty," this authority being premised on the claims of

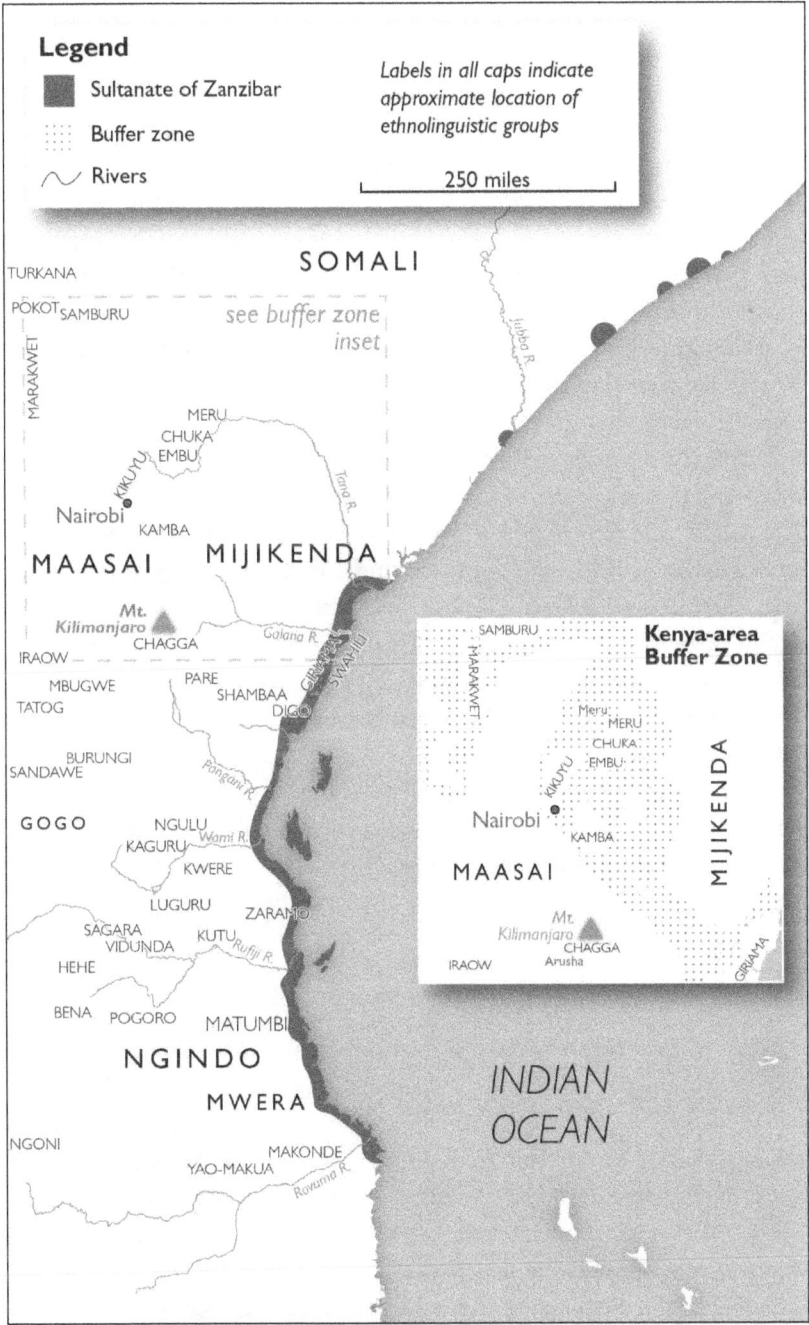

Legend

▪ Sultanate of Zanzibar

⋮ Buffer zone

∿ Rivers

Labels in all caps indicate approximate location of ethnolinguistic groups

250 miles

TURKANA

SOMALI

POKOT SAMBURU

see buffer zone inset

MARAKWET

MERU
CHUKA
EMBU

KIKUYU

Nairobi
KAMBA

MAASAI

MIJIKENDA

Mt.
Kilimanjaro
CHAGGA

IRAOW

MBUGWE
TATOG

PARE
SHAMBAA
DIGO

Golana R.

GIRYAMA

SWAHILI

BURUNGI
SANDAWE

Pongara R.

GOGO

NGULU
KAGURU

Wami R.

KWERE

LUGURU

ZARAMO

SAGARA
VIDUNDA

KUTU

Rufiji R.

HEHE

BENA POGORO

MATUMBI

NGINDO

MWERA

NGONI

MAKONDE

YAO-MAKUA

Rovuma R.

INDIAN
OCEAN

Jubba R.

Tana R.

Kenya-area Buffer Zone

SAMBURU

MARAKWET

MERU
CHUKA
EMBU

KIKUYU

Nairobi
KAMBA

MAASAI

MIJIKENDA

Mt.
Kilimanjaro
CHAGGA
Arusha

IRAOW

GIRYAMA

Map of colonial-era social buffer zones

"seigniories along the ridges," each of whom demanded "tribute."[112] The Kamba similarly demanded *koti* in a bid to keep aspiring traders out of their corridors.[113] By contrast to the IBEA, this conception of sovereign authority was not grounded in property, for people did not "so much consider the land as belonging to them, as that they belong to the land."[114]

Negotiating the reallocation of sovereign prerogatives from these local "sultans"—rarely an easy task—was thus critical to the labors of the corporate-state.[115] And IBEA administrators often failed to convince local leaders that the Sultan had indeed vested them with his authority. Absent the Sultan's letter, local authorities often demanded "a lot of money" in exchange for allowing the IBEA to cut roads through their territories, arguing that it was "strange" hubris that officials came to their regions with "no letter" in hand.[116] Other leaders were critical of the company for arriving with "no representative of the Sultan."[117] Others still received the letter, but "ignored it altogether."[118] Indeed, the Sultan's authority was not recognized by many on the mainland, it being largely circumscribed to the "10-mile zone" along the shore and extending "only a comparatively short distance inland."[119] The Sultan's authorization, while a crucial first step in "materially . . . dealing with these chiefs," was evidently not sufficient, forcing company administrators to concede to the demands of local sovereigns.

If company men were not received as bearers of the Sultan's authority and negotiations appeared on the verge of foundering, company officials turned to the threat of violence in the face of what they claimed were extortionate demands.[120] Making recourse to violence was not desirable, however. The IBEA was a quasi-sovereign, but not one with a monopoly on the use of force, a claim which would have been tantamount to "usurping the Sultan's authority."[121]

Debt and Deference in Two Registers

In pursuing sovereign authority, and in keeping with long-standing expectations of debt and deference, the IBEA routinely distributed "gifts" to local leaders. To secure the good favor of the Sultanate, the IBEA presented bin Said's successor, Sayyid Khalifa I bin Said Al-Busaid, with a maxim gun delivered "on behalf of the Directors." The provenance of these items was important, with George Mackenzie recommending a gunmaker on Edgware Road who had "much experience in doing up guns . . . for the Persian Chiefs," knowing the "peculiarity of fixing the sling as they like it."[122] The IBEA was routinely forced to conform to local expectations of quality and taste.[123] Novel but ambiguous relations that tethered the company to the Sultanate were marked through a simple inscription: "from the I.B.E.A Company."

These dynamics shaped relationships from the top of the political hierarchy to the distributed and overlapping tiers of authority structuring space and social life in the countryside. The Sultanate maintained a vast web of clients to whom it regularly distributed favors, gifts, and cash.[124] As bin Said extended his authority along the coast in the 1880s, this group of clients grew, requiring the disbursement of payments to local rulers in exchange for their fealty.[125] In distributing gifts, company officials hoped to build on extant practices that characterized the transfer of power, as it was the custom when "a new Sultan comes to the throne that all the Liwalis, chiefs and Elders accept certain presents from him as a mark of submission to his rule." In mimicking the Sultan's customs, the IBEA hoped to "more clearly impress upon the people and tribes the nature of their relations [to the Company which] . . . had assumed the Govt. of their country."[126] In making these offerings, the IBEA was forced to confront existing frameworks of sovereignty that laid out in their own terms a vision of the relationship between space and authority, between debt and deference, that was itself patchwork and layered.[127]

For roads to materialize corporate sovereignty, however, gifts would have to function as infrastructural attachments, tethering people and their futures to the IBEA and its networks of marketization. And the Sultan was not alone in demonstrating a desire for roads, with some reports cheerfully recounting that "the natives . . . seem very pleased about the prospect of the road passing through their country."[128] But people were discriminating, some "chiefs" making clear that they did not "want the road nor us to come near there."[129] Company officials were forced to acknowledge that people brought their own ideas to bear as they evaluated the relative merits of infrastructures that company officials proposed to cut through their lands.

The region was indeed a busy and competitive world of valuation, requiring company men to discern people's tastes in each settlement through which they planned to lay new roads. As one report recommended: "You should carefully ascertain and advise me of the basis on which hongo [tribute or honorific] is charged at each point by the various tribes and the amount and description of each article levied."[130] This practice mimicked historic expectations that caravans journeying inland pay tribute to leaders as an acknowledgement of their authority.[131] If roads were to be built, the men of the IBEA would have to conform to local expectations as they worked to "secure the good will of the native chiefs who claim[ed] to rule the country."[132]

Company men struggled to understand the different metrics of value that extended throughout the territory. In some periods, gold watches, guns, money, and silver were in high demand.[133] In other instances, beads of specified size and color, and cotton cloth were at a premium.[134] Tribute and gifts were often augmented by the mixture of bodily fluids, with blood brotherhood between company officials and local "chiefs" accepted as rites critical to securing relations of trust and mobility.[135] In this, too, men of the IBEA built on practices that governed the caravan trade, blood brotherhood historically ensuring travelers' movement through the region without incident.[136] Territorial boundaries, the limits of bodies, and ideas of sovereignty were porous. Distributed sovereignty, itself the logic underwriting the IBEA's incremental expansion in the region, was the standard form of action.[137]

The Politics of Payment in Two Registers

Company officials struggled to establish a monopoly over the meaning of the material transfers requisite for the operations of the corporate-state. In entering into these relations, company officials worked to ensure that gifts were not read by local "sultans" as inaugurating a dynamic of routine payment, that they did not constitute the company acceding to regimes of sovereign seizure—namely the payment of tribute or *hongo*—operative in the countryside. These material offerings were *Ada*, or "customary gifts," that would be paid out once. And for these *Ada*, local leaders ought to be grateful, for the IBEA's real gift came in the form of money the company deployed for the "good of the country."[138] Put simply, company officials worked to impress on leaders that these exchanges were not evidence of their deference to local sovereigns, that company officials were patrons rather than clients. These were, at base, struggles over valuation.

Gifts were not freely given, then, but were to act as infrastructural attachments, debts tethering leaders to the company and its roads. In exchange, local interlocutors were to begin repayment by commencing the "preliminary work of clearing the jungle" in preparation for "road-making." The return of labor was critical, it being a means of avoiding such payments being viewed as local leaders levying "blackmail" on the company.[139] Incrementally, company officials forecasted, their sovereign authority would be recognized in spatial terms, radiating out from the nascent infrastructural networks that they hoped local leaders would take a lead in conscripting people to construct.

Word-of-mouth guarantees were not sufficient for the men of the IBEA. Face-to-face agreements were backed by treaties which company officials claimed were "written out in the various native dialects."[140] Those who were willing to accept the IBEA's status as delegated sovereign were to mark this transition by flying company flags—visual texts that semiotically matched the treaties used to underwrite them.[141] Pursuing emulation as a tactic of centralization bore fruit for the IBEA. By July 1887, Mackenzie had secured twenty-one "treaties of protection" from local leaders from Taita country up to Taveta.[142] General Matthews, for his part, was credited with negotiating twenty-one treaties on behalf of the Sultan and the firm along the coast and inland.[143] These treaties, too, functioned as infrastructural attachments mobilized "for the specific purpose of opening up a route through which caravans might safely pass."[144] This process was repeated myriad times as IBEA officials moved inland.

The exchange of currencies and gifts for treaties in the countryside marked an extension of the peculiar logic that saw the company exchange rents for the concession. In the hands of the IBEA, treaties operated as commercial instruments signaling the marketization of sovereignty.[145] But treaties were geared toward marketization of a much more thoroughgoing kind. In the minds of company men, gifts of *Ada* were exchanged for land, transforming the topographical landscape of the region and undermining the layered sovereignties that had historically governed spatial practices. As one report noted, the sum spent on gifts was "not extravagant" given the "enormous tracts of land . . . it practically gives the Company . . . freehold rights over." This, too, was part and parcel of the civilizing arm of the commerce and civilization dyad, with this official confidently writing that there was no "doubt of the beneficial influence" this process of expropriation "exercised over the entire native mind."[146]

The world of contract that administrators sought to impose contravened historical dynamics of debt and deference in a region where a "debt on land . . . [could] never be finished." Instead of "hard and fast arrangements," the exchange of goods and livestock for tillage rights had historically kept "alive the sense of mutual obligation" among the parties, with the seller "entitled to receive goats" from the buyer, a privilege that extended to his descendants. These arrangements were geared toward temporal extensions that bound the futures of participants in ongoing dynamics of debt and deference.[147]

If these expectations mitigated against the permanent transfer of land, they also offered a vision of "public" and "use" that was at odds with the vision IBEA officials brought to bear as they engaged in this process of expropriation. As one report noted, "All the lands not under cultivations are public lands and are the property of the Company the Chiefs having already entered into Treaties with them to sell." Gifts of *Ada* were essential to these supposed transfers of land. Other visions of the relationship between authority and use were, of course, at play. One such vision guided historical spatial practices in the unpeopled territories dividing the Kamba from the Kikuyu. According to this epistemology of space and social life, the absence of cultivation did not signal these as unused spaces, but indicated that no group held proprietary claims over these corridors. For the company, by contrast, the absence of cultivation marked these as "public" spaces and thus spaces subject to expropriation by the nominally private firm.

The exchange of money and goods for sovereign prerogatives and land was part and parcel of these labors of marketization, and roadbuilding was a key weapon in the company's arsenal. As one report read, "If the Sultan . . . would grant us land on each side of the road . . . we might be able to do much in the way of spreading civilising influences by getting people to settle on these strips . . . who would probably begin to cultivate the land where they knew they could count on protection and felt that they had a good road to a ready market."[148] Officials, evidently, did not conceive of roads simply as disenchanted routes of commerce. They were capable of "spreading civilising influences," both in their very material form and as vectors of marketization.[149] New road networks were freighted with ideology, their very materiality capable of serving both a critical pedagogic, and an essential material function—civilizational uplift and market-making came together along the road.[150]

Confident though it was in the wisdom of corporate sovereignty, from the outset the IBEA was under considerable strain. These tensions were augmented as the IBEA faced competition from the sovereigns that sanctioned its operations. These conflicts, too, centered on contested understandings of the relationship between debt and deference, and tournaments over the relationship between sovereignty and seizure sat at the heart of the disputes.

The Limits of Divisible Sovereignty and Battles over Sovereign Seizure

By the late 1880s, it became clear that other distributions of sovereignty subtending emergent visions of territorial authority were undermining the IBEA's status as a corporate sovereign. The ambiguous place of the coast in Britain's imperial portfolio, specifically its connections to the Great Lakes region, was at the center of these problems. By the 1880s, Britain's position in Uganda was under threat from German operatives. The British "nation" turned to the IBEA "as the agency whose duty it was to guard the national interests in Uganda."[151] In 1888, the Crown tasked the Consul-General of Zanzibar, Charles Euan-Smith, with opening "up friendly relations with Uganda," which he did, sending a letter to Kabaka Mwanga (1884–1888; 1889–1897). This letter assured the Kabaka that the operations of the IBEA were sanctioned by the Sultan. Company officials hoped that emphasizing their role as the Sultan's "plenipotentiaries" would enable the company to be looked on favorably by Mwanga, who harbored a healthy suspicion of Christian missionaries who had made inroads by the 1870s.[152] In mobilizing the authority of the Sultan, the British also hoped to win over "Arab traders" in the interlacustrine region. In this world of divisible sovereignty, this was sensible, the Sultan being both "their Sovereign and co-religionist."[153]

This was a tumultuous period in the Kingdom, and the Kabaka came to view outside forces—both Christian and Muslim—as a threat to his authority.[154] British intervention seemed wise, leading Mwanga to appeal to the IBEA for help. Critics of the IBEA claimed that this responsibility was implicit in the terms of the charter. Aghast, the IBEA worked to separate its role as a corporation from its role as a sovereign. As the company historian wrote, the "fallacy underlying all the arguments and assumptions as to the responsibility of the . . . Company . . . was that the immediate interests of the Company were identical with those of the nation. Nothing could be more specious."[155] Nevertheless, and under the pressure of public opinion, in 1890 Lugard traveled to Buganda and secured a treaty with Kabaka Mwanga, which guaranteed British "protection" over the Kingdom.[156]

Enacting British power in Uganda was an expensive undertaking, the IBEA claiming that securing the "national interest" had eaten up the lion's share of its initial capital of £425,000, costing the company £220,000.[157] If the company was being asked to enact administration through the mobilization of private capital, they argued, should the treasury not subsidize private capital investments to secure the Crown's "sphere of influence" when investors' commercial interests were not obviously at stake? The line dividing the sphere of the political from the sphere of the economic was, evidently, very thin.

Infrastructures occupied center stage in this debate. After years of celebrating the merits of roads, it had become clear that a railway was required to consolidate British dominance.[158] Mackinnon thus requested assistance from the government, "either in the shape of a guarantee of 3 per cent or a grant to the extent of 30,000 to 40,000 pounds a year for 25 years," to fund the construction of a railway from the coast to Lake Victoria.[159] With this security guaranteed, the cash-strapped firm "would have no difficulty in finding as much additional Capital as may be necessary for the general purposes of Administration and the development of an Enterprise of National importance, largely advantageous to Imperial interests."[160] Parliament rejected this request, which, the company claimed, made it impossible for the IBEA to raise capital from the "investing public" who read Parliament's refusal as a sign that the government did not intend to "stand by the company."[161]

While the IBEA had worked to establish itself as the dominant sovereign in the region, these commitments were malleable. If the prospect of profit-making was undermined, so too was the company's commitment to territorial sovereignty. This was an issue of no mean importance. In August 1889, three-quarters of a million worth of shares were issued to the public, only one-third of which were purchased.[162] The government's refusal to offer its support as a political security backing financial investments had evidently eroded investors' confidence. Without government intervention, the IBEA threatened, the company would be forced to withdraw from Uganda. Relations with the government were poised to get worse. In 1890, Britain and Germany signed the Anglo-German Agreement, which stipulated that Germany relinquish its claims on Zanzibar.[163] Britain promptly declared Zanzibar a protectorate, hoping to use this new position as a vantage from which to link the coast to the Great Lakes region. These new arrangements swiftly shifted the balance of power away from the IBEA and its headquarters in Mombasa.[164]

If the IBEA's status as a corporate sovereign was being undermined from the metropole, so too was its status as a territorial sovereign being challenged on the coast, as disputes between the company and the Sultanate over the status of sovereign seizure made clear. In 1891, the company appealed to Consul General Gerald Portal, charging that Sultan Ali bin Said, successor to Sultan Bargash bin Said, owed the company Rs. 89,801.90.[165] This claim was founded on the accusation that the Sultan's officials had collected duties on goods exported from ports leased by the company. In undertaking these collections and "depriving the Company of one of the sources of revenue from which the Company's annual payment to the Sultan" was derived, the Sultan had "broken his agreement with the Company." In turn, the IBEA withheld rents owed as collateral securitizing the Sultan's debts.[166] In doing so, the company overturned historical relations of debt and deference that had characterized the dynamic between the Sultanate and its Indian financiers. This the Sultan could not accept.

The Sultan countered that the company had no right to claim drawbacks and demanded that the company pay the Sultanate outstanding rents. While Portal conceded that the company was entitled to drawbacks, he concluded that it was the company, not the Sultan, that was the debtor, with the IBEA owing the Sultan an outstanding Rs. 50,410.11. Portal castigated the IBEA for "prejudging their claims in their own favour" by withholding rents it owed the Sultan. This action, he wrote, was in "danger of being misinterpreted as implying a distrust of the Sultan and his government, even though the Company are aware that, since the declaration of the English Protectorate ... the Sultan, his Government, and his finances, have been under the immediate and direct control of the English Consul General."[167] In declaring Zanzibar a protectorate, the government had effectively shunted the IBEA aside, asserting itself as the *de jure* political and fiscal authority in the region.

The longer-term territorial goals of the British government increasingly came into conflict with the short-term, returns-based logic of the company. So long as the IBEA retained the right to administer the coastal strip, it retained a monopoly over not only transport and communications connecting inland regions to the coast, but also over import and export duties on goods passing through its ports. Once the government had declared Zanzibar a protectorate, the corporate-state posed a direct threat to the prospective profitability of formal colonial occupation. Conflicts over the status of sovereign seizure were central to these debates. As Portal wrote, "If Zanzibar takes over the interior and Uganda . . . and the English Company continues to hold the coast ports under their concession from the late Sultan, one of the two must starve. We, i.e. Zanzibar could not consent to spend money in the interior . . . simply in order to increase the customs receipts of the Company—that system would run the Zanzibar exchequer dry very soon."[168] This battle pitted an imperial against a corporate sovereign, and central to these conflicts was the prospect of profit.

For home audiences, at issue was how the IBEA's status as a corporate sovereign related to its right to engage in sovereign seizure. As one vitriolic article in *The Economist* read, the IBEA's threat to withdraw from Uganda was tantamount to proposing to "keep the port and the coast, but surrender the far interior to the British taxpayer."[169] The article continued, roundly challenging the merits of company rule. In the event of the IBEA's withdrawal, a British capitalist wanting "to do business there . . . must cut the roads. If he desires to prevent massacres he must keep an armed force. If he wishes to hold the Lake he must provide the steamers. And then . . . the Company will graciously take the profit of them in the shape of duties levied on all goods at the port of import."[170]

The IBEA was shrugging off its commitment to the national interest—or more accurately, its commitment to the interests of British capital. Worst of all, these peculiar arrangements were designed to secure the IBEA's profits alone, with the "responsibility of the Government" being "as if the Company did not exist, but the means out of which that responsibility ought to be met are to be reserved for the Company's dividends."[171] Ironically, however, and missed by this critic, the company's request that the government mobilize the public hoard to shore up the corporate-state was the result of the IBEA's increasingly circumscribed ability to exact sovereign seizure within eastern Africa itself.

The IBEA's already compromised position was facing challenges from all directions. In 1892, the company charged the Crown with putting its relations with France and Germany ahead of its commitments to the company. The Foreign Office (FO) had, it claimed, "in an act of unjustifiable aggression" placed the company's possessions under the provisions of the Berlin Act, which proclaimed the entire region a free trade zone.[172] This deal had been struck without the consultation of the IBEA. While the Sultan had accepted the terms of the agreement, the IBEA continued, he had only done so on the promise that the principle of commercial liberty would not apply in his domains, with the Sultanate retaining "fiscal independence." The "principles of free trade," it claimed, simply did not apply to the Sultan's territories.[173] The company argued that this "fiscal independence" had passed to the company through the Concessions of 1887 and 1888, which granted the firm the right to levy customs and taxes, prerogatives the Sultan had "ample right to assign."[174] In signing a "contract" with the IBEA, the Sultan had effectively ceded his sovereignty to the company, abrogating his right to sign treaties without the company's approval.[175]

Mobilizing the law, the FO argued that the concession was not a "contract," but merely authorized the "delegation" of the Sultan's sovereign authority to the IBEA.[176] No legally binding agreement had been breached. The company's relationship to the Crown that granted its charter, while ambiguous in theory, was proving to be less ambiguous in practice.[177] At stake here was the status of sovereign seizure. Were the free zone recognized in all the territories over which the IBEA had delegated authority, the company would lose the 5 percent import duty and tariff, as well as export duties.[178]

The company angrily enumerated the services it had provided as corporate sovereign and vector of imperial power. The screed sought to remind "Lord Salisbury that the Company has really been doing Imperial work with private funds; that it has facilitated the introduction of a British Sphere of Influence in Eastern Africa which otherwise might have fallen wholly to the Germans; that it has provided an instrument by which Government may, through a Chartered Company give effect to . . . their anti-slavery policy . . . [At] considerable outlay, [it had] explored, mapped, and reported on wide regions in the interior . . . and that it has done all this without any pecuniary aid from Government." Seizure was central to the vision of sovereignty mobilized by the company to bolster its claims against the government. Without being permitted to raise taxes, "universally recognized as essential to the maintenance of any and every form of administration," the government had fundamentally undermined both the sovereign authority of the IBEA and its solvency as a corporation.[179]

The findings of the solicitors were conclusive. Legal action could only be taken against the British government in the form of a "petition of right." In this case, they argued, there was "not any contract between the Crown and the Company for breach of which a petition of right could be successfully maintained."[180] As for the Sultan, the solicitors conceded that in "depriving the Company of one of the sources of revenue from which the Company's annual payment to the Sultan" was to be made, the Sultan had "broken his agreement with the Company." While this was an offense for which the Sultan could be sued in the United Kingdom, this was not feasible in the present case. The issues were jurisdictional, reflecting the profound uncertainties generated by divisible sovereignty in this period. The Sultan had created a court consisting of the British Consul and an assessor; however, this "instrument" did not "have any jurisdiction over the Sultan."[181]

The IBEA charged that debts were owed, for "in performing the services for the Empire," the company had established itself as "a creditor of the nation." The IBEA's occupation of Uganda was central to this argument, which, the directors claimed, had been undertaken not in their own interest, but in response to pressure from the government and public opinion. Indemnities were owed. Contesting the finer points, the government acknowledged that the directors' claim was fair. Compensation "on the basis of ten shillings and sixpence in the pound" would be paid to the IBEA, equivalent as it was to roughly two-thirds of shareholders' investments.[182] In total, the British bought the company, including its assets, for £250,000.[183] And so, in 1895, the company was recompensed for the surrender of its concession and charter.

In a move that will sound familiar to readers today, these debts were socialized, as tax revenues were mobilized to offset the costs required to purchase the assets of the faltering corporation. Rather than turn to British taxpayers, the government seized money from the Zanzibar Treasury, £200,000 being paid by the Sultan for the purchase of the concession and the company's property—roads included—with Parliament paying £50,000 as compensation for its charter rights and for the treaties the IBEA had secured with "native Chiefs."[184] It had been the Sultan who had granted the company his sovereign prerogatives, so too would it be the Sultan, whose territories were now formally under the control of the British, who would pay to buy back his increasingly nominal sovereign authority.

Conclusion

By the 1890s, both territorial sovereigns had evidently reached their limits. While they had apportioned their sovereign prerogatives to the company in the form of the charter and the concession, each came to see corporate sovereignty, specifically the right to engage in sovereign seizure, as a threat to their own political and fiscal aspirations. Without the right to tax, the corporate-state simply could not fund its operations—road-building among these—and it certainly was unable to deliver promised returns to investors. If limitations in the state fiscus had enabled the emergence of the corporate-state, it was those same limits, enforced on the company by the British government and the Sultan, that heralded the IBEA's unraveling.

But the corporation's failures ran deeper than this. Indeed, the infrastructure firm's claims notwithstanding, the IBEA was in truth not very successful in bringing what it termed "commerce and civilisation" to the region by constructing road networks. As a British parliamentarian wrote of the region in 1897: "The only road, in the true sense of the term, in the East African [sic] Protectorate is the great caravan road to Uganda." "Uganda Road" consisted of "Mackinnon Road," from Mazeras to Kibwezi, and "Sclater Road," from Kibwezi to Kedong and from there to Lake Victoria.[185] Those roads that the IBEA did construct were perennially on the brink of disappearing. Frequently, the region's flora simply "killed the road," as a supposedly common saying put it.[186] As for "Mackinnon Road," the road with which we began, the British Government purchased the route, which had largely been paid for by Mackinnon himself, from his heirs for £3,431—about KSH 210 per kilometer or twenty cents per meter, "something of a bargain"—on the eve of the company's withdrawal from the region.[187] Why this road was so cheap had everything to do with the IBEA's labor regime, as the next chapter explores.

The collapse of the IBEA led to the formal takeover of the administration of the region by the Foreign Office. While "sound finance" had prompted the Crown's sanctioning of the firm from the outset, the government's takeover did not mark an end to imperial austerity, but rather its deepening, as the Crown worked to recover monies sunk into the region's infrastructures from African communities in the form of direct taxation (see chapter 2). Put simply, this framework of austere governance would be formalized and scaled up with the onset of formal colonial occupation, as the colonial state pursued the imperial corollary of "sound finance," colonial "self-sufficiency," and set to work constructing the routes of "commerce and civilisation."

2

The Politics of Valuation

Building Attachments, "Taxing"
Infrastructures, and Transforming
Expert Work into Labor

In 1892, Ernest Berkeley, an official working for the Imperial British East Africa Company (IBEA), wrote to his superiors regarding the relationship among infrastructures, sovereign authority, and seizure, or what he trepidatiously referred to as "tax." "I am not prepared at present," he wrote, "to actually lay down a positive opinion regarding the levying of taxes on shambas [farms] situated along the road . . . The Natives already inhabiting the track through which the road is intended to run would probably in their present undeveloped state, fail to recognize in it such advantages as would warrant in their eyes the levying of a tax from them, and it appears doubtful whether the presence of the road would induce others to settle in its vicinity when they became aware that this situation entailed taxation."[1]

As chapter 1 explored, gaining the sovereign right to seize taxes was at the very center of the corporate-state's mission. As this chapter shows, however, getting East African communities to accede to the company's *de jure*, let alone *de facto*, claims to have gained a monopoly over administrative and fiscal authority proved to be no easy task. These transformations would require a revolution not simply in regimes of sovereignty, but in regimes of work as well—processes which turned on the company and later the colonial state struggling to gain a monopoly over valuation.[2]

These conflicts over valuation largely played out on the road, the site where the dual projects of commerce and civilization, profit-making and securing the commonweal, came into most direct conflict. These contests were the outgrowth of a central paradox structuring the place of roads in the imperial imaginary. Africans living in the region were to not only benefit from the existence of new infrastructures but also from the new regimes of labor and taxation required to build them, both of which were held to have a civilizing effect on people and the landscapes they called home.

Confident proclamations of roads as bearers of "civilisation" notwithstanding, roadwork was difficult work. And in the early years, metropolitan surveyors and engineers leaned heavily on African topographical expertise—namely African knowledge-workers' intimate understanding of the natural-, animal-, and human-made characteristics of the region. This prosthetic infrastructural work was transformative, critically shaping the emergent colonial transportation network. Administrators, though, effaced the central role played by African interlocutors, shoring up a pernicious but durable vision of Africa as a place without infrastructural expertise.[3] However, people who called the region home could not easily be induced to labor on these new routes of "commerce and civilisation," which demanded a form of exertion that was at odds with prevailing theories of both civilization and valuable work. While these epistemologies of space and social life shared a vision of culture as nature reformed, their goals in culturing space, and the spaces subject to this transformative work, were often out of sync. And people brought their own ideas regarding the relationship between the production of cultured spaces and work to bear, often rejecting the new labor regime that roadwork portended.

At this conjecture, company administrators tried to mobilize new fiscal technologies: a company currency, on the one hand, and new regimes of sovereign seizure—variously referred to as a "toll," a "tax," and a "levy"—on the other. A company currency, they hoped, could consolidate the IBEA's status as *the* sovereign authority in the region, while new taxation regimes would increase company revenues and form the basis of corporate profits. Working in tandem, they forecasted that these fiscal and administrative forms would operate as infrastructural attachments, creating a need for currency in what had been a cash-scarce economy, and binding people as labor to the road and the wage as they struggled to "find their tax."[4] Put simply, as the corporate-state mobilized African topographical expertise as the basis for accumulation, it concurrently tried to transform this value-generating work into physically taxing and rote infrastructural labor.[5]

These contradictions would not end when the Foreign Office took over the administration of the region, but instead became the foundational logic of the early colonial state. In what would become the East Africa Protectorate, infrastructural development was increasingly underwritten by economies of coercion, violence, and ecological disruption that forced people to take to the road, entering the wage-labor economy—often as laborers for the Public Works Department—in order to raise their taxes.[6] Thus, infrastructural labor was both the prerequisite for the generation of new markets, creating the networks on which commodity export depended, *and* infrastructural labor was a means of generating a revenue pool through taxation. And taxes seized from Africans—in both money and labor time—functioned as subsidies to the infrastructural needs of the white settler class, materially shoring up the racial hierarchies that structured the Kenya colony. Evidently, securing the commonweal turned on mobilizing new forms of coercion, *while* delivering African communities "commerce and civilisation."[7] These lofty ambitions notwithstanding, the colonial state continued to struggle to gain a monopoly over valuation, finding many uninterested in entering the money economy, preferring instead to retain other media of value and alternative notions of self-actualization. Extant grammars, administrators found, were remarkably durable.

Expert Distributions in Two Registers

European merchant adventurers arrived in eastern Africa armed with a hubristic vision of their technopolitical prowess. These technofetishistic fantasies underwrote the dream of infrastructural connectivity widely celebrated by the IBEA. Roads, administrators claimed, were not only essential to "open up this unknown region" to "legitimate commerce," but the labor regimes required to build them would move people forward in civilizational time.[8] The mobilization of the latest tools in civil engineering were key to this arsenal. If the road—both as a material form and as a sign—was to have civilizing effects, trigonometrically aided visions were required to construct right-angled roads with wide road reserves boasting clean lines, all "properly" graded and surfaced.[9] It is no surprise, then, that on the heels of gaining the concession, George Mackenzie proclaimed that "one of the first important essentials for the Company" was the appointment of a "thoroughly competent Engineer who should bring out with him instruments suitable" for constructing "the line[s] of the route[s]," for "roadmaking" was the *"sine qua non* of development."[10]

While company men were confident that the construction of new roads would be generative of civilizational uplift and the creation of new markets, they were forced to concede that they were confronting a historically constituted topography boasting extensive networks of mobility.[11] As Frederick Lugard reported as he moved inland, there was "always a path," whether cut by "game" or by "man."[12] Those "network[s] of paths" that were the product of human intervention tended to wend their way across the territory "in every direction," accommodating themselves to the undulating hills and valleys that characterized inland areas, and offering protection against dangerous others.[13]

The very form of "native paths" subjected them to scrutiny, with company officials routinely dismissing them as inefficiently "circuitous"—while "narrow path[s]" were well-suited "for one" or caravans marching single file, they argued that such paths were poorly suited to new technologies of mobility that would traverse their surfaces.[14] What was needed were roads that offered the most direct connection between two points, roads that cut "straight through the forest."[15] The rationalization of space that underpinned these universalist pretensions of infrastructures were difficult to enact in practice. Indeed, the knowledge of savvy engineers trained in the metropole and armed with technologies of surveying, often proved to be of limited utility in eastern Africa; material conditions on the ground portending different infrastructural futures.

Engineers and surveyors' limited knowledge of eastern Africa's topography was a perennial issue for metropolitan experts. The Chief Engineer, Lt. Swayne R. S., routinely reported that his survey party had been forced to abandon routes that initially appeared promising. In some instances, grading was too difficult across the steep hills and deep valleys given the quality of the soil.[16] In other instances, prospective routes were obstructed by waterways. Engineers also lamented discovering chains of "spurs," "numerous swamps," and "intervening valleys" that marred prospective routes.[17] It was not that Europeans had come unprepared. As Lugard recounted, the savvy surveyor moved through the countryside with a "haversack" filled with "a small file, a spare foresight, a bit of bee's-wax, a measuring tape, the road-book for surveying . . . a bit of whip-cord, a tiny bit of chalk, [and] a small screw-driver."[18] However, eastern Africa's topographical and social characteristics demanded expertise of a different sort, requiring the mobilization of African infrastructural workers whose knowledge of the region often proved more valuable than the expertise of metropolitan engineers and surveyors.

European surveyors and engineers routinely cited the role played by "native guides"; however, they framed this work as being of negligible import. When guides abandoned survey parties, engineers confidently reported that they were able to manage "perfectly without them."[19] When the importance of guides was acknowledged, it was mainly for the role they played as social lubricants. As one report noted, the absence of a guide could render marches "useless," as parties were unable to pass through "principal villages where we should have endeavored to make friends."[20] On this framing, the work of these men was the social and cultural work of translation. As Lugard wrote: "Since Swahilis are of various tribes of the interior there is never any difficulty in finding interpreters in the various countries passed through, and even though you may not have a man of the tribe in your caravan, there are sure to be several who have picked up a little of the language on some former trip."[21] These linguistic skills were born of people's experiences working on the caravans that moved between the coast and inland regions.

The social work of guides was critical to securing the mobility of survey parties as they negotiated the terms of the IBEA's presence in the region with "local sultans."[22] Consider the work of Dualla, "a Somal [sic] by birth," who Lugard selected as a guide due to his access to vast social networks.[23] As Lugard argued, there was "probably no living man who has travelled so much in Africa."[24] Dualla was routinely recognized by "village chiefs" who thus "escorted" the group "on its way"—social work that enabled Lugard's movement toward Uganda.[25] It was likely Dualla, an avid polyglot, who advised Lugard as to which "symbols of friendship" were appropriate in which regions, this competency being critical to securing intermediary guides for the group as they "moved through the belt of the forest."[26]

Administrators framed this work not as material, and certainly not as technoscientific, but as merely rote labor.[27] Absent guides, parties could always turn to a key technology—the "compass"—which, some claimed, obviated the need for the mobilization of Africans' topographical expertise. However, these tools were often of limited utility, with Lugard admitting that his compass only lent him "a very, very vague and hazy idea of the general direction we wished to pursue."[28] When company officials acknowledged Africans' topographical knowledge, they routinely denigrated it, claiming that guides offered only "falsehoods," revealing their "incompetence."[29] Indeed, administrators across the colonial period downplayed the critically important role played by African knowledge and expertise when it came to technologies and infrastructures.

The work of African topographical experts, many of them acting or former porters, was much more material and transformative than such representations acknowledged. Engineers and surveyors were dependent on the input of these men to fill in the vast blank spots in their knowledge of the territories and its people. It was essential that African guides know where and at what frequency parties could access drinking water, a task for which the compass was of little use.[30] In other instances, guides explained geological features, such as the presence of volcanoes, to company men.[31] This knowledge, born of experience, was augmented by the mobilization of technologies, the "gun-bearer" being dressed up "like an (African) Christmas tree," decorated with a "haversack, the aneroid, and the prismatic compass."[32]

The IBEA engineers' reliance on "native information" ensured that new roads routinely followed "native paths" and existing caravan routes, fetishistic hopes for straight lines notwithstanding.[33] Though old paths were followed, new

Photo of clearing of Jelori Road, constructed between present-day Malindi and the interior of Kenya in the 1890s, file 49, box 72, PP MS I IBEA/Mackinnon, SOAS

technologies of mobility required widened routes that were differently graded and surfaced. As John Ainsworth recalled, his party "had been clearing a track suitable for animal and wheel transport from Kiwezi on towards Uganda. . . . The new road lay . . . to the West of the Nzawi on the old safari track which was not possible for wheel transport."[34]

When new routes were "fixed upon" they were "marked off roughly," a process administrators hoped would displace existing paths, transforming company roads into "acknowledged track[s]."[35] Overhanging bushes were cut back, grasses weeded out, and bamboo uprooted.[36] Earthwork and dynamite enabled the construction of roads across the mountainous region.[37] Gradients were not to exceed 1:20. Trestle bridges fortified with "road-bearers" were built over rivers.[38] Earth beaten by feet was to be modified by unmetalled surfacing to accommodate bullock-carts.[39] These new roads were often sited at a distance from, though adjacent to, their older siblings.

In undertaking this prosthetic work, African interlocutors demonstrated their preference for retaining existing spatializations of social and material life. As William Beardall reported to Gerald Waller: "All the natives here are anxious that the road should follow the Wanyamioze route to the Kinagani . . . so they are all prepared to swear that every other route is bad or impassible."[40] Beardall resisted complying with the recommendations of these men, arguing that it would take the road down toward the soft, "muddy" soils of the Kingani river.[41] Other routes were proposed, but "from all native accounts," Beardall reported to Mackinnon, "a course between these two routes would be imprac-ticable."[42] And so Mackinnon instructed the party to concede and survey the Kingani river route, cutting a road adjacent to the "native path," which wound "about a good deal."[43] Where possible, company roads would reflect a prefer-ence for angularity, with Beardall reporting that he would follow the route of the path, but would "keep a straighter course with the road."[44]

These spatial preferences extended beyond African topographical experts' insistence that new roads follow old paths. At Nzoi, "natives" explained[that there was "one possible track" along which to proceed if the party did not want to make a "detour of one day's journey north or east." While willing to point out the route, they refused to undertake the journey themselves, as a "spirit of exceptional powers" resided on the "lofty peak."[45] Porters, for their part, abandoned caravans proceeding along routes that were "unpopular."[46] Indeed, people routinely threatened to refuse cutting company roads that contravened reigning organizations of space and social life. In the 1890s, George Wilson wrote to Mackinnon on the progress of a track that African workers were con-structing through Ukambani west toward Maasai territory. "This is to give you early intelligence," he wrote, "that 22 miles of road were finished on the 23rd . . . tomorrow I will cross the Masai [sic] road so much referred to." Some days later, Wilson sent an update: "I greatly fear that the Masai [sic] Road is an insuper-able bar to the further progress of the Wakamba with us. The dread of this road becomes magnified as we approach it. . . . They see both myself and my men al-most daily in the vicinity of the road but they're not to be influenced by this."[47] This "route" was used by the Maasai—a "warlike tribe"—"when raiding coast-wards."[48] While not disclosed in the company archive, this "Masai [sic] Road" was likely the terminus of the semi-permeable zones that crisscrossed the terri-tory (see chapter 1). In this instance, and no doubt under duress, these Kamba men crossed the road. As Wilson wrote some days later: "I have to report having carried the New Road across the Masai [sic] Road without unusual incident. It is most satisfactory to state that the Wakamba took their share of the work."[49]

European surveyors iteratively built on knowledge of the region's topographi-cal features that they gained through "native information."[50] One survey party followed a path documented by Lugard based on the knowledge of guides as "far as Nakuro [sic]," but changed course on confronting an "old Masai [sic] road," which led them to Kavirondo. Subsequent parties were able to chart the same course based on "rough sketches," inscriptions that, in tandem with the employment of "Masai [sic] guides," ensured that the "survey party had no difficulty in finding the way."[51]

As Africans' knowledge of the region's topography was inscribed cartograph-ically, it formed the basis of company maps—"rough pen & ink sketch[es] . . .

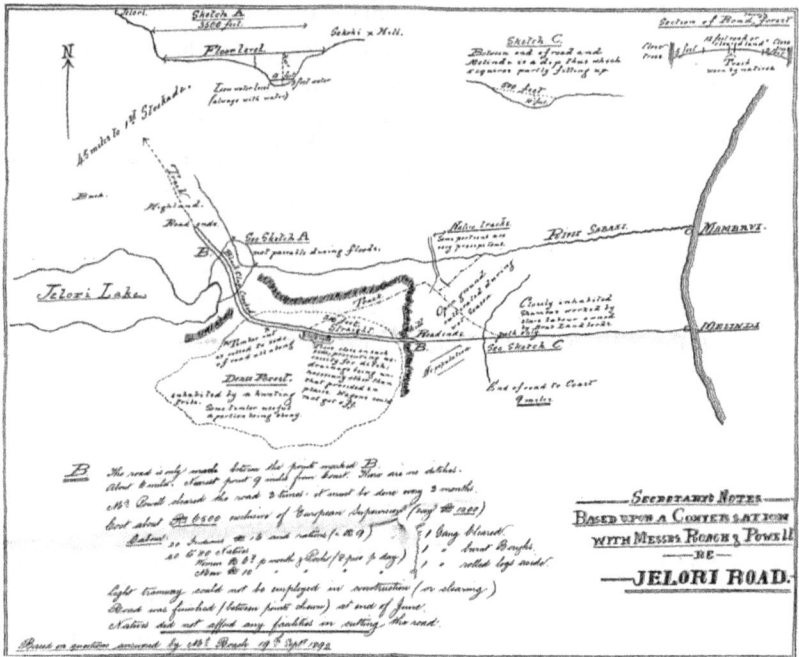

Map of Jelori, with road construction notes, 1892, file 49, box 72, PP MS 1 IBEA/Mackin-non, SOAS

[which showed] the approximate position of this station & route taken by the Company's Caravan to date."

By enclosing African topographical expertise and fixing it on the page, administrators created maps that obscured their origins. As Africans' knowledge was inscribed, it was rendered partially invisible—embodied expertise of the terrain was flattened as it was translated, appearing instead to be the result of the work of British engineers and surveyors armed with techniques of mapping and civil engineering learned in the metropole. Consider the "Wakamba highway." It, like so many other paths and caravan routes, formed the basis of the emergent colonial infrastructural network, with the line becoming the basis of "Mackinnon Road" and later being chosen as a link in the Uganda Railway.[52]

As "native knowledge" was appropriated and subsumed, communication networks were rebranded: from "native path" to IBEA road, from Safari track to Sclater's Road, from "Wakamba Highway" to "Mackinnon Road," from "footpath" to route of "commerce and civilisation." In the process, the critical roles played by African interlocutors, like the paths themselves, were effaced. This was part of the process of reducing African topographical expertise to infrastructural labor. These displacements notwithstanding, such discursive in-

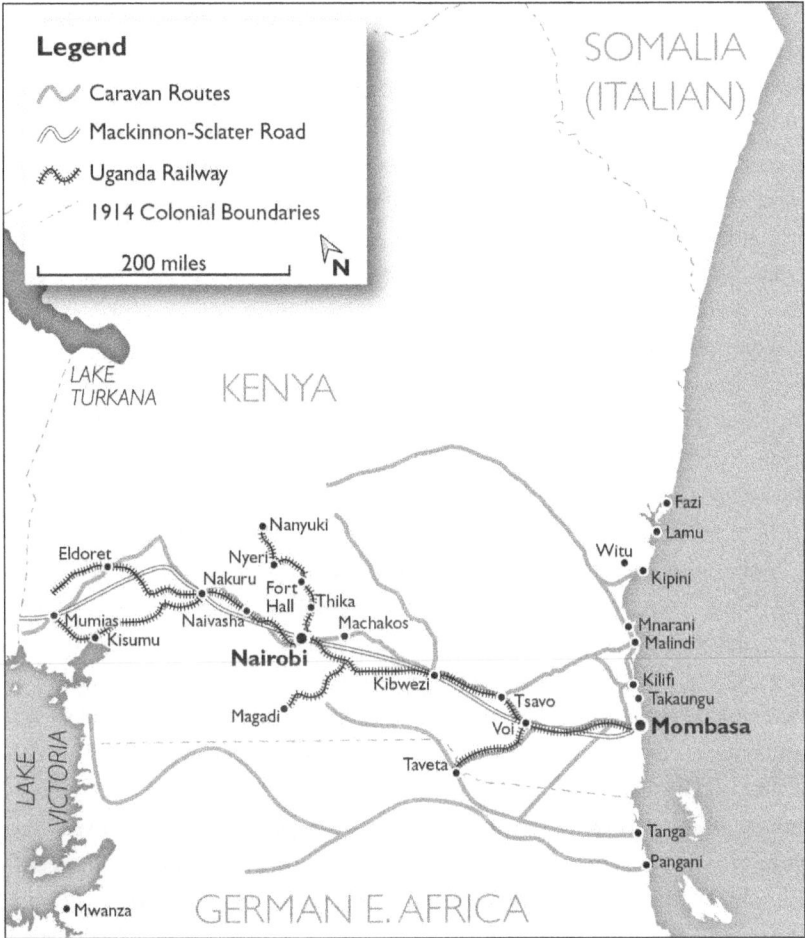

Map of caravan trade routes overlaid by colonial transportation infrastructures through present-day Kenya

scriptions formed the basis of material interventions on the land, with African topographical experts acting as infrastructural prosthetics whose knowledge critically shaped the emergent infrastructural topography of the region.

The Virtues of Work in Two Registers

"After some persuasion we got the native Elders, their headmen and people to take an interest in clearing tracks . . . We laid out the lines and the natives did the work," recalled Ainsworth.[53] Administrators unreflexively linked a desire for roads and a willingness to engage in roadwork to the relative civilizational status of people they tried to induce to labor on these routes, and held that the labor itself would be generative of civilizational "uplift." As one update read: "With regard to the Wakamba, as workmen, I have the honor to report satisfactorily . . . They are cheerful, and can be led."[54] This capacity to "be led" was linked to the future, with this administrator confidently stating that with "more extended experience they should become a useful factor in the native labor question."[55] Administrators framed the materiality of road labor itself as having pedagogic value, remarking confidently of the "beneficial influence on the natives of the discipline of the work."[56] This epistemology would not have been altogether illegible to many of the communities living in the region, for whom both maturity and culture were defined by skilled interventions in the natural world. However, not all work was considered equal by people living in eastern Africa. Realizing the pedagogic value of roadwork would take time and the cultivation of new rhythms—rhythms that would transform work into labor.

IBEA administrators were forced to contend with existing epistemologies regarding work and status which African communities brought to bear as they debated the merits of laboring on company roads. African leaders seeking "wealth in people" sought retinues composed of people with multiple and diverse skill sets.[57] Generic quantities, according to this epistemology of wealth and power, were always modified by personal qualities. These insights are critical for understanding the value placed on categories of particularization—fame, honor, and expertise. Kamba visions of value generating work, and self- and collective-actualization similarly placed a premium on singularity and specialization. As Myles Osborne argues, at the pinnacle of social life "were the *andu anene* ('big men' or 'prominent individuals')." These people "exhibited virtues that were most highly prized: bravery in battle or the hunt; specialized or secret knowledge about the martial disciplines and the wider world."[58] By the mid-nineteenth century, Kamba of the warrior class were well-integrated into global markets as they worked on caravan routes that took them from the coast inland. For these men, work on the caravans was a means of securing adult masculinity, analogous as it was to large-game hunting. Like hunting, time spent with the caravan not only held out the possibility for travel but also promised large, consolidated returns and esoteric knowledge of the "wider world."[59]

And this work was anything but rote. As J. R. Cummings notes, among the Kamba historically, there was little to distinguish the caravan porter from the trader. Kamba men involved in the caravan trade were "producers and . . . transporters, and upon their arrival at their trading destination . . . [they acted as] traders." And the trade itself was anything but homogeneous. For the Kamba, "trade," or *kutandithya*, only took place on their return to the *kyalo*, or home area, or at other sites within the regional trading system. Transactions at the coast, by contrast, were understood to be merely "economic intercourse." For Kamba communities, then, trade was a socially embedded phenomenon, a far stretch from the abstraction of transactional relations mediated by money that company men hoped to enact.[60] The goal of Kamba men, by contrast, was to assemble more wives, dependents, and cattle—gaining "internal societal prestige and authority" that resulted from "widening of one's sphere of influence."[61]

By the time the company began surveying inland regions, the power of Kamba trading networks had eroded, leading some among their ranks to sell their labor to European traders. This was a boon to the IBEA, who put the knowledge of these men to the task of mapping the territory and constructing its roads. But the rhythm of roadwork that sat at the heart of the IBEA's new infrastructural labor regime was radically different from the rhythm of the caravan. Unlike the excitement and prestige of long-distance work along the caravan routes—where men were able to demand "an exorbitant figure" as they mobilized their skills as warriors—labor on roads was banal, routine, without interruption, and remuneration was incremental.[62] And so, people were resistant to these patterns of labor "for hire" that took them to parts of the territory that were great distances from their communities for long periods of time and which offered small returns.[63] As Beardall wrote to Waller in 1879: "You must understand" that delays in construction have been considerable, "owing to our having been unable to get day labourers . . . All the men that I had got together last September many working continuously and regularly, are now scattered up and down the country, some busy on their own shambas, others india-rubber hunting, or gone on journeys, or at the coast and as yet they don't seem inclined to return to work."[64] Kamba men brought their own systems of valuation—systems which held that, for men, it was mobility and hunting that brought prestige—to bear as they considered working on company roads.

If infrastructural work was to strip people of expertise and create a pool of laborers, it would require reformulating people's apprehension of time and authority. As George Wilson wrote, Kamba road laborers were now "working miles away from their homes." While this brought them "under the useful discipline of the Camp," Wilson conceded that it "caused no small measure of resistance." "Naturally" this new labor regime, he wrote, caused "extreme anxiety," the Kamba "being like so many children; each waft of wind turns them from point to point like a weather-cock. Their free and carefulless [*sic*] life has unfitted them for steady toil and makes them impatient of control."[65] Like the work itself, the time of infrastructural labor was generic and homogenous, and the shifts embedded in this new labor region entailed a radical departure from the past. Indeed, as the work of Kamba porters was increasingly circumscribed to the transactional realm, the distributed nature of caravan work—which had historically included both women and "seers"—was undermined.[66] As the expertise of women and elders was sidelined, caravan work was masculinized. These processes, which narrowed the circle of workers and circumscribed the nature of work, accelerated as male caravan workers contracted their labor to the IBEA. Coming under the "useful discipline of the Camp," then, was part and parcel of administrative efforts to strip people of particularity, thus transforming expert work into infrastructural labor. This was no easy sell for people who, administrators claimed, had "no idea of the value of time."[67]

Administrators' condescensions were premised on a series of misrecognitions. Kamba men's ethos surrounding time and work reflected a historically constituted sensibility that infrastructural labor contravened in fundamental ways. In the past, the caravan trade had been structured to mirror the rhythm of the seasons, ensuring men were available to help in the work of planting and harvesting.[68] Sensitivity to seasonal fluctuations would have been augmented by experiences of the nineteenth century, which had seen the outbreak of periodic drought, famine, and disease.[69] People drew on their recollections of the famines of the 1870s and 1880s when they considered engaging in roadwork. These experiences placed a premium on people being attentive to the capriciousness of the rains. Being attentive to the "waft of wind" was essential to ensuring people's *own* capacities to cut culture out of nature. Like the land itself, time was neither empty nor homogenous, and attending to its lively character marked the pivot between harvest and dearth, between life and death.

It is no surprise, then, that the moments of greatest frustration coincided with the planting seasons.[70] "During the past month," one report noted, "I could not get as many daily laborers as I could wish . . . at this season of the year the people are busily engaged in cleaning and planting their own Shambas."[71] Such correspondences were written in February, immediately preceding the long rains that, ideally, would run from March through June. These months were the time of planting. While both epistemologies of civilization were premised on the need to expend human energy to turn nature into culture, the rhythm of the road was out of sync with the rhythm of the seasons. Missing the time of planting could prove to be disastrous not only in the short-term but in the long-term as well.

Regimes of Valuation in Two Registers

Just as eastern Africa's spatial order was crisscrossed with competing visions of sovereignty and cultured space (see chapter 1), so too was it a busy and competitive realm when it came to units of account, stores of value, and media of exchange.[72] In the early days, wrote Ainsworth, "money as a medium of exchange was unknown among the natives of the interior, the buying of food, any payment for services etc. was done by way of barter."[73] The limited use of money required company men cultivate knowledge of people's extant grammars of valuation as they solicited them for favors, namely, requesting permission to cut new roads and asking local leaders to call people up for work.[74]

"Runners" were routinely sent back to camp with information "detailing the beads, cloth &c. that are necessary."[75] Generalities were not sufficient; instead, these men were to report on the "article most in favour with the natives for barter."[76] Brass wire, beads, muskets, and *merekani* cloth were all in high demand by mid-century.[77] The exchange of goods for food was essential to sustaining caravans as they moved inland. While there was routine to this—with each "vendor" filling a vessel "till it overflows," in exchange receiving "his or her equivalent in beads"—the value of the beads was subject to variation, increasing as one moved away from the coast. And people's tastes varied widely. As Lugard reported: "Each tribe has its particular fancy—the Wakamba will take only red and white beads, the Wasoga only dark blue, the Wakavirondo only pink. The Masai [*sic*] . . . only . . . iron wire."[78] Elsewhere, "natives . . . prefer[red] cash (pici) to either cloth or beads."[79] "Trade goods" had to be carefully selected "to satisfy [these] varying needs."[80] Ascertaining people's tastes, though, was also of the utmost importance if company men were to secure pools of road laborers. As one report noted: "I had a very pleasing shauri [counsel] with [Kamba men] at which they fell in at once with my wishes for their cooperation on the road. If my supplies could have stood the strain I could have engaged hundreds here without trouble."[81]

Non-currency media were not the only forms of value with which the IBEA had to contend. In the mid-nineteenth century, the Swahili Coast was a zone of competition characterized in no small measure by long-standing battles among a plethora of national, sub-national, and commercial currencies. While in theory the "Zanzibar dollar" (the Maria Theresa thaler) was "the unit of account," the American dollar, as well as the Indian rupee, had made inroads by the 1870s.[82] By the 1880s, the Indian rupee, followed by the Indian copper pice, had gained ascendance; the former becoming the coin company men had to "deal with," it having "completely supplanted all other coins."[83] The rupee's ascendance reflected the dominance of the South Asian merchant class, the coin being the primary medium through which regional economies articulated with global circuits of exchange and international markets.[84] In the early years, the company worked within this regime of valuation, using the "coin most in circulation" for the "payment of labour."[85]

While relying on the Indian rupee was a pragmatic decision, the circulation of the coins presented problems for the sovereign aspirations of the corporate-state. From the perspective of company officials, the multiplicity of currency and non-currency forms was dizzying; there was simply too much variation in units of account, too many stores of value jostling for authority. What was needed was a medium over which the IBEA held a monopoly. A company currency, administrators hoped, could dislodge the myriad currencies, as well as non-currency units of value—livestock, rhino horns, hippo teeth, beads, and cloth—circulating in the region.[86]

A company-issued currency would not only consolidate the sovereign prerogatives of the corporate-state but would also, by contrast to Indian coins, offer the prospect for profit through seigniorage.[87] In asserting this sovereign prerogative, the company proceeded with caution, producing company coins as near facsimile copies of circulating tokens. As one report noted, the company coin should "adhere as closely as possible to the coins current, and perhaps that of the East India Company with the scales on the reverse side [but] adapted to the [Imperial British East Africa] Company would be the most suitable."[88] If the emulation of the East India Company's inscription dominated one side of the coin, on the other the company would emulate the "style of the latest coin minted by the Sultan," with the word Mombasa replacing the word "Zanzibar as appears on his coin." This change in location was significant, marking the shift of the administrative center of gravity from Zanzibar to Mombasa that attended company rule. By 1888, an "experimental shipment" of these coins—a hodgepodge of mimicry—which bore the "name of the Imperial British East Africa Company" was made to Mombasa, entering the protracted struggles over valuation increasingly making themselves felt inland.[89]

The company was insistent that administrators working in the countryside confirm that IBEA coins were operating as messengers of a new sovereign order.[90] To their relief, while they remained marginal, the circulation of company coins was proceeding apace. This was a battle which company officials viewed as being central to enacting corporate sovereignty, but a form of sovereignty that was, in practice, divisible, distributed among the Sultan, the IBEA, and the Crown.[91]

Company officials hoped IBEA coins would act as infrastructural attachments for paying wages in a standardized unit of value was, among other things,

Coins minted by the Imperial British East Africa Company

designed to subsume diverse lifeways and forms of expertise by rendering both time and labor equivalent to a single unit of account.[92] Administrators saw in a company currency one such prosaic attachment. But as company officials were quick to learn, ascertaining people's preferences in forms of payment was of limited utility when it came to reformulating people's ideas regarding value-generating work. Through the intermediation of Kamba porters who had transformed themselves into labor recruiters for Swahili and European traders, the exchange of work for coins was introduced to inland communities. Many outright rejected efforts to leverage cash as a means of transforming work into labor. As Cummings writes, various stories recounted Kamba "laborers receiving the rupee given to them only long enough to hear of their proposed purposes before throwing them either away or back to their employers." Others spoke of "individuals who kept the new 'money' and buried it on returning to their homes vowing never again to work for the Kamba agents who were attempting to cheat them by paying their own people with the worthless coins provided by the Afro-Arab traders."[93]

Kamba men were not alone in drawing on historical expectations of value and work as they considered committing themselves to the IBEA's road-labor regime, and company employees continued lamenting the difficulty of recruiting and retaining labor. As one report noted, on "the third day in the second week" the men indicated they were "tired of the work," arguing that they "came to be soldiers and [so] would not work any more."[94] Many others rejected the IBEA's terms of labor outright. Some demanded advances to "send home to their wives."[95] Others wanted payments on the day, fearing they would forget "how long they had worked."[96] Still others rejected the IBEA's paltry wages, demanding "20 pice now instead of 15" pice on the day.[97] Work was by no means generic, and people tried to leverage the company's need for their labor as they negotiated the terms of their employment. Such labor unrest was often met with violence—flogging, irons, and imprisonment were all made use of.[98]

Part of the problem from the perspective of company men was that there was no material incentive for people to sell their labor. As one report recounted of people living around Kavirondo, "Their only want being a few beads, they have no adequate inducement to work."[99] This problem was endemic. As another report read, there is a "great scarcity of labour and the few people obtainable are difficult to keep," obliging us "to pay high wages." Labor scarcity was ubiquitous and, company officials argued, was due to the "abundance of food in the country this year."[100] This observation was prescient. Destructive dearth would be the pivot that would ultimately force people to take to the road as they entered the wage economy.

Sovereign Seizure in Two Registers

It soon became clear that the IBEA-issued currency was not a sufficiently robust infrastructural attachment. The coin could not, on its own, tie people to the road. What was needed was a more coercive fiscal technology. In taxation, company administrators saw one such possible attachment that, they hoped, could bind people to the road. As one report noted, "When public roads are constructed the principle of levying tolls upon the traffic which they carry should be regularly put in force." It was hoped that the "revenue in tolls," in tandem with the "leasing of lands adjacent to the road," would "defray the outlay in construction."[101] Put simply, these new infrastructures were part and parcel of administrative efforts to create new "fiscal subjects" by enforcing new regimes of taxation.[102]

These forms of extraction signaled broader transformations. Indeed, taxation was conceived of as a means of tethering people as labor to the road, all the while generating a revenue pool for the company to fuel further infrastructural expansion and line the pockets of shareholders. If consolidating the authority of the company currency was critical to creating a labor market, taxation was the imagined infrastructural attachment that would force people to turn to road labor in order to "find their tax."[103] But we would miss much if we saw in taxation a simple economic instrumentalism.

The rationalization of space in the form of planned roads was matched by the rationalization of sovereign seizure as both were centralized in the hands of the corporate-state. This discursive reframing was part and parcel of the ideological claims of the IBEA. In the hands of the company, *hongo* (tribute) was rebranded "blackmail" and other forms of value capture were deemed "barbaric." These forms of seizure were contrasted against "civilised" forms of value capture—the "tax," "toll," "duty"—that would be undertaken by the corporate-state. As one article in *The Economist* read: "Roads . . . cannot be made or maintained without some regular authority, nor except under a European flag is it possible to prevent divided and often hostile chiefs from levying the inland transit duties, which, better than any other barbarian contrivance, throttle trade."[104] In European hands, arbitrary forms of seizure would be routinized and rationalized; rather than acting as an obstacle to trade, they would facilitate its expansion.[105] For administrators, then, these new regimes of seizure were ideologically dense, core components of the civilizational value of the road and road labor itself.

Administrators were well aware that inaugurating novel regimes of sovereign seizure would take work and time. People would need to be eased into new arrangements, which would see claims to *hongo* move from their hands to those of the corporate-state.[106] As Berkeley, with whom we began, concluded: "In regard to the toll to be charged over the road . . . this should not be imposed until the road is somewhat actively used by the natives, who having once appreciated its utility [as a route to market], would be more likely to pay the toll cordially."[107] Here, Berkeley not only recognized the constructed nature of markets, and the work that went into enacting and naturalizing them, but in so doing, acknowledged the existence of incommensurate systems of valuation—grammars that the corporate-state was unable to dislodge.

The IBEA failed in its efforts to fund its operations through exacting novel regimes of seizure, not least because it was unable to assert a monopoly over valuation, with multiple currencies and non-currency stores of value—the Indian rupee among them—jostling for authority.[108] This was, in part, because company administrators did not imagine the IBEA currency simply as a unit of exchange. As Wambui Mwangi persuasively argues, they were as concerned with company-minted coins' potential in a hegemonic project geared toward establishing social, cultural, and economic domination. The Indian rupee, by contrast, was a flexible unit. It was the instrument of "trading concerns . . . [and] did not have the same stringent requirement of a conceptual value consensus as those required by, for example, taxation or wage labor employment," both of which sat at the heart of the IBEA's project.[109]

These failures were consequential. By 1892, the company was functionally bankrupt, and retrenchment in investments began as soon as the Foreign Office took over in 1895 (see chapter 1). The new administration really did not know how to administer the territory and was largely dependent on company officials, many of whom, like Ainsworth, were absorbed into the nascent apparatus of the state.[110] It was not simply people that trafficked across the boundary dividing the (private) corporation from the (public) state. When the Foreign Office took over the administration of the region, the policies of "sound finance"—that both led the government to grant the IBEA its charter *and* structured the IBEA's operations—not only persisted but became more pronounced, with African communities left to offset the costs of this ideologically dense but materially thin infrastructural state. Indeed, while the IBEA largely failed in its efforts to fund its operations through taxation, these techniques would have a second lease on life as they were taken up and extended by the colonial state. Consolidating this method of fiscal governance would be experienced by communities living in the region as an acute reversal. Indeed, these administrative changes dovetailed with a period of calamitous ruin for Kenyan communities, and this ruin was firmly tethered to colonial infrastructures.

The Politics of Immobility

The first deaths on the road emerged at a troubling conjuncture. Over the course of the nineteenth century the region in general, and what would come to be called Kenya's Central Province—the location to which this story now pivots—in particular, had been wracked by periodic drought and epizootic epidemics. In 1892, rinderpest claimed the lives of approximately 50 percent of the herds on the land. Then, again, in 1897 came drought, on the heels of which came famine.

Materiality here mattered in at least two interrelated ways. Not all were affected equally, with southern Kikuyuland significantly worse off than northern Kikuyuland. This had less to do with topography or variations in rainfall specific to the region than to the place of the south in relation to the increasingly robust infrastructural and administrative reach of the nascent colonial state. Southern Kikuyuland had been the main grain store for passing caravans heading west toward Uganda, surpassing the role of the Kamba, which peaked in the 1870s.[111] In periods of bounty, people reported large markets, where women arrived "laden with the superfluous produce of their fields," which they were, apparently, "very glad to dispose of."[112] However, these emergent markets had drained communities, leaving them vulnerable when the rains failed in 1897 and the following year.[113] It was not death by famine that caused the bulk of the devastation, however, but smallpox.[114] Unlike famine, which emerged at the intersection of drought and the consolidation of new markets, smallpox is a communicable disease and, while "naturally occurring," it is a disease of empire and trade.[115] And its mobility was linked to new infrastructures in critical ways, the contagion spreading along the "profusion of [new] routes."[116]

Communities living in the region were savvy in connecting the devastation of the famine years to external forces—infrastructures among them—and the "white man" who brought them.[117] As the railway replaced roads as vectors of commerce and civilization in the colonial imagination, "coolie" laborers, many of them indentured, were imported from India to begin the work of laying the tracks.[118] The railway hungrily gobbled up land so that what Maasai *laibons*, or seers, referred to as the "great serpent," and what Kikuyu seers referred to as the "iron snake," might make its way from the port at Mombasa to its terminus at Lake Victoria.[119] The railway was not the only symbol of the exogenous forces that led to disaster. "Traditional accounts of the famine years in Ulu," writes Charles Ambler, "described the period as Yua ya Panuga, 'the famine of rice,' or Yua ya Maguina, 'the Gunny sack famine.'" These patterns of naming firmly linked the devastation to external forces. "The rice that was imported and distributed to the hungry by Europeans," notes Ambler, "was then largely unfamiliar in central Kenya, as were the bags used to transport it. In a deeper sense, these and other new products were seen to have engendered the disequilibrium that led inexorably to catastrophe."[120] For the communities of Central Kenya, in other words, there was nothing natural about this calamity—its loci were evident. And it was a calamity which had grave moral as well as material consequences.

Missionaries and government agents alike reported on the scale of the death, which according to American missionaries claimed the lives of 49,000 people in the south.[121] It was impossible to ignore the devastation, with corpses lying on paths large and small, as well as new roads and the rail line. One resident recalled seeing "hundreds of poor wretches dead or dying on the road."[122] Missionaries of the Africa Inland Mission (AIM) reported that "shambas and paths are literally strewn with corpses."[123] These observers were not alone in remarking on the conspicuous visibility of these deaths. A British settler recalled that "no matter where one went corpses [were] strewed [on] the tracks," both "native" and new.[124] As another settler frankly noted: "The railway line was a mass of corpses."[125]

This archive does not disclose how these bodies made their way to these new arteries of "civilisation and commerce." Perhaps people hoped that it would be along these networks that they might gain reprieve. In this, perhaps, they would not have been wrong. In 1898, the colonial government initiated food relief programs at the coast, which it later extended inland.[126] And yet, the placement of the bodies of the dead also mirrored popular understandings of the loci of death, and firmly situated IBEA and colonial infrastructures as uncultured, wild spaces.

Kikuyu ideas regarding individual and collective well-being were intimately wrapped up in the painstaking work of clearing the forest, of cutting culture out of nature, of civilizing uncivilized zones. For these communities, the untamed forest was a cultureless and potentially corrupting space. Historically, the bodies of the dead had been placed in the forest, cordoning off the polluting qualities of dead bodies from the confines of cultured life. The placement of the bodies of the dead during the famine suggests that, in this time of calamity, the road joined the forest as a space outside the bounds of civilization— roads as spaces of corruption, cultureless places contrasted against the securely bounded *ithaka* or cultured land that householders sought to protect against death. If there was a tragic poetics to how and where these deaths were made visible, they also offered a powerful indictment of the respatialization of social and material life engendered by external forces.

The status of new roads as uncivilized spaces would have gained conceptual power in contrast to existing networks of mobility. Kikuyu conceptions of the paths of movement, like prevailing conceptions of space in general, were not generic but closely linked to their function. There were *njira ya agendi*, a path visitors trod on foot from one part of the country to another; *njira ya mukira*, side paths or diversions; *njira ya rûtîmo*, through-paths, the way one sends a child to do an errand; and *njira ya thama*, the way of migration, the path trod when one takes down one's home and moves to another place.[127] These patterns of naming suggest that travel was vocational, part of a task—the paths themselves named to reflect their social function. These routes were the means by which the people engaged in the everyday work of maintaining social and material life, not generic means to some abstract end.

At this conjuncture, new infrastructures emerged as powerful symbols not of civilization but of destruction. They were sources of scarcity and vectors of communicable diseases. It is no surprise, then, that many associated the devastating famine of 1897 with new infrastructures.[128] And people worked to protect themselves against these external forces, with families and communities building barriers against disease. With appropriate rites, they purified the gates to their homesteads and the paths leading into their communities.[129] Older paths gained renewed salience as both networks of illness and sites of potential fortification. Properly medicated sticks were hammered "down across all the footpaths leading" to villages, protecting communities against the outsiders they saw as carriers. Boundaries of identification became less permeable as people worked to insulate themselves from various others.[130]

The material devastations of the famine years were matched in scale by the moral crisis they engendered and the generational politics through which they were framed. Deprivation brought the abrogation of social responsibility as elders refused their obligations to dependents, being forced to cast them off the land in order to provision for ever more intimate circles of people. Starving wives and children were pawned to wealthy men of Nyeri to the north. This was a social calamity, a "terrible failure of responsibility for fathers and husbands." This marked a critical rupture in generational relations, as social juniors used evidence of their fathers' moral failings during the famine to claim for themselves rightful authority. As noted by Derek Peterson, they called this new peaceable epoch *utheri*, and it was ushered in through *ituika*, a period of transition and cleansing, a way of "restoring social discipline."[131]

As for the colonial state, the period suggested that new relations of dependence would enable the spread of the wage labor economy as some who had avoided waged work arrived at sites of administrative control to provision for themselves and their families.[132] Missionaries, too, remarked that those "who never . . . would have heard the glad news" emerged from the calamity "hungry for the Word of Life."[133] There was, of course, nothing natural about this labor market or this market for souls, nor the starvation and disease that made them visible. If the famine had enabled new relations of dependency, fiscal reforms ensured the thickening of these infrastructural attachments when the rains eventually returned in 1899.

Taxing Infrastructures in Two Registers

In the early years of colonial occupation, administrators worked to assemble a system of production that would ensure "sound finance's" imperial other: colonial "self-sufficiency." Strapped for cash, they pinned grand hopes on the fiscal possibilities of white settlement. In 1897, the government opened land for white settlement, with settlers holding 99-year leases. By 1903, this was taking place on a large scale.[134] This was accompanied by the partial enclosure of lands. Building on IBEA practices (see chapter 1), the Land Acquisition Act of 1896 "allowed the administration to acquire land compulsorily for . . . *public* purposes."[135] In 1915, the Crown Lands Ordinance rendered Africans tenants of the Crown. But it was debt—debt in the form of taxation—and not the simple fact of lands' partial enclosure that, colonial administrators hoped, would ensure people's participation in the wage labor economy.[136]

The first Commissioner of East Africa, Arthur Hardinge, lost no time in working to stiffen these infrastructural attachments. In 1899, he inaugurated direct taxation of Kenya's African majority to shore up the state's revenue regime. Building on the schemes of the IBEA, taxes would first be seized along the infrastructural lines of the state, beginning with the railway center at Machakos.[137] Some Britons conceded that critics might find this arrangement predatory.[138] "It may be argued," Lord Hindlip wrote, "that it is not right to tax natives living in their own land, but it must be remembered that if it were not for the Pax Britannica the tribes who pay tribute would be subject to raids (as they very often are in spite of it) from their neighbours, and perhaps have to pay the equivalent of taxes to another and more powerful tribe."[139] Like its corporate forebear, the colonial state labored to mimic the extant forms of seizure it was working to displace. This emulation did not detract from the claim that the colonial state's variant of sovereign seizure was decidedly more civilized.

This nascent taxation regime was extended with the Hut Tax Regulations of 1901. As Mwangi has argued, in taking the hut as the site of extraction, the colonial state worked to impress on Africans their status as subjects within this new colonial dispensation.[140] Their subordination cast them as a class of debtors who owed payment to the supposedly benevolent and paternalist colonial state. Africans protested the Hut Tax, not least because it disproportionately affected polygamous unions. Within the colonial order of things, having a large retinue of dependents was not a measure of wealth as it was in the precolonial period, but rather subjected polygamous men to more aggressive forms of dispossession, with the Hut Tax obliging men to pay a tax on each wife's hut.[141] The Hut Tax, however, did not apply to mobile men, those "loafers and general riff-raff" who dodged administrators throughout the colonial period.[142] To remedy this, the state inaugurated the Poll Tax in 1910, which could be extracted from all able-bodied men sixteen and over.[143]

The primacy placed on the taxation of the African majority reflected the priorities that guided imperial austerity in the age of colonial "self-sufficiency." While in the early years of formal colonial occupation, the Treasury loosened its purse strings, the prudent logic of "sound finance" soon prevailed. Funding steadily declined and by 1913–14 was halted altogether. The allocation of funds such as they did exist required making a set of decisions regarding distribution. The result: The development of road networks that were unevenly spread over the colonial topography, with the state constructing roads in the areas of white settlement in the Highlands, as Britons and South African whites began flooding the region. And these routes were mainly devised to ease the movement of crops being exported by white farmers, rather than for the benefit of the African majority. And as with its corporate forebear, the colonial state continued to, by turns, exploit and expropriate African infrastructural labor to service the needs of white accumulation, in this case not of the private firm but of white settlers. The 1906 and 1910 Masters and Servants Ordinances authorized the government's use of coercion to raise unremunerated labor for infrastructural projects.[144] Unfree labor was back.[145]

The railroad was used to justify even more predatory forms of extraction. Following the collapse of the IBEA, the Treasury had loaned the Protectorate government £5.5 million to construct the Uganda Railway, which largely followed the path first laid out by the Kamba caravan route and which was subsequently overlaid by Mackinnon-Sclater, or Uganda Road.[146] This investment, too, was a subsidy to white farmers for whom the line would facilitate commodity export. As the logic of colonial self-sufficiency came to dominate administrative practice, the government set its sights on recovering monies it had sunk into the railway over the past decade.[147]

Like the IBEA before it, in new regimes of taxation, administrators saw a possible infrastructural attachment—one that could tether people to the wage labor economy, all the while funding the expansion of the infrastructural state and paying down its debts. As Governor Sir Percy Girouard speculated in 1910, Africans paid up to 40 percent of total revenues in tax and import duties. Settlers, for their part, paid only 20 percent.[148] In Kikuyu Province alone, African tax receipts comprised around 95 percent of the colonial state's revenue by the 1920s.[149] And these taxes were hugely burdensome, with reports recounting that after "paying their taxes (which they are not always able to do), money for food and clothing" was "practically non-existent."[150]

If African taxation was to fund the expansion of infrastructures designed to serve a growing settler population, much of this tax-generating labor was itself infrastructural. On the eve of the First World War, over 10 percent of those employed in wage labor were working for either the railroad or for the Public Works Department (PWD).[151] In Central Province in 1915, acting District Commissioner of Fort Hall, E. L. Pearson reported that the majority of laborers were employed on public works projects, specifically road-building, for which they labored for KSH 5 per month and rations.[152]

Infrastructural labor was the prerequisite for new markets—creating the circulatory system on which commodity export depended—*and* infrastructural labor was a means of generating a revenue pool through taxation. And the balance between these two imperatives was rarely clear-cut. In regions where labor was not readily available, local administrators were often granted permission for road construction projects on the promise that the funds spent on wages would return to the colonial state in the form of tax revenues.[153] Those who were unable to pay their taxes were to pay with labor, often roadwork. This was a tax in labor time.[154] Taxation, then, was not simply critical to raising revenues, but was a coercive technology designed to consolidate a labor regime.[155]

In Britain's colonial holdings in Africa, by contrast to the English countryside from whence Marx narrated his critique of capitalism, wholly alienating people from their lands was not the modality through which capitalism's "primitive accumulation" proceeded. Leaving people in partial control over the means of production allowed settlers and the state to employ Africans at very low rates of pay, women at home taking up the affective and material slack of social reproduction. But if people retained a measure of control over the means of production, how were they to be forced into emergent labor markets? The question was how to make people dependent on cash. Taxation was a means of doing this, but it was essential that it take the form of direct taxation. Had administrators settled on income tax or indirect taxes, for example, or the toll, as had been experimented with by the IBEA, only those already implicated in the money economy would have been tapped. Forms of direct taxation like the Poll Tax, by contrast, were to be extracted from all able-bodied men.

Theoretically exacted in colonial currencies, direct taxation, then, was to operate as a form of primitive accumulation, obviating the need for wholescale dispossession, all the while forcing people to submit to the "freedom" of wage labor.[156] As Hindlip wrote, if "natives" were allowed "to pay their taxes in grain, etc., the supply of labour will visibly decrease." To that end, he argued that a "demand for cash should be created among the natives, who would then have to obtain coin . . . to pay their taxes."[157] This was a more pronounced form of seizure than dreamed of by the men of the IBEA, who had proposed taxing people for their proximity to roads and the access to markets they ostensibly held out. The fiscal and administrative regimes of the early colonial state thus operated as critical infrastructural attachments, firmly binding people and their futures to the road—new paths taken as young men struck out to "find their tax."[158]

But administrators understood taxation to be *both* a fiscal and a cultural technology, with Under-Secretary of State Winston Churchill writing that not only did the "existence of the tax . . . not impose any harsh or severe burden," but that it was a "stimulus to habits of industry and the use of coin."[159] It was at this intersection of revenue-generation and pedagogy that commerce and civilization appeared to be two sides of the same coin. Coins which people continued to have little interest in holding.

Coining Conquest in Two Registers

Administrators had hoped that direct taxes would be paid in rupees, which were established as legal tender of the protectorate in 1898. In truth, though, the spread of the money economy was proceeding slowly and unevenly across the protectorate. And administrators, like their corporate forebears, found themselves confronting competing regimes of valuation that proved to be remarkably robust.

Concerned with taxation's role in an ideological project designed to cement the sovereign status of the colonial state and the status of Africans as subjects, the early administration accepted taxes in kind—"sheep, goats, ivory, grain, [and] labour" were all made use of.[160] Entering into this world of multiple regimes of valuation, early administrators dispensed with the need to establish commensurability, ascribing seemingly arbitrary values to objects used in the stead of currency. The Hut Tax of Rs. 1 could be satisfied by paying one sheep, three hoes, or an unspecified number of chickens.[161] In other districts, two hoes were rendered equivalent to Rs. 3. In the region surrounding Lake Victoria, crocodile eggs were accepted as payment, administrators hoping that this would lead to the destruction of the dangerous animals.[162]

By 1906, however, the administration's labors to gain a monopoly over the media of exchange and units of account were seeing some success, with the rupee circulating more broadly across the region.[163] As Ainsworth recalled: "Money came to be readily accepted by the natives living in the areas adjacent to the Administration centres and the Railway line, and so there came about the change from the barter system to one of cash payment."[164] There was a networked spatiality to this shift, which radiated out from nascent infrastructural networks and the emergent disciplinary regimes with which they were associated. Grand hopes notwithstanding, the circulation of the colonial currency did not supplant extant regimes of valuation. As the Provincial Commissioner of Central Province wrote to the Chief Secretary as late as 1939: "[I]t is still not uncommon to hear the native use the expression when going to get tax money by the sale of cattle that he is going to buy money. This implies that his view ... [that] money is cattle and he just purchased something with his money whereby he can pay his tax."[165] Money had not won out as the final arbiter of value in either social or economic terms.

If taxation appeared a benign and bureaucratic expression of notional sovereignty, methods of collection were backed by the potential use of force. Those who resisted the colonial state's nascent taxation regime found their livestock seized and, in some instances, their villages razed.[166] People sought to protect themselves from colonial seizure, again working to fortify themselves against the dangerous incursions of outsiders. People took to the road, blocking paths, "even the smallest track[s]" with trees, and digging "pitfalls" boasting "sharp stakes" that were "cunningly covered over with loose earth."[167]

These strategies of fortification notwithstanding, the efficacy of this regime that placed seizure at the center of infrastructural expansion is evident in the numbers. In 1923, government revenue was 1.8 million pounds sterling, with half a million pounds sterling, or about a third, being garnered through the taxation of African men. Only 10 percent of the taxes collected in African areas was reinvested locally, and this was predominantly devoted to paying the costs of future tax collection and administration.[168] As for the other 90 percent, it was taken out of African districts and redirected to other locations, namely those of white settlers, making it "quite clear that the severity of African taxation was a direct consequence of the financial needs of . . . European infrastructure."[169] Taxation, in this period, was a means of extracting value without making any contribution to the "public good" of the African majority. Settlers, for their part, successfully lobbied to ensure that they were subject to very light indirect taxes.[170] As the state translated "sound finance" into colonial "self-sufficiency," it established a revenue regime premised on racialized rentierism.

Infrastructural Labor in Two Registers

These arrangements quickly produced contradictions. As the District Commissioner of Machakos reflected in a rare moment of clarity: "The native labourer look[s] . . . on the road, through his particular portion of the country, as a never ending source of hard work which seldom grows less but rather tends to increase."[171] Labor was not the dominant frustration. Rather, people highlighted the difference in the quality and number of roads serving settled farms versus those serving African areas. Conceding the point, he concluded: "It seems very unfair that quite large sums of money are voted annually for roads which often only serve a very few occupied farms, when not a penny is available for one which taps the majority of a Native Reserve."[172] People were well aware of this as a kind of technopolitics premised not on technical intervention but on paltry infrastructural investments that relied, by turns, on the exploitation and expropriation of African infrastructural labor. And these contradictions became something over which people could argue.

If, for administrators, roads were to be the arteries of "commerce and civilisation," for many they continued to be popularly experienced as spaces of blockage and civilization interrupted.[173] Those called up to undertake unremunerated roadwork were tied together and forced to march from one district to another along the administrative routes of the colony.[174] These forced marches would have been humiliating for people for whom the self-possessed labor of establishing a homestead was the most valorized form of work. The violence of this labor regime was not simply symbolic—sometimes, labor conditions led to the end of biological life. In the mid-1910s, settlers living in Central Province wrote a flurry of complaints to the District Commissioner (DC), requesting "police authorities . . . look into the matter and the manner of death of Kikuyu who are continually to be seen dying or dead in the ditches by the Nairobi Road between the Chania Falls and Nairobi."[175] The DC admitted that those dying were engaged in forced labor; their deaths a consequence of malnutrition. The road again emerged in this period not as a symbol of civilization but of debasement, not as a route of commerce but of dearth, not as a sign of mobility but of irrevocable immobility, of death. In each of these registers, these deaths on the road would have chillingly echoed the devastations of the 1890s. Both conceptually and in material terms, then, for many African communities, colonial roads were not spaces of movement but of stasis. If this was a form of (infra) structural violence, the nature of the violence was anything but invisible.

The politics of immobility were only to become more pronounced. The Natives Ordinance of 1915, implemented in 1918, required every African over the age of 15 to be registered and fingerprinted. These registration certificates, or *kipandes*, were to be worn around the neck in metal frames. Within two years, 194,750 certificates had been issued. By 1924, this number had climbed to 519,056, rising to 1,197,467 by 1931.[176] Any African found outside their reserve who was unable to produce the document was punished with hard labor—often road labor. By 1937, people reported that police were demanding money as well as the *kipande* when they stopped people, arresting their movements along the road.[177] If taxation was a means of ensuring the movement of men as labor, the *kipande* was a means of controlling where people moved and to what ends.

This was the central contradiction of mobility in the region. Maintaining roads as the arteries of "civilisation and commerce" turned on truncating the movement of Africans. And people were well aware of how the state deployed the techniques of strategic immobility to shore up this highly stratified racial order. The evidence was everywhere, from the uneven distribution of road networks across the landscape of the Kenya colony, to the labor regimes used to construct the arteries, to the different materials used to build roads in settler areas versus reserves, to the everyday experiences of being stopped along the routes. The colonial road network, for many, operated as a core symbol of the highly racialized social and economic order of the Kenya colony; it was a metonym for the African majority's marginalization within the colonial order of things.

Through an impressive feat of ideological acrobatics, this violent regime of infrastructural labor did not detract from a vision of roads as vectors of civilization. One of the main arguments justifying the use of forced labor on roadworks hinged on the instructional value such labor afforded, with colonial officials arguing that coerced labor—particularly on the extension of road networks which were for "the common good"—would lead to the cultivation of patriotic colonial subjects.[178]

Conclusion

While it is perhaps true, as John Lonsdale argues, that by 1905: "'Kenya'—if such a retrospective concept may be permitted" had been "transformed from a footpath of 1000 km (600 miles) long into a colonial administration," by the late 1920s, administrators had still not secured a monopoly over valuation.[179] This led to a shift in tactics, which turned on renewed efforts at marketization. If taxation was not sufficient to induce people to enter the wage labor economy, administrators speculated, perhaps the desire for consumer goods would be. The most forceful advocate for the cultivation of a new market of consumers was none other than Ainsworth, who had gotten his start working for the IBEA, becoming the colony's first Chief Native Commissioner in 1918. While many administrators understood that African taxation was essential to ensuring colonial "self-sufficiency," for Ainsworth it was Africans' desires for material goods—affected through "educating them to the wants of Western civilisation"—that would lead Africans to seek out employment, rendering coercion unnecessary and ensuring the reproduction of a laboring force, and thus the state's revenue regime.[180] He was emphatic on this point, advising that each "and every native in a Reserve must be encouraged to want something which he does not produce himself, and in order to obtain it he must be taught to earn money."[181] Commerce as civilization was the argument here.

This was the framework against which colonial officials (despite their anxieties) viewed the desire for technologies of mobility—bicycles, cars, and lorries—alongside demands for roads servicing native reserves as evidence of both the relative development of individuals and communities as well as colonial officials' relative success as civilizers. By 1924, colonial officers were reporting African vehicle ownership, which, when in the right hands, was reflexively linked to the civilizational status of individual owners. Technologies of mobility and emergent markets would join currencies in generating infrastructural attachments, administrators hoped—attachments that would bind people to the road and tether them to the wage.

And yet, while by the 1920s, Africans were indeed consuming things that they did not produce as they entered the cash economy, the types of subjectivities and the forms of self-fashioning that these modes of consumption enabled could not be controlled. And gaining a monopoly over what people apprehended as value-generating work continued to be an uphill battle.[182] In 1925, the DC of Fort Hall reported with anxiety that mission education had produced a generation of young men who fancied themselves "literary types."[183] The following year he remarked on the threat these mission-educated men posed to colonial authority. This population of "readers," he argued, "constitute a potential danger as, being averse to unskilled manual labour and lacking employment, they . . . tend to sell the ranks of the discontented and seditious. Some of them . . . become teachers at Missions School or traders but a large number of them rest on their cars and do nothing."[184]

Colonial officials were unable to monitor the purchasing power of Africans whose status as consumers was, in part, a consequence of the state's desire to implicate them in the wage-labor economy. As Africans in increasing numbers set their sights on purchasing technologies of mobility, they contravened the aspirations of colonial officials, putting vehicles to uses both unanticipated and undesired by the administration. At issue, however, was not simply that these men refused to participate in the wage labor economy, but the fact that this rejection indexed the continued proliferation of economic spaces, social practices, and ideas of value-generating work premised on alternative grammars of valuation, grammars that the colonial state remained unable to dislodge.

"Tropicalising" Technologies

Cable and Wireless Ltd. and
Making Broadcasting "Work"

One morning in January 1944, a mobile information unit arrived at the compound of a Church Missionary Society (CMS) school, located in Maseno, western Kenya. Samwel Okwako Libuko put to page the excitement he felt that day. When the van stopped, "many boys rushed out of their classrooms." Others "peeped through the windows and doors." The reason for these students' interest was clear to Samwel. Such "a motor car had not been seen by us for [a] long [time]."[1] This vehicle was storied, rumors of its existence circulating through the countryside.[2] Some were less excited by the arrival of the unit. These young men, one student wrote, "pretended that they never cared for that motor car." This student did not reflect on the reasons for their disinterest, but perhaps, like Ambrose Wakaria (to be discussed below), these young people had been taught by parents and grandparents at home to view European novelties with circumspection.

Students were among the targeted audiences of the many viewing and listening parties staged by the information arm of the colonial state during the Second World War. Initially, the administration imagined these units serving small audiences. Arthur Champion, the first mobile operator, had changed all this. By obtaining "a non transparent screen which he erected some 12' in height by means of a series of poles . . . by choosing natural amphitheatres and by marshalling the audience into a narrow segment so that they did not stray away from the reflection of the light he was able to show to audiences as large as 5,000."[3] To use the language of the time, in rejigging the equipment such that it worked in the Kenya colony, Champion had "tropicalised" the unit and its technologies. The work of "tropicalisation" took many forms.[4] Sometimes the unit was set up on hillsides, or "little slopes," at other times in the "big hall[s]" of schools, in "godowns" of local businesses, or in "amphitheatres," which made "first-class outdoor . . .

hall[s]."[5] What the crew sought was "favourable ground [which] enabled . . . a large number to see well."[6] Broadcasts were accompanied by silent films over which an African commentator spoke, pedantic developmentalist lectures and educative performances by African knowledge-workers, and the distribution of broadsheets.[7]

But the mobile information unit was never to be a permanent feature of the state's broadcasting apparatus. In the beginning, administrators imagined radio listening would be a domestic affair because "private listening on personally owned sets in the house is . . . the best form of listening."[8] This was a universalist vision of technologies. In this networked future, government development initiatives would travel over the airwaves, rather than the roadways, with centrally produced broadcasts from Nairobi reaching into the lives of communities living throughout the territory.

This was in keeping with an emergent developmentalist ethos emanating from the Colonial Office (CO) that, officials hoped, would see the centralization of administrative authority, with experts in education, agronomy, and health fanning out across Britain's colonies following the war. For many, radio had a central role to play in this changing landscape. Indeed, it was precisely in the absence of funding for larger and "harder" infrastructures of development— roads, health clinics, piped water, electricity—that many came to see radio as a suture, the government's messages for social and material transformation suffusing people's consciousnesses over the airwaves. The medium, they claimed, could literally broadcast development.[9]

Historically an agricultural term, to broadcast meant to widely scatter seeds for harvest. For administrators, broadcasting held out just such a promise—the medium itself capable of spreading development in both its ideational and material forms.[10] As noted in 1936 by a committee appointed to report on broadcasting in the colonies, "We envisage the development of Colonial Broadcasting—and its justification—not as an instrument of entertainment . . . but . . . as an instrument of advanced administration . . . for the enlightenment and education of the more backward sections of the population, and for their instruction in public health, agriculture etc."[11] Radio, many believed, could usher both people and landscapes into a developed future. As E. R. Davies proclaimed over the airwaves in 1941, radio was to "ventilate and explain to the African this plan of development in all its aspects and the methods by which it is proposed to carry it out."[12]

As this chapter shows, even this parsimonious aspiration was stymied from the outset, and proponents of radio were forced to reckon with arrangements born of an earlier moment of developmentalist thinking. While the British government had debated constructing an empire-wide communications network since the early twentieth century, Parliament's commitment to "sound finance" led the body to repeatedly vote down such proposals.[13] The extension of broadcasting in the colonies was thus a frugal affair; the dearth in metropolitan funding requiring that prospective stations be "self-supporting" in keeping with the imperial analog of "sound finance," namely colonial "self-sufficiency."[14] Absent funding from the Treasury, the government had thus entered into early agreements with what would become the British monopoly, Cable and Wireless Ltd. (C&W), to service white settler and South Asian listeners. These early decisions—driven by the pragmatics of fiscal constraint and a logic that gave primacy of place to the needs of white settlers—functionally attached the information arm of the state to infrastructural networks funded and established by a corporation. In the absence of funding, the corporation, the state hoped, could function as an infrastructural prosthetic. But C&W's profit prerogative, in tandem with Kenya's unique topographic, atmospheric, and demographic conditions ensured that, over the course of the colonial period, broadcasts from Nairobi failed to provide "satisfactory reception over considerable areas."[15] Aspirations notwithstanding, administrators acknowledged that it would be "a very long time before one can hope for . . . [the] ideal [of domestic listening] to be reached in Africa"[16]

Administrators, then, realized that material constraints would make domestic listening over personally owned sets impossible. Seamless radio broadcasting—like brick-and-mortar schools, electrification, well-resourced health clinics, and paved roads—was not a possibility in the immediate or near future. In the meantime, the government hoped that "through the medium of a wireless receiving set attached to the van," the problems thrown up by the lumpy reach of the state could be mitigated, as the van brought government messages and entertainment "to audiences which would not otherwise be reached by the vernacular broadcasts from the Nairobi Station."[17] For all intents and purposes, then, the mobile information van *was* radio in this period. It *was* broadcasting. This van was a tropicalized technology designed to bridge the gap between an emergent developmentalist ideology and the particular material world of the Kenya colony shaped as it was by Kenya's unique geography, racial hierarchies, and fiscal policies that saw the state outsource broadcasting to a corporation, Cable and Wireless Ltd. At one level, then, efforts to enact radio in this period both indexed the change in direction of colonial policies and exemplified colonialism on the cheap, particularly as plans for centralized administration devolved to "community development." This was colonial austerity in the age of social welfare.

While the work that went into orchestrating these events was occluded in information officers' reports, these visits were carefully staged and depended on the input and practiced labor of African knowledge-workers and experts. And this work was critical. Not only were the technical components on which the unit depended routinely in a state of disrepair, but the ideological and developmentalist ambitions wrapped up in new media forms were perennially on the brink of failing. In the minds of administrators, these registers of breakdown were linked—the breakdown of technics would lead to a breakdown of the messages on offer. The maintenance and repair of both imperfectly tropicalized technologies and the messages that rode on their rails were essential. The Information Office called on African technologists and knowledge-workers, colonial "middle figures" and experts of both technics and culture, to the task of themselves acting as infrastructural prosthetics—critical maintainers of both technology and ideology.[18] Central to this maintenance and repair was the work of translation, which required content be not only translated in linguistic terms but in conceptual terms as well.

The importance of these contributions would be tacitly acknowledged by the colonial state during the Mau Mau uprising of the 1950s, when the Kenya Information Office (KIO) tasked African knowledge-workers with generating content that would resonate with communities in the countryside and with the young men and women fighting in the forest. During the uprising, this cultural expertise was appropriated and subsumed by the state, and put to the ongoing task of "tropicalising" Kenya's media infrastructures. It was precisely this critical work of translation that rendered African knowledge-workers powerful brokers and authors of culture. Like "ethnic patrios" writing for the vernacular press, and those putting to page "timeless" ethnic histories, these radio men sought to establish a monopoly over the past to manage the turbulence of the present.[19] As they fulfilled their vocations, they mobilized their status as infrastructural prosthetics, engaging in their own projects of culture-building and generating new attachments with "their listeners."[20]

Forging Infrastructural Attachments

While Kenya was not the only colony to use information vans as the primary medium of broadcast, it did not do this as a response to the absence of a radio network, but rather in spite of the presence of an infrastructure of broadcast. To understand why this was the case, we must first explore the history of this network and the interests guiding its consolidation and operations. This history began in the 1920s, and drew together the pecuniary concerns of the British state, the desires of white settlers, and the interests of a British corporation, Cable and Wireless Ltd.

The British government's interest in the role radio communications could play in colonial administration was born of technological transformations, namely Guglielmo Marconi's successful use of shortwave beam transmission.[21] The development of beam technology led to fierce competition between cable and wireless companies, prompting the British government to organize the Imperial Wireless and Cable Conference in 1928.[22] It was agreed that Sir John Denison-Pender's Eastern Telegraph Group, and Marconi's Wireless Company would stage a merger, bringing together the interests of both wired and wireless industries, and establishing a major monopoly firm.[23] As *The Economist* reported, the rationale for the merger was threefold: to secure all the advantages to be derived from unification of direction and operation; to preserve for the Governments concerned control over any unified undertaking which may be created, in order to safeguard the interests of both the public and of the cable and wireless users; and to secure these *desiderata* at a minimum cost to the Governments.[24] And so, Cable and Wireless Ltd. was born, with Imperial and International Communications Ltd. (IIC) operating as its subsidiary, an arrangement that largely benefited shareholders—insurance companies, pension funds, and investment trusts—located in London.[25]

While IIC was a private firm, communication infrastructures were inextricably linked to the prerogatives of the state. The IIC was deemed a "public utility," mobilized to secure "the public interest," a designation that subjected the firm to government oversight.[26] The chairman and one member of the board were appointed by Her Majesty's Government. Dividends were limited so as to avoid the company being guided solely by a returns-based logic. And the IIC was required to consult the Imperial Communications Advisory Committee, which was composed of representatives of the governments present at the conference. The future of radio broadcasting in the Kenya colony would be shaped by these infrastructural attachments, which bound the colonial government to the corporation and its shareholders.

For many observers, though, the novelty of the IIC undertaking the work of building and managing imperial communication infrastructures lay not so much in its status as a private company, but in its use of the newly developed beam technology.[27] From the perspective of the CO, the materiality of beam technology mattered in three ways. Propagated through the air, beamed messages could not only be sent at three times the speed of those traveling by cable, but required a mere one-fiftieth of the power, and operated at a twentieth of the cost.[28]

If the British government's interests in wireless were guided by pecuniary considerations, the aspirations of white radio enthusiasts in Kenya—people for whom the colony was a "white man's country"—were similarly shaped by questions of economy, namely the prospect of accumulation. Between 1922 and 1928, Kenya experienced a period of economic prosperity. Export revenues from coffee and sisal poured into the colony.[29] For settlers working as commercial farmers, the Nairobi-London link would allow for the near-immediate coordination of prices, prospectively giving them a significant advantage over regional competitors in Uganda, whose messages would have to be routed through Nairobi.[30] It was likely with these material transformations in mind that in 1927, the year before the Conference, the government of Kenya granted the white settler-run British East African Broadcasting Co. (BEABC) call signal VQ7LO.[31]

Over the course of negotiations, the BEABC secured two licenses. The first authorized the company to maintain a beam-type telegraph for direct communication between Kenya and England, which would be built by the British Post Office.[32] Some government negotiators appear to have been savvy as to what the future might hold, insisting the Kenyan government offered this deal "only on the understanding" that the BEABC would "inaugurate a local broadcasting service to cover the whole of the Colony."[33] And so the second license permitted the company to provide shortwave services to the colony.[34] While this new service—financed by "prominent local business men,"—was to provision for what was referred to as "the public," this public was from the outset limited in both space and by race, with radio broadcasting mainly servicing white settlers living in Nairobi and in the countryside.[35] As in Angola, Kenya's settlers, too, hoped to use radio to broadcast "whiteness, stretching it out sonically, [and] hailing listeners to orientate themselves around and to the sounds of settler whiteness."[36]

While C&W's services were racially exclusive, the agreement granted the Postmaster the right to rent the licensed apparatus for "public purposes"—namely broadcasts for the African majority—at the rate of KSH 20 per half hour.[37] In negotiating this agreement, the colonial state functionally attached the future of its information arm to the infrastructures of the company. In 1928, the station opened with much fanfare, making Kenya the second British colony, only after India, to have regular, daily wireless broadcasts.[38] As it was a business venture, listening would not be free, with radio licenses costing KSH 50 per annum.[39] In exchange for offering these services, the company was to receive 50 percent of all licensing fees collected by the Kenya Post Office.[40]

Provisioning a colony-wide service was no easy task in material terms. What was then called the Kenya colony is a large territory with a highly variable topography and few people on the ground. Technological developments in radio broadcasting were no match for this novel environment. As settlers and radio hobbyists Norman E. Walter and Allan S. Ker recalled: "The medium wave transmitter was not powerful enough to cover adequately the required area." Medium wave, while desirable over short distances, is a sensitive medium, susceptible to atmospheric and electrical interference, both of which subject broadcasts to distortion. The frequencies at which medium waves travel are also variable in a 24-hour cycle, messages moving better during the night, but struggling to travel during daylight hours.[41] As for shortwave, the technology was out of reach in the early days of the BEABC, it still being "in its infancy."[42] Financial constraints, namely the absence of government funding, further challenged the team. As Ker and Walter noted: "To begin with, all the equipment used for . . . broadcasting services, was designed and constructed locally to cut costs."[43] Kenya's early radio network was thus born of government policies that outsourced the cost of constructing and administering the network to this rag-tag operation—it was, from the outset, a "tropicalised" technology.[44]

The life of the BEABC was short-lived, and within fifteen months of launching its services the company was absorbed by the massive London-based IIC.[45] When the IIC purchased the firm in 1930, it begrudgingly agreed to take over the BEABC's broadcasting obligations.[46] While the contract allowed for a government buyout, the cash strapped colonial state had little interest in taking over the unremunerative service from IIC during the first decades of broadcasting—though some administrators, perhaps not realizing their prescience, disputed this decision.[47]

While the absorption of the BEABC by C&W largely led to the replacement of local hobbyists, amateurs, and hastily constructed technologies with C&W's state of the art manufactures, there was nothing seamless about the services offered by C&W.[48] Radio, while seemingly a stand-alone infrastructure, relies on the presence of other networks for its smooth operations. These conditions could not be counted on in Kenya. The daily news often arrived at the studio late as messengers struggled to bicycle from Nairobi to the station along a non-macadamized road that was near impassible during the rainy seasons.[49]

But things appeared poised to change. In 1934, the IIC rebranded itself Cable and Wireless, and by the late 1930s, the station began broadcasting on shortwave. This technological fix, many hoped, would render issues of topography and atmosphere a thing of the past—shortwave not only having a more expansive reach but also being less liable to atmospheric and electrical interventions and variations over the course of a 24-hour cycle.[50] Despite this technological transformation, and the new affordances it portended, the reach of the broadcasts was by turns more expansive and more limited than the group had hoped.[51] The uneven reach of broadcasts emanating from Nairobi was a perennial problem with which administrators had to reckon. And these limits, like those resulting from the state having attached itself to the private firm, would routinely undermine colonial aspirations to use radio to enact development at a distance and on the cheap.

In 1938, C&W constructed a new studio in Kabete. The timing of this expansion was felicitous, coming as it did in the months leading to the outbreak of the Second World War. C&W worked with the newly established KIO to facilitate a "large increase in broadcasting" to Africans.[52] In preparation for this expansion, the Kenyan government initiated an "experiment" designed to gauge "Africans'" responses to broadcasting in Kiambu, a district in Kenya's Central Province, in August 1938.[53] The purpose of the experiment was "to discover how broadcasting . . . [could] best be used for the educational development of Africans" and, as "a social service," it was financed by the government.[54] The spatially constrained nature of the experiment, though, was primarily guided by fiscal restraint—there were simply not funds available to "conduct listener research on any but the smallest scale."[55] This was likely in part due to government concerns regarding the powerful white settler lobby, which routinely denigrated funding for what it deemed to be overly indulgent investments in the lives of the colony's African majority.[56]

Running on a "shoestring," the KIO relied on myriad non-state actors to subsidize its labors. District Commissioner D. H. Williams assembled a team in possession of various registers of expertise in devising programming for the experiment, which included not only representatives from the "Medical, Education, and Agricultural Departments" but also Louis Leakey, of whom we shall hear more, and "an educated Kikuyu, Mr. Harry Thuku."[57] Fiscal constraints required that the government mobilize these extra-state entities as it considered incrementally extending infrastructural connectivity. The administration simply could not have pulled off the experiment on its own. These constraints notwithstanding, administrators were confident in the results. Radio was truly a miraculous tool of "advanced administration," they believed, which "proved quite definitely that the African is 'Wireless' minded."[58]

The exigencies of the war heightened the CO's perception that imperial communication infrastructures would need to be expanded, leading the government to abandon the experimental model in 1939.[59] Soon thereafter, the government initiated broadcasts in Kikuyu, Dholuo, Kikamba, Luhya, Luganda, Kinandi, Kisii, and Kiswahili on transmitters rented from C&W.[60] But C&W's disinterest in provisioning for the largely unremunerative African majority, in tandem with Kenya's unique topographic and atmospheric conditions, continued to ensure that broadcasts emanating from Nairobi did not reach the bulk of prospective African listeners. In light of these limitations, for many in the 1940s, radio arrived in their lives not in the home on personally owned sets, but over the colony's troublesome road networks and in a van. In the seams of this lumpy infrastructural network, African knowledge-workers and technical experts asserted their autonomy. It is the story of one such knowledge-worker, Ambrose Wakaria, to which we now turn.

The Making of Colonial Social Categories

I met Ambrose Wakaria in Ngong Town, a large peri-urban city on the out-skirts of Nairobi in 2014. I was meeting Wakaria because he was rumored to be the oldest serving broadcaster in Kenya. I soon learned he was also revered as an expert in Kikuyu culture. Wakaria was born in 1928, the year after the BEABC got its start, in an area of Nairobi that is today called Ridgeway. "These were all Kikuyu lands then," he told me, gesturing out the window. His parents were early Protestant converts, joining the CMS in the early 1920s. Before Ambrose's birth, they had a "Christian wedding," paid for largely by his father who worked as a cook in Nairobi.[61] When Europeans stole his family's land, he and his family moved to Kiambu. Like many, his father's decision to join the church led to conflict within the family.[62] Wakaria's grandfather was par-ticularly angered by his son's decision to send Wakaria to mission school. "My father was very much interested in schools . . . [but] my grandfather could not allow me to go to school," Wakaria recalled. The two were at an impasse. Wa-karia grew up in the midst of this tension, but walked the line with equipoise, working to suture the breach.[63]

His father's conversion and relative affluence brought Wakaria first to a CMS mission school and then, in 1942, to boarding school. Wakaria's relation-ship with his education was fraught. While he enjoyed history and agricultural courses, the teachers were harsh, he told me, "beating us all the time." Follow-ing high school, Wakaria studied pharmacology at King George Hospital. This unusual trajectory landed him a position at the KIO in 1950. And it was through this work that he gained his expertise (and celebrity status) as an authority on all things Kikuyu. Men like Wakaria were both the voices and the ears of the information arm of the colonial state. They were researchers and experts of "culture," lay ethnographers whose knowledge was routinely mobilized by the colonial state as it fine-tuned its information apparatus. Men like Wakaria—critical translators and professional knowledge-workers—occupied positions of relative power. The colonial state viewed these men, and the knowledge they bore, with apprehension.⁶⁴ Wakaria, like many African knowledge-workers, got his start working for the mobile information unit.

If Wakaria's itinerary was relatively unique, the same was true of the British information workers under whom men like Wakaria worked. Arthur Champion, like Wakaria, was an avid observer of people as well as place, coupling his administrative duties with a penchant for lay ethnography. In 1944, following his retirement, he penned "Native Welfare in Kenya."⁶⁵ In it, he laid out his ambivalent feelings regarding Christian education, arguing that conversion had led to undue rifts with the mission-educated constituting a separate community within a "sea of paganism."⁶⁶ While Champion would have had a good deal to say about the conflicts that structured Wakaria's upbringing, the unit and the information arm of the state were utterly dependent on men like Wakaria.

Champion appreciated the possibilities for "development" that many at the time located as being the special purview of the mobile information unit. Unlike the Church, the unit's reach, though nodal, was theoretically sociologically all-embracing. In 1940, Champion became the first mobile unit operator, crisscrossing the colony in the single van serving the territory from which he broadcast "news" (with African broadcasters providing translation), screened films (with African knowledge-workers providing commentary) and, with his crew, organized live shows (with African assistants executing performances). According to observers, Champion was attached to "his staff," which consisted of "the Machakos driver a Kamba called Katele and his assistant and ... two interpreters." Moses, the Kikuyu interpreter, spoke Kikamba as well as Kiswahili fluently, while John, "the Luo," spoke Kiswahili as well as his own "mother tongue Dholuo."⁶⁷ The linguistic specialization of these knowledge-workers, as we shall see, was a critical component of their expertise. The KIO was thoroughly dependent on the skills of these men as it worked to convince communities to embrace its largely unpopular developmental interventions.

A Fragile "Development"

The mobile unit traveled widely, with every district "anxious to have a prolonged visit."⁶⁸ Staged in largely unelectrified regions of the colony, the glow of the screen, the projection of voices, and the large audiences marked these occasions as exciting events. These shows not only drew crowds from far afield but were heterogeneous and lively. In Nakuru, one 1940 show drew a crowd of "between 2,000 and 3,000 natives ... of mixed tribes." This linguistically diverse group demanded that "Kiswahili, Jaluo and Kikuyu languages were all made use of." This linguistic flexibility, the report noted, "appeared to give general satisfaction."[69] Not only did heterogeneous populations often watch and listen side by side in this period—generating a multilingual and polyphonic space—but the content was anything but exclusively "local" in character, "the keenest interest and curiosity" being "aroused in new scenes of unfamiliar peoples, lands and customs, giving rise to a desire for reading and a wider knowledge of the world."[70] In the minds of administrators, this interest in the broader world was intimately linked to the developmentalist mandate of the tours. As one wrote: "There can be little doubt that the use of selected films . . . would rapidly make the minds of the Africans in the Reserves sympathetically disposed towards the acceptance of improved conditions of life."[71]

These spectacles followed a fixed temporal regime. As Champion wrote, "The performances . . . start at 6.30 p.m. preceded by a few items consisting of band music, songs in English, Kikuyu and Kikamba on the loudspeaker and the reading of the introductions in the appropriate language of two or three of the films takes place . . . After that one or two local films follow with running commentary only . . . We conclude [at] about 8 p.m. as a rule."[72] The regimented nature of these visits was in line with the developmentalist-cum-disciplinary effect that administrators hoped these events would inculcate in audiences. Not only were attendees to be "taught" how to reform everyday life and so achieve development, but the administration hoped to cultivate spectators of a particular sort: quiet, attentive, interested.

The unit was not only regimented by the clock, but the timing of these trips was shaped by the seasons, with the unit staying "in Nairobi during the month of April when the long rains [arrived]" and in "November for the short rains." For African knowledge-workers and experts, these were periods of reprieve, enabling them to tend to their *shambas*, take holiday, and visit family.[73] During these periods, British officials working with the mobile unit "would travel around on the tarmac roads in the vicinity of Nairobi showing to various institutions such as the two breweries at Ruaraka, [and] the schools at Pumwani."[74] These men were only able to do this work during these months by virtue of the presence of tarmacked roads, a luxury limited to Nairobi and its immediate hinterlands. The unit, in other words, was dependent on another infrastructure whose lumpy reach was similarly born of policies of austerity: roads (see chapter 2).

For audiences, the unit's arrival bore with it the possibility of temporarily traveling to various global elsewheres. As Bartholomew Orwa recalled, this brought pleasure and excitement. Orwa wrote, the "first thing which made me interested was [the] Gramophone. This made me stand and dance, and I said in my heart that this was the only thing which I should hear till I shall leave the place."[75] But music was merely the initial draw. It would have been followed by propaganda films designed to shore up popular support for Britain's war effort, developmentalist films, and on many occasions, crew members tuned into broadcasts from the Nairobi station.

On one occasion, as J. E. Obala recounted, music was followed by a film featuring "Bukura Farm," which boasted a "garden" with "many crops," and "good milking cows," one of the best of which was milked "by a brown boy." Obala continued his composition, writing: "This boy . . . used all the methods he learnt at the Veterinary School like using a clean bucket[,] washing his hands before milking, washing the cows udder and rinsing the bucket with hot water. All of these made it sound to correlate the crops work and the animals work."[76] This developmentalist vision reflected an emergent rationality that, following the war, would see the rise of new cadres of experts and new genre of expertise that sought to reshape the colonial landscape and effect "rationalized" development. This vision got its first lease on life with media technologies, tools of both "development" and "advanced administration."[77] These aspirations notwithstanding, during the war and thereafter limitations in financing ensured that these messages were spread in nodal form.

The content of the mobile unit's shows offered audiences a vision of a seamless, well-integrated development apparatus. In this instance, agricultural and veterinary work were presented as intimately connected—now rationalized, they unfolded together as an integrated and coordinated totality reflecting the emergent developmentalist ethos. But the fiscal policies of the colonial state rendered these visions chimerical—they were spectral and experimental at best. This was not a "real" shamba, but the "Bukura Farm Institute," a "12 acre model holding" to which "married couples were invited to live" so as to get a "good idea of how they should look after their own land." This, then, was a representation of an ideal developmentalist future, rather than a reflection of existing conditions. As one administrator wrote: "This made an excellent propaganda film," but it "did not work out in practice." The couples living on the farm complained that "it was all very well for governments to have these farms because there was no risk attached if the crops failed." By contrast, if the couples used the proposed agricultural methods on their own plots and the crops were to fail, they "faced starvation."[78]

Just as word of the mobile unit preceded its arrival, so too did rumors of failed agricultural experiments and development initiatives gone awry. Popular responses to government broadcasts and films were filtered through experiences wrought by colonial developmentalist interventions into the landscape and people's lifeworlds.[79] This raised the stakes for government-initiated information work. The unit was to mitigate not only against the uneven infrastructural reach of the state, but its contents were to convince and cajole audiences in the face of conceptual breakdown. The unit's operations had to offer proof of concept. In both material and conceptual terms, African experts would be asked to fill the breach—to act as critical maintainers, repairers, and translators of both technics and culture. Their labor—largely occluded in the colonial archive—was essential prosthetic work, which concurrently sutured together the lumpy infrastructural state while transforming its shape.

Medium as Message: Mechanical Breakdown, Prosthetic Work, and the Cultivation of Technical Expertise

In 1943, Champion retired and was replaced by W. F. P. Kelly.[80] For Kelly, this new role joined together his professional ambitions and his desires for administrative intimacy in a moment of flux as the new emphasis on centralized development saw "men on the spot" being replaced by metropolitan experts.[81] "The Officer-in-charge of the Mobile Cinema has one of the most interesting jobs in the Colony," he wrote. "He sees more of the country and the people than anyone else, and has the most leisure to digest his experiences. Nothing could be better for an officer who having spent years in the Secretariat, feels he has lost contact with reality, than to spend a few months with a Mobile Cinema."[82]

For Kelly, the success of the mobile unit required producing a sense of awe in audiences.[83] While this had been an expressed goal of British information workers from the outset, Kelly worked to strengthen this effect when he replaced Champion. Kelly's "improvements" to the unit, importantly, did not pertain to content, but to form. He had raised the height of the screen, thereby enabling the projector to be located some distance from the screen itself. This reconfiguration was undertaken to better conceal the mechanics of the unit by minimizing the "noise of the generator." This was critical. "From this eminence," the show could "be heard for several miles, the crowd . . . [could] be silenced at will, and, when silent, the commentator . . . [could] introduce all kinds of modulations into his voice."[84] As was the case with the "Bukura Farm," the "success" of these shows, administrators believed, required minimizing the fact that they were, in fact, curated and staged. New media technologies were supposed to awe, yes, but they were also to be received as transparent lenses onto reality.[85]

Kelly's interventions were an improvement on the old system of harder, more interventionist discipline. Rejigging technologies arriving from the United Kingdom had made it possible to allow spectators to "laugh, talk, and enjoy themselves within limits without that continual blowing of whistles by . . . Tribal Police . . . an irritating feature of the old shows."[86] This was an attempt to move from discipline to ideology. And these shows, Kelly assured the administration, were a great success, with audiences reaching 2,300.[87] The system remained unwieldy, he admitted. He fantasized about a unit in the future that was "less elephantine and more mobile."[88] In reconfiguring the unit, Kelly followed in the footsteps of Champion, who impressed on Kelly the importance of actively tropicalizing this imperfectly tropicalized unit by navigating its imperfections, responding to its breakdowns, and improving on its design. Backgrounding the mechanics of technologies required labor and vigilance, responsibilities the KIO devolved to African technological experts.

The KIO mobilized African technological experts to mitigate against the possibility of conceptual breakdown that would be the result of visible or audible disjunctures between the state's aspirational developmentalist future and the material present. These men were responsible for ensuring that imperfectly tropicalized technologies worked in the novel material setting of the Kenya colony. European information officers expended few words on the African men staffing the unit, routinely downplaying their technological expertise. Consider the following 1944 report: "The work of the African staff has been satisfactory considering they are not qualified electricians or mechanics."[89] Far from being "middle figures," these men were not even represented as having been cogs in the unwieldy operation that was the mobile information unit.[90] This supposed lack of expertise would need to guide the selection of technologies, administrators argued, which would need to "operate over long periods *without skilled servicing*."[91]

The long distances traveled over poorly cut roads was hard on bodies, hard on the van, and hard on equipment.[92] It is precisely in the context of "indiscipline" and "breakdown" that these experts and their expertise become visible. The archival traces emerging in the context of indiscipline tell us something of the strictures under which these men performed their work, and their emergent status as a social category. Archival traces emerging in the context of breakdown reveal how critical their work was to the operations of the Kenya colony's lumpy information apparatus.[93] And breakdown was routine. One monthly report recounted that "performances have been somewhat marred by a defect in the loud speakers which has begun to assert itself," leading to a "complete breakdown of our arrangements for amplification of the human voice." This defect, it continued, "would not assert itself or perhaps be a defect at all under conditions in England."[94] Administrators routinely attributed breakdown to imperfectly "tropicalised" technologies. And this was not limited to mechanical technologies. Film reels, too, had a different shelf-life in the "tropics."[95] Degradation, noted one report, was the result of "many factors."[96] Climatic conditions were partially to blame, but so too were the "[b]ad and dusty roads," which led to "heavy wear and tear on the films,"[97] causing "a considerable amount of slacking" to take place "on reels of film . . . [leading] the adjoining surfaces . . . rub against each other at every jolt on the road."[98] These media technologies were simply not designed for these conditions, requiring the intervention of African experts who routinely took on "duties" that were supposedly beyond what "their education and experiences warrants."[99] But the work of maintenance and repair was both critical and creative.[100]

These imperfectly tropicalized technologies made developing a series of work-arounds essential. Crews devised "various devices for securing films tightly on their reels to stop them from rubbing in transit. An efficient and cheap one" was the use of "a thin cross section of an old motor-car inner tube. A loop of tape on this home-made elastic band . . . [making] removal easy."[101] The need for such work-arounds were both generated and shaped by policies of colonial austerity. Innovations, both "efficient and cheap," were part and parcel of the work of maintenance, but that did not render them unstudied.[102] Technical interventions of this sort were essential to making broadcasting work, and the perceived stakes of failure were high.

Administrators were convinced that mechanical mishaps would shape the messages received, that an interruption in the medium of conveyance would undermine the messages on offer. Such interruptions would have durable effects, disrupting efforts to attach people to the world of colonial information and retarding the "development of broadcasting in East Africa for many years."[103] As F. H. Knight reported, absent the dedicated maintenance of imperfectly tropicalized technologies, deterioration was inevitable, the results of which were "obvious to the audience," who were quick to point out any interruption in the flow of broadcasts and live shows.[104] Audiences' evaluations of breakdown were a major concern, for they rarely "blame[d] the projector." Instead, audiences left with "a poor opinion of either the film, the projectionist, or the mobile cinema generally."[105] Anxiety over audience responses rendered the daily work of African crew members critically important.

These technical experts routinely repaired the Megavox, changed worn out tires, tinkered with imperfectly tropicalized loudspeakers and amplifiers, tuned the wireless receiving set, managed the van-engine generator and the Bell and Howell 16 mm projector, and spliced together films when they broke. Officials' reports notwithstanding, these men were technologists who fine-tuned their practice on the road. And this practiced labor had rendered the crew expert maintainers.[106] The "responsibility of maintaining the equipment in good condition," wrote W. Sellers, "falls upon the operator . . . [who] will . . . set up a regular routine which will enable the equipment to be maintained. . . . In this way, he will become familiar with the entire equipment and be able to replace or readjust minor parts without causing delay in an itinerary or the expense and miles of travel by a skilled mechanic or electrician."[107]

The disjuncture between the actual daily work of African crew members and colonial representational practices was rooted in a deep-seated ideology that maintained there was something incommensurate between Africa and technologies, between Africans and technological expertise. This impeded administrators' capacity to see technological expertise where and when it was happening. And yet, evidently, African operators were routinely saddled with the task of maintaining, repairing, and translating persnickety technologies.

While the Bell and Howell was widely viewed as being an ideal projector for the region, working "well under tropical conditions," it was a complicated technical object.[108] As Knight wrote: "The portable film projector is a highly complex precision instrument, and consequently requires . . . careful handling and maintenance."[109] Without this work of maintenance, the equipment would be subject to the particularities of the "tropics," the "very roughest treatment" allowing "dust and dirt" to "accumulate."[110] These conditions demanded the mobilization of specialized knowledge, without which the "projector's steady journey on the downward path" was assured. The expert work of these men, far from being unexceptional, was practiced.

If African experts and technologists were responsible for maintaining and rejigging imperfectly "tropicalised" technologies, they were also responsible for negotiating worn-out parts—the material stuff of policies of colonial self-sufficiency. After complaining about a series of mechanical upsets, Champion noted that not a single show had been delayed, admitting that credit was due to his "native driver who . . . [had] jumped into the breach." This act of jumping into the breach was routine. In this instance, the "driver/operator,"—one of the few technical experts I can track in this partial archive—was Katele, who had worked with Champion since the earliest days of the unit's operation—navigating the colony's treacherous roads, as well as operating, maintaining, repairing, and translating the generator, projector, and loudspeaker along the way. Conditions in the field and the habitual breakdown of non-tropicalized technologies demanded this daily creative work of innovation, maintenance, and repair.[111]

Message as Medium: Conceptual Breakdown

Savvy administrators betrayed their understanding of the centrally important role played by knowledge-workers. As one noted, "There can be no evangelism without evangelists—no propaganda without propagandists. A good African propagandist is as rare as a good European one."[112] The comparison of African information workers to African evangelists was not incidental. The success and spread of Protestantism in the region turned on the knowledge, the work, and the commitment of these critical intermediaries.[113] Perhaps it is no surprise, then, that when issues of translation came up in relation to media, missionaries had a lot to say.

Reverend Leonard Beecher of the CMS was centrally involved in orchestrating the production of broadcasts to African communities, seeing the "value of the Christian contribution to African Broadcasting."[114] In 1941, an ad hoc Translation Bureau was established as a joint effort between the administration and the CMS.[115] Five African translators were engaged under the direction of Beecher to undertake translations from English into Kikuyu, Kikamba, Bantu-Kavirondo (likely Luhya), Kiswahili, and Nandi.[116] Miss Clarke of the Union of Seventh Day Adventists supervised Dholuo translations.[117] The government launched two half-hour programmes weekly in a handful of the vernaculars, with an additional fifteen-minute programme daily in Kiswahili, using transmitters rented from C&W.[118] Gusii and Taita were each given one program per week.

In mobilizing the labor of missionaries, the government built on long-standing responses to historical limitations in financing, conjuring up missionaries' sense of purpose as they tethered missionaries to state-directed projects of development. Yet this was not merely colonialism on the cheap. Administrators saw missionaries' participation as being essential to generating new attachments between communities and new media, a capacity that turned on missionaries' "knowledge of the African."[119] These women and men brought their knowledge born of this supposed intimacy to bear as they entered the world of colonial information.

While for all involved, the language of broadcast was a matter of concern, the contours of this concern varied widely.[120] Administrators, for their part, imagined that broadcasting in the future would be undertaken in Kiswahili, itself a stop-gap solution, with English eventually replacing Kiswahili as the single language of broadcast (see chapter 4). Missionaries, by contrast, emphasized the importance of reforming and standardizing vernacular languages. This belief was born of missionaries' conception of language, which posited a direct correlation between language and culture.[121] As a member of the Advisory Committee on Education in the Colonies wrote, "The mother tongue is the true vehicle of mother wit. Another medium of speech may bring with it, as English brings with it, a current of new ideas. But . . . [it] is through the vernacular . . . that the new conceptions of the mind should press their way to birth in speech."[122] Language was not simply the vehicle for expressing reality. Instead, it registered changes in, and in turn reshaped, reality.

A Kikuyu Anglican representing African interests was in partial agreement with this position, arguing: "It is difficult to see how a child can acquire any real understanding and command of English if he has never been trained to express himself with ease and freedom in his mother tongue."[123] For this representative, though, vernacular-language acquisition was not incommensurate with the learning of other languages. He continued, noting that Africans' second language of choice was English, comprehension of which was an "African's right," it being the "only door . . . that is precious in the common inheritance of the peoples of the Empire."[124]

Not all agreed with this position. Many viewed Kiswahili, not English, as being a key *lingua franca*. As an anonymous writer argued, "Swahili is a necessary part of a country such as this, containing so many tribes and tongues, with people at so many different stages of development."[125] For this letter writer, Kiswahili would be the language of a new cosmopolitanism. He continued: "If people will only realize that township life and consequent mixing of tribes and tongues is going to spread, they would realize too that in Swahili we have a priceless means of inter-communication and of imparting information and instruction."[126]

The position of this writer was born of an emergent broadcasting infra-structure, with Kiswahili gaining popularity "due to the increasing use of Swahili in broadcasting . . . and to the wider world brought to the notice of the tribes by the war" who saw "the value of Swahili in their contacts with fellow soldiers."[127] This was not an argument for parochialism. Theirs was a politics of linguistic and conceptual multiplicity, a world of potential wherein multiple scales of belonging—in linguistic, social, and political terms—were simultaneously at play (see chapter 4). It was the state and missionaries, though operating from different vantage points, that maintained a parochial vision of languages as constraining structures that would dictate the desirable scale— here in the singular—of social and political inclusion. If, for administrators, vernacular-language broadcasting was a stop-gap solution in preparation for the days of monolingualism for some Africans, Wakaria among them, register switching was not a stop-gap position but a desired state.

The problem was fundamental: direct translation was impossible. Nor was translation simply a linguistic feat—a consideration which itself required "unremitting attention to every detail between the preparation of the original English script . . . and the eventual delivery to the microphone"—but an achievement that required the painstaking labor of translating concepts.[128] This was completely uncharted terrain that required "constant trial and error," for "the task of broadcasting to Africans . . . [was] a new one."[129] The easiest and least threatening fix would have been to employ Europeans fluent in the vernaculars. However, not only was media that was obvious "government propaganda" dismissed out of hand, with listeners responding with "only scepticism and suspicion," but native English speakers' handle on the vernaculars was the subject of much derision.[130]

African Knowledge-Workers and Translating "Development"

In pursuit of conceptual translation, the KIO recruited African knowledge-workers. These men, like Wakaria, came from the ranks of the mission-educated, often pursuing higher education following high school, and spoke a minimum of two languages.[131] Most importantly, perhaps, they displayed a sensitivity for the local. And these men were relatively autonomous in their work. Beyond being largely in charge of the unit and responsible for the execution of the shows, they translated government scripts into vernacular languages and explained government policies and the rationale behind development campaigns.

Indeed, the central importance of these knowledge-workers was directly related to their work of translation, which was required in both linguistic and conceptual terms. Just as the unit's technical components were perennially subject to breakdown, so too were the messages the unit bore. Just as the unit's technologies required rejigging, so too did the developmentalist vision on offer. Administrators leaned heavily on the cultural expertise of these men, whose work of conceptual suturing rendered translation an object of administrative and missionary concern.

The skilled work of translation required shuttling meaning across what state officials and knowledge-workers alike perceived to be divergent epistemological fields. This work of conceptual translation was difficult to achieve, and ensuring quality translations took time and demanded vigilance.[132] As Mervyn Hill, an information officer, noted in 1941: "[T]he preparation of . . . English script[s] was often unsatisfactory and frequently lacking in [an] appreciation [of] . . . the form and style essential to effective translation into the vernacular language[s]."[133] This work of conceptual translation required mobilizing local idioms and salient metaphors, "allusion[s]" that appealed to the "history and the inherited wisdom of the tribe."[134] While some reported that "translator-announcers" took "to broadcasting like a duck to water," others acknowledged that being an effective broadcaster was not a given but a skill.[135] As Champion recounted, "The absence of competent personnel for the technical work . . . combined with a particularly incompetent announcer . . . who has neither the voice nor the mentality for the job" would destroy a tour.[136] These criticisms of personnel brought into full relief "how dependent the success of . . . demonstrations" was on "well read introductions and intelligent and well timed commentary."[137] What was required, in other words, were knowledge-workers capable of making British developmentalist messages legible and meaningful to African communities. Carbon copies across scripts were impossible; loose conveyance was as necessary as it was threatening.

In the 1940s, the KIO tasked African knowledge-workers with not only translating government broadcasts but with generating new knowledge. This took many forms, ranging from conducting research on local conditions to reporting back on local responses to the information disseminated through the mobile information unit.[138] The information gathered and interpreted by these men was subsequently cycled up and integrated into the state's expanding information apparatus, aiding the state in its ongoing efforts to tropicalize colonial media.

The recruitment of African knowledge-workers was undertaken in a bid to firmly root broadcasting in the lifeworlds of listeners.[139] Broadcasts were to "'feature' local news," and to ensure that "local news was of interest to Africans it had to be collected *by* Africans," argued E. R. Davies, head of the African Section of the KIO.[140] Knowledge-workers were "expected to make as many personal contacts as possible."[141] Some acted as "local correspondents, with their homes widely scattered." By 1941, approximately fifteen correspondents, who were tasked with sending news from the districts to Nairobi, had been appointed.[142] The goal was to attach African listeners to broadcasting by creating something unique, "something that was real *African* propaganda and not just an adaptation of mechanical European propaganda."[143] Vernacular-language broadcasts were to function as infrastructural attachments, tethering people and their futures to the developmentalist aspirations of the colonial state.

African knowledge-workers were thus critical intermediaries and maintainers of ideology as well as technics. This placed African knowledge-workers in a place of threatening power, and the colonial state looked on anxiously at evidence of emergent socialities born of the world of colonial media, all the while relying on such intimacies to make colonial broadcasting work. So thoroughgoing was this issue that knowledge-workers and broadcasters, researchers and translators, were always viewed with suspicion. If this work of conceptual translation required a kind of poetics, it also brought into full relief the politics of the infrastructural world of which broadcasting was a part.[144]

Missionaries, in particular, viewed radio as being an unsteady and volatile medium, leading those among their ranks to be even more anxious than the administration about mobilizing the expertise of African knowledge-workers. In coming to their conclusions, missionaries likely drew on their experience working with African catechists, and their knowledge of the importance of innuendo, proverbial forms, and idiomatic sayings in shaping communication.[145] They understood the complexity and the Janus-faced nature of language; the ability for language to communicate cultural facts that might be missed by European listeners.[146] Mobilizing local knowledge, all were beginning to realize, was a requisite, but it did not come without risks.

Such critiques raised fundamental questions about African knowledge-workers as a social category, and the medium of broadcasting as a form of communication. Programmes were broadcast live, and "African entertainment"—being "essentially impromptu in character"—bore a number of threats.[147] As Davies proclaimed in a 1941 broadcast targeting European settlers and employers: You "may rehearse them as often as you like and find that on the actual occasion of the broadcast they may give quite a different version to what they have rehearsed."[148] Thus, it was of the utmost importance that administrators locate knowledge-workers they could trust to faithfully broadcast the material supplied.

Colonial anxieties indexed a fear that knowledge-workers used their prosthetic work as a means of engaging in their own political and cultural projects. The threat: They were transforming these new mediated forms into their own prosthetics, extensions by which they could call to order new audiences, spread new messages, and engage in their own political contests over the mediated worlds of people they hoped to secure as *their* listeners. And administrators were not wrong. Deftly moving between subject positions—of loyal intermediary and dissenter, of soil expert and oath-taker—these men cycled through the colonial hierarchy, leveraging their positions to their own advantage, sometimes undermining the colonial authority that underwrote their own, while concurrently augmenting their power and prestige in domains largely illegible to the state.

This is best apprehended not as a movement between discrete categories of belonging and practice but rather as evidencing a social, political, and cultural world that was rife with code-switching and multiplicity. For knowledge-workers, in other words, their work as intermediaries was not incommensurate with their work as dissenters, their work as state-employed knowledge-workers was not at odds with their efforts to leverage new infrastructures to forward their own political projects. By 1946, likely in response to these concerns, African knowledge-workers were limited to "men of high educational standing." This hiring policy was matched by an expansion of responsibility, these men being tasked with "advising of the form in which the material should be put over to their own people."[149]

Despite the central role African knowledge-workers and technical experts played, not only was their import perennially elided in official correspondence, but they were poorly remunerated. In 1940, African translators recruited from other government departments to "ensure a satisfactory degree of reliability and responsibility" were not compensated for their work.[150] The frugal logic of "community development" had evidently taken hold. This generated no small measure of conflict, with the Information Department complaining that the seriousness of its mandate was being undermined by the reliance on "voluntary workers."[151]

Extending Prosthetic Work

By the 1950s, the KIO began actively courting dedicated African information workers. Given administrative anxieties, these men were subject to stringent oversight and scrutiny, but their intermediary status also opened up opportunities for creative forms of action. As the recruitment pamphlet detailed, those traveling with information vans were guaranteed attentive audiences, and would-be information workers were promised space to exercise their creativity and expertise. "Staff," would-be recruits were reminded, performed "interesting and creative work," which included "the writing of Press stories, the production of newspapers and magazines, broadcasting, the making of films and the running of cinema information vans." But like technical experts, these recruits would not be recognized as shapers or producers of knowledge, nor as creative innovators or designers. The African author of this pamphlet remained unnamed.[152]

Nevertheless, African knowledge-workers retained an unprecedented degree of autonomy. Each van was staffed by an information assistant, who was largely in charge of the unit and responsible for the conduct of the show. These knowledge-workers were accompanied by experts of other kinds—a few of whom we have met, such as Katele—"driver/operator[s]" who were to "drive the vehicle" and assist "with the operation of the generator and projector."[153] The peripatetic itineraries of these men often placed them beyond the reach of the state.

Wakaria entered this world of information, acting not only as the voice of broadcast, but also as a researcher. Later he became the editor of "Mugambo wa Kiambu," a newspaper directed to Kikuyu communities, after which he opened "the first information office" in Muranga. In 1956, he was tasked with opening the Mount Kenya Broadcasting Station, where he started another newspaper called "Matemo."

While the value of the knowledge generated by these men turned precisely on their ability to cultivate dynamics of intimacy with would-be listeners, the colonial state anxiously worked to modulate the emergent infrastructural attachments these men generated as they moved through the countryside. African information workers were to comport themselves with ascetic discipline. Screening was strict, many being "rejected on political grounds eg because they are members of associations, the activities of which are likely to make them unreliable."[154] People's schedules both on and off the clock were closely monitored, as were their movements. The administration put a whole host of regulations in place to ensure African information workers did not put Information Office technologies to their own uses. The administration desired the maintenance of a firm hierarchy and a strict division of labor. "Drivers are to drive. Operators and Officers in Charge of vehicles will on no account take the wheel."[155] But access to a vehicle, and the mobilities it afforded, registered these knowledge-workers as men of high status. Wakaria recalled with pleasure the prestige he gained driving around the countryside in the information department's vehicle. Prohibitions on unsanctioned movements were put in place to limit these emergent socialities.[156] These regulations were not simply preventative, then, but were a response to unsanctioned though widely documented practices. "It is known," one report noted, "that it is the custom of many drivers, operators, etc., to use the Departments' vehicles for the purpose of visiting friends who live near to places where entertainments etc., are given. This practice must cease forthwith."[157] Failure to conform to these regulations—even to be perceived to not conform to them—often led to "instant dismissal."[158] All of these prohibitions notwithstanding, overhearers could never tell, really, what knowledge-workers were communicating nor how their messages were being received. The use of the van enabled knowledge-workers and technologists to transform the unit and media infrastructures into their own prosthetics—technologies critical to the formation of new socialities and forms of attachment.

Vernacularizing "Development" and Mau Mau

The KIO's expansion in these years was a direct response to the specter of "disorder." Central Province, the region from which Wakaria hailed, was in crisis. Soil exhaustion, increasingly stark class differentiation within Kikuyu communities, and misguided government development initiatives proved to be a heady brew. While white settlers remained the dominant class in Kenya, those loyal to the government and people of high educational status, primarily among the Kikuyu, had amassed considerable wealth, constituting an African bourgeoisie.[159] Mobilizing a Kikuyu epistemology, these men claimed that they had attained *wiathi*, or self-mastery. With access to land came access to capital, both of which they could translate into "wealth in people."[160] Squatters and *ahoi*, or tenants, for their part, found the possibility of attaining *wiathi* functionally blocked.[161] Rumors of widespread oathing spread throughout the Kenya colony in 1950, as these social juniors gathered in the Aberdares Forest and those of Mount Kenya. Some 15,000 Africans, mainly Kikuyu, but also Meru, Embu, and fewer Kamba and Maasai, joined the movement.[162] The colonial state panicked, declaring a state of emergency in 1952.

What the colonial state called Mau Mau, otherwise known as the Land and Freedom Army, scholars have variously described as an anti-colonial struggle, a civil war, or a class struggle which pit young, marginalized Kikuyu men and women fighting to achieve "self-mastery" against an older generation of wealthy men who had forgotten their obligations to social juniors.[163] For some, this required constraining the behaviors of young women and men in the city.[164] For others it involved shoring up a putative "Kikuyu" center perceived to be under threat from ethnolinguistic and racial others.[165] For others still, it was a struggle to draw connections among class formation, land scarcity, and social adulthood.[166] By all accounts, though, Mau Mau was a project of social and moral reform wherein various players sought to assert themselves as the legitimate protectors and guarantors of contested notions of community and contentious visions of the future.

In this war of words, the colonial state deployed "development" as a key tactic designed to convince forest fighters of the Land and Freedom Army of the wisdom of surrender.[167] There was nothing novel in this. Indeed, the push for social welfare via radio broadcasting was a direct response to the labor unrest of the 1930s and 1940s.[168] It is no surprise, then, that efforts to mobilize development as an antidote were equally prevalent in the fight against Mau Mau. If development's failures were a product of Mau Mau's actions, the administration argued, future development along welfarist lines was the promised return for loyalty.

But globally circulating visions of development were not the only ones the state sought to harness in its fight against Mau Mau. Simultaneously, the KIO mobilized another, and deeper, set of ideas regarding social transformation— what we might think of as vernacular conceptions of development and the social good. If the history of radio broadcasting in colonial Kenya had taught the KIO anything, it was the importance of tropicalizing both the technologies of media and the messages they bore. During the Emergency, the colonial state mobilized African knowledge-workers and vernacular notions of development as they sought to convince fighters to leave the Aberdares Forest and those of Mount Kenya and return to "civilisation." In pursuit of this project, the Information Office actively combined notions of development premised on a logic of social welfare with Kikuyu notions of complete personhood, emblematized in the concept of *wiathi* or self-mastery. These twin notions of development, broadcast over the airwaves, were to reach into the cultural consciousnesses of Kikuyu forest fighters and their allies in the reserves as the colonial state worked to transform radio into an infrastructure of counter-insurgency.[169]

Kikuyu social thought holds self-possessed labor as the most valorized form of work. Through active labor on the land, boys became men, people became moral persons. According to this ethic of labor, culture was cut out of the wild. It was through the painstaking work of removing trunks, stumps, and roots that the land was made tillable and social reproduction was secured.[170] These homesteaders literally "sweated culture from nature."[171]

Labor as the means of self-actualization was historically reflected in Kikuyu structures of authority. An acephalous society, the homestead was the foundational unit of social organization.[172] The highest responsibility of a man was the successful management of his family. On being circumcised, a man would seek to separate himself from his father's *shamba*, or garden, cutting his own bit of soil and laboring to produce a tillable plot. This piece of cultured land was the first step toward establishing a household. It was at this stage that adult life was pursued (and pursuable)—a man marrying his first wife, who, after the initial clearing of the land, was largely responsible for tilling the soil. A man's worth was defined as *wiathi*. The designation of *wiathi* constituted recognition of the social value of a man's labor.[173]

Wiathi is a profoundly future-oriented theory of value. It was only by virtue of pursuing land, marriage, and children that a man would be remembered beyond his death; his legacy being sustained through his children, their mothers, and the generations that succeeded them. Adulthood was the promise in the immediate future. The status of venerated ancestor and protector of the line was the promise in the long-term. Failure meant adulthood was not realized. But *wiathi* is also a profoundly classed and gendered vision of maturation. Male adult status was premised on access: first to land, then to wives, and children thereafter. Poverty was a moral failure that merited social exclusion. Here, then, we have a Kikuyu epistemology of both individual and collective development. Mau Mau was, in part, an argument over this vision of maturation. Social and material conditions had been revolutionized by the 1950s. Land was scarce and thus the possibility of gaining the requisite materials for maturation was functionally obstructed. This was not the only time such disruptions had happened (see chapter 2), but it was the most extreme.

Within this epistemology, the uncut forest was the place of the wild, the forest standing in for the troubling forms of disorder against which Kikuyu civilization gained its definitional strength. The forest was a place of anti-culture; its creatures the opposite of man and civilization. Kikuyu enclosures maintained strict boundaries between the uncultured world of the bush and the cultured spaces of domestic life.[174] This spatialization of culture was reflected in foodways, with Kikuyu almost uniformly maintaining prohibitions against the consumption of wild animals.[175] These frameworks of both self and society were at the center of the arguments that formed the basis of what the colonial state called Mau Mau.

Over the airwaves, people who joined the government—so-called "Loyalists"—were promised a future marked by development along social welfarist lines. As one Kiswahili broadcast detailed: "The government . . . [knows] that black people are not satisfied by houses that they live in." It was in response to this discontent that one broadcaster promised, as he called to order his audiences in Kikuyu, a future wherein Africans would be "given food and houses to live in, and after that, we'll find a way of letting their wives and kids go live with them [*nitūre tabarira ūria atūmi na ciana ciao*].[176] People were promised "other things so that they can live happily and [be] satisfied." These included "a football pitch, and places they can watch movies [*nyūmba cia kuonanagīria thinema*] and also places where their kids can play."[177] All these developments, listeners were promised, would be provided by the government, which had "spared funds for that to happen." Together, these reforms would ensure that people had houses to live in, and places where they could "go for meetings and where they . . . [could] be reading books."[178] Broadcasts situated these futures as a kind of return wherein people could "start having the fun that we used to have where people will dance and sing." Promises of new developmentalist initiatives were routinely grafted onto visions of the good life that tapped into longer histories.

If the future was represented as the realization of this augmented vision of the past, Mau Mau was presented as an atavistic throwback, the inverse of both development and progress—it was an anathema to civilization. The notions of civilization conjured up by the colonial state explicitly drew on Kikuyu notions of complete personhood. As one European—that secondary evidence suggests was none other than Louis Leakey—argued over the airwaves in Kikuyu: "[T]hose who are like wild animals [*nyamū cia gīthaka*] should think of their people and respect them, their children and their wives [*andū ao, ciana ciao, atumia ao*]."[179] Leakey framed Mau Mau forest fighters as having abandoned their families. In so doing, they had abrogated their central responsibility as effective fathers, husbands, and homesteaders. For those unrepentant, this decision was irreversible. As Leakey continued, the "already captured Mau Mau will never go back to their families [*makīria ya ūguo, ndīna wītekio, andū a Mau Mau arīa mahīngeretwo matire hīndi mangeītekerio gūcoka kwa andū ao*]."[180] The stakes of this threat would not have been lost on listeners, so central was the connection between the cultivation of a homestead and the realization of social maturity.

But Leakey sought to cut even deeper than this. Mau Mau fighters, in rejecting this central tenet of Kikuyu masculinity, had rendered themselves non-human. Not simply boys—arguably a framing that led them to the forest in the first place—but "wild animals." They were the inverse of humanity, the opposite of civilization, the wild animal, a symbol of anti-culture, and the opposite of intellect.[181] As one broadcaster provoked those in the forest: "If you know you're not animals, then think."[182] What these men were being called to think about was the way in which their decisions to become like "wild animals" had placed them beyond the realm of culture.

If Mau Mau fighters were wild animals, it so followed that they had no claims over the gardens they had cut. Those who refused to confess would lose their "*Ithaka* [lands]" and these would "never be given back either to their children or their children's children, forever and ever." This was not simply a matter of property. For Kikuyu listeners, the distinction between life and land was not so neatly drawn. Cultivated land was not simply symbolic of culture, it was the materialization of culture, central as it was to Kikuyu notions of individual and collective actualization. Land, in other words, was not an addendum to life—it was life.

The threat, then, was existential, not only calling into question the present and immediate future, but the futures of generations to come. The very viability of the Kikuyu as a people was on the line. These men would not be mourned, one broadcast warned, but would "die like animals of the forest." Broadcasters' reduction of men to the status of wild animals was tethered to a serious set of existential uncertainties. "Now, are you among those that will be killed like animals? Without a wife, who will cry for you?"[183] Without family, there was no future. "Posterity, not burial, conferred remembrance and reincarnation," writes John Lonsdale. The threat here was not simply death or the loss of land, but a social death that would extend through the generations.

This social and moral universe was profoundly classed. Though framed in generational terms, for those marrying "late and often dying young, the poor were at greater risk of oblivion."[184] Land reforms of the past many years, but most recently the Swynnerton Plan (see chapter 4), which formalized land tenure, had exacerbated this state of affairs, attenuating relations between elders and juniors in critical ways. From the perspective of Mau Mau—youth by

dint of class rather than age—the wealth and status of these elders was illegitimate; they, in collusion with the government, had disrupted the moral economy. Poverty was now a permanent position, rather than a temporary status.
This class immobility and emergent inflexibility led many to the forest. It was
in the domain of vernacular, not social welfarist visions of development, that
they found their paths blocked. The threat: forever remaining children. Without land or family, these "boys" would simply fade into oblivion. Here, the
forest fighters were framed as having sacrificed their land, the critical hinge
between past, present, and future. This threat fundamentally undermined the
animus underwriting Kikuyu social life, anchored as it was in material interventions on the land.

These broadcasts actively weaponized this existential threat. As one Kikuyu broadcaster stated: "Any person who submits easily [and] shows that he
is ready to change . . . it's possible that they will be given a new bush [githaka]
where they will live."[185] Here, Mau Mau forest fighters were being promised
that for which they fought: A piece of nature out of which they could cut
culture. With this provision in place, they could become self-actualized adults,
respectable men and homesteaders—the legitimate bearers of wiathi.

Cultural Prosthetics and the "Bush Telegraph"

Understanding the premium administrators put on mobilizing vernacular notions of development and the cultural insights of African knowledge-workers
requires understanding what the state perceived itself to be up against: other
highly elaborated networks of communication. The information arm of the
colonial state was, indeed, designed to combat what was routinely referred to
as the "bush telegraph."[186] During Mau Mau, these other communication networks would have called men and women to the task of joining fighters in the
forest along many of the same registers as those vernacular notions of development the state mobilized in an effort to combat Mau Mau.

This was a fight that pit "young" men against "elders"—social categories
defined as much by class as by age. These "youth" were called to arms to regain
ithaka na wiathi, or cultured land and self-mastery. It was a fight over how, in
the context of ongoing material changes, these putative youths could achieve
wiathi—the measure of social adulthood and complete personhood. This was
a battle over both the future and the capacity to establish futurity. And these
messages would have made their way into the homes and locations of people
along the inscrutable lines of the "bush telegraph."[187] Of this, the state was
well aware. "Don't accept to be lied to by the voice [twanya toa ohoro] of those

who call themselves leaders of the opposition [*atongoria a ūtīti*]," one Kikuyu broadcast advised.[188]

Evidently, the colonial state's communications network was not the only one in effect. And often the same people generating cultural knowledge for the state's use were engaged in other, crosscutting communication networks. Walking this fine line required careful calibration, subtle movements to convince audiences on both sides. During Mau Mau, these men, as employees of the Information Department, particularly as Kikuyu men, were vulnerable. As broadcaster John Gitonga noted, he "received several letters from Mau Mau [stating] that they wanted [his] head badly."[189]

During our discussions in 2014, Wakaria also emphasized this awkward and dangerous middle status. "Many of us were killed by Mau Mau," he told me, but many "of us were [also] chased away by [the] British." As Wakaria ruminated, "You had to be very careful" to "make sure both sides . . . [understood] you well." This required articulating a seemingly singular message that could be interpreted by each side according to their reigning epistemology and visions for the future. The British needed to hear the broadcasts and interpret them as loyal. As for Mau Mau, Wakaria continued: "We could say proverbs or idioms" that communicated messages illegible to the British. In this way, broadcasters could ensure that "Mau Mau was happy, [and that] the British were happy." But broadcasters "had to be very careful . . . During the emergency you had to be very careful. Because you have to please the government, and also you have to please the Mau Mau. The government can kill you. Mau Mau can kill you." This might have been a skill rooted in wordplay, but it was dangerous. It was this work of pivoting between two epistemologies that, for Wakaria, made him an expert "propagandist." I asked Wakaria if he remembered any of the proverbs he mobilized during this period. "Of course," he told me, "I remember. I use them even today in broadcasting." In this, Wakaria signaled how his work as a "propagandist" was part and parcel of the "return to culture" that has enabled him to become an authority on all things Kikuyu.

Wakaria no longer self-identifies as a propagandist, but during Mau Mau, the "tension was high" and information workers "had to play it very well." "To be a spy is not easy," Wakaria explained. "It's not easy."[190] Ears were everywhere. Savvy British listeners were a source of great anxiety for knowledge-workers, particularly those with a facility in vernacular languages. In this, Leakey loomed large. "Our big man was Louis," Wakaria recounted. He "knew Kikuyu very well . . . [so you] had to be very careful" to ensure that you were not found out. "We were followed, every word we said," Wakaria explained. And people occupying this doubled position were, on occasion, found out.[191] The colonial state's efforts to secure a monopoly over communications in the region were routinely thwarted, however. Robust parallel networks were rife. "So there were many communications networks?" I asked Wakaria. "There were many communications. We didn't have telephones but we used to communicate."

Indeed, according to Wakaria, all Kikuyu, information workers included, were on the side of Mau Mau. For his part, not only did Wakaria mobilize proverbial forms to communicate with Kikuyu listeners—"to the Mau Mau I will only give a single sentence, but it says a lot"—but he used his connections as a one-time would-be pharmacist to get medicine to the Mau Mau forest fighters. "They knew very much that I was in pharmacy. And, therefore, I was given a mission to supply them with medicine." In this, too, he mobilized parallel networks, relying on women to move goods to the forest.

Conclusion

Despite all these dangers, Wakaria was emphatic that he enjoyed his role as a knowledge-worker before independence in 1963: "I enjoyed very much. I enjoyed broadcasting." From this position, Wakaria discovered and pursued his vocation as the preeminent authority on "Kikuyu culture." "When people first saw the unit," he told me, they "didn't like it, not from the beginning." As he explained: "It was a new thing to us. We were not coming across those things. They were done by Europeans." His opinion changed in 1950 when he joined the unit and "saw it was a good thing."

What he found the most satisfying was knowledge-work. "I enjoyed looking for news." For this, he had to do "a lot of research for my language, because it was a Kikuyu broadcast." This allowed him to "interact with the people . . . [and go] to meetings." This work granted him access. "After I joined the information," he told me, "especially when people liked me so much, [the elders] called me to sit with them." For Wakaria, this allowed a return from the path of deviance pursued by his father. His research, he explained, allowed him to go "into the culture." This ethnographic work he then broadcast first over the loudspeakers, and later over the airwaves, calling together a community bounded by language and culture through *his* "Kikuyu broadcast."

Undertaking this work clarified for him his capacity to mobilize broadcasting as his own prosthetic, enabling him to call to order a new community of listeners. For this work, he made KSH 142 per month, a wage that placed him just above the semi-skilled workforce employed in Nairobi.[192] His work, theoretically, was to close the gap between the desired reach of this ideologically rich but materially poor developmentalist state and the arterial nature of its presence in the lives of the bulk of the population living in Kenya's rural areas. From this vantage, though, he claimed for himself the position of authority on all things Kikuyu.

Knowledge-workers like Wakaria, then, routinely turned their positions within the information arm of the colonial state into ones from which they could pursue their vocations as purveyors and authors of culture. In this, they reformulated their prosthetic work, appropriating the form of these new media for themselves as their own prosthetics, reformatting radio broadcasting into a tool they could put to work in calling to order new audiences—audiences bound together by language. And this position brought Wakaria fame. "I was a big man by the way. I was a boss. An Information Assistant. ABS. African Broadcasting Service. Driving in that car. Oooh! I loved it."

For Wakaria, this work of cultural rejuvenation persists into the present. When we spoke, Wakaria explained that he was mobilizing his cultural expertise in what he referred to as a "campaign." "The culture," he told me, "it's coming back now, it's coming back." It will require time, "another 50 years," but the cultural inheritances of Kikuyu people would be restored to them. This required that men and women, boys and girls, be reeducated in Kikuyu culture. "I teach men, I teach women," he told me. "I bring them together." Much of this happened over the radio, where listeners would call in to consult Wakaria regarding "correct" Kikuyu practice. As though untouched by time, it was over the radio that Wakaria explained to them their cultural inheritance. Over the airwaves, these inheritances will be restored and the Kikuyu will "find themselves correct," find themselves "standing straight." When we spoke in 2014, by contrast to the past, Wakaria was well-compensated for his labors. "And they pay me well!" he laughed. "No joke!"

The "bush telegraph" was not the only alternate communications network with which the administration was forced to contend in the 1950s. While of a different shape, scale, and intensity, the state's protracted media war against Mau Mau was matched by other efforts to gain a monopoly over the informational worlds to which Kenya's colonial subjects had access in these years. Here, the Egyptian Revolution and the broadcasting network of Radio Cairo loomed large.

By contrast to the inscrutable lines of the "bush telegraph," which wended its way from hilltop to hilltop and over *panya routes* (rat routes) crisscrossing the territory, Radio Cairo made good use of the airwaves, its shortwave broadcasts touching down at nodal sites across the region. These vernacular-language broadcasts were heavy with messages that exceeded the boundaries of the colony, inviting listeners to envision an emancipated future premised on tethering together nodes of discontent and linking them to a single cause: colonial occupation. It was within this busy and competitive informational world that the colonial state set for itself the task of developing a robust, state-run broadcasting network as it worked to displace C&W, whose monopoly, once a boon, was now a dangerous burden. It is to these stories that the next chapter turns.

4

Broadcasting the Future
Airwaves and the Politics of Affinity

A 1954 report authored by the Kenya Broadcasting Commission reflected on the future directions of radio broadcasting in the colony. With no small measure of disappointment, the report read, "A complex society like that of Kenya, with its three major groups [white, Asian, and African] and its many languages, creates special problems for broadcasting; and these problems can only be solved by a complex, and therefore expensive, system of broadcasting. It was clear from the evidence put before us that, as far as Africans are concerned, while Swahili is understood in most areas, there are important tribal groups which would prefer, and, would only derive full benefit from, broadcasting if it were given in their own vernaculars."[1] The Commission was not exaggerating the problems it faced. With many dozens of languages spoken by communities scattered across the undulating topography, establishing a broadcasting network that could accommodate all vernacular languages would be a proposition near impossible to execute in practice. Nevertheless, crucial here was the admission that popular demands for vernacular-language programming would need to shape the broadcasting network of the future.

As I showed in chapter 3, administrators had imagined broadcasting as a key infrastructural prosthetic, shoring up the developmentalist aspirations of the colonial state since the 1930s. This chapter turns to the postwar period which saw the Treasury, for the first time, make monies available to assemble a state-run infrastructure of broadcast. This was a direct response to labor organizing across the British empire in the 1930s and 1940s. Within Kenya, these movements culminated in the general strike of 1947. These actions clarified that unless the British were going to provision "social services" for colonial communities to increase their "welfare," administrators risked forms of disorder that might herald the end of empire.[2]

These circumstances convinced the administration that developing its own broadcasting network was a matter of urgency. This position marked an about-face in colonial broadcasting policy, and reflected administrative anxieties in a period of profound flux. In this twilight moment, the colonial government hoped to use broadcasting to guide, constrain, and mold the social and political worlds available to Kenyans, by generating subjects of an "intermediate" scale—beyond "tribe" but before "all-embracing citizenship." This turned on efforts to engender a new and more robust parochialism through Kiswahili-language broadcasting. Administrators hoped that this technopolitical fix could secure a future that was bounded by the borders of the colony.

These goals were complicated in practice. From the perspective of the Kenya Broadcasting Commission, there were three considerations with which technologists and administrators would have to contend as they set to work imagining a state-run broadcasting network. Not only was Kenya a unique atmospheric and topographic zone, but problems of financing remained. While metropolitan monies were available for the design and construction of the network, demand outpaced resources. Despite the hopes of some administrators, broadcasting "development" would continue to rely on the remediation of the "market," not least for the design and construction of tuning receivers. People who were to be the subjects of this developmentalism would need to be brought into the fold as consumers of both media and the technologies of broadcast.

Material issues could not be easily divorced from the question of politics. Indeed, as the opening vignette suggests, compounding these concerns was the question of the language of broadcast. While the government had hoped to use Kiswahili, and later English, as the language of the future, they were confronting a highly variable ethnolinguistic field. The stakes of these linguistic politics intensified over the course of the 1950s, particularly in light of the Mau Mau uprising and the Egyptian Revolution, both of which appeared to threaten the integrity of the political unit that was the Kenya colony.

Indeed, while one of the promises of radio from the perspective of administrators was that it was a medium that knew "no frontiers," this came to be seen as a dangerous liability.[3] As people started tuning in to broadcasts from various global elsewheres—namely Egypt's Radio Cairo—on their shortwave receivers, this affordance presented an acute threat. Foreign broadcasts in a number of languages bleeding into Kenyan airspace invited people to generate forms of affinity that were by turns anti-imperial, regional, religious, and ethnolinguistic— alliances that in some instances turned on creating new modes of attachment that transcended colonial boundaries, while in others hardened ethnic constituencies into being.

These external concerns articulated with internal transformations, as administrative reforms placed downward pressure on communities to increasingly view a landed and ethnolinguistically bounded mode of identification as the surest means of securing their futures. Taken together, these crosscutting projects convinced administrators of the need to gain control over the informational worlds of communities, who thought of the Kenya colony in rather ambivalent terms when they discussed the boundaries of home. These dynamics clarified that C&W's monopoly over broadcast was more than an inconvenience; this infrastructural attachment was a dangerous liability that the colonial state was eager to shake off. Here, the state found itself playing catch-up as it worked to deploy radio to shape the boundaries of social and political belonging thinkable to colonized communities.[4] Evidently, by the mid-1950s, the invocation of preference, of desire as demand, had decidedly political overtones.

Listeners seized on this opportunity, conjuring up the anxieties of the colonial state as they called the government to task, demanding that the state recognize their linguistic difference and provision them with vernacular-language broadcasts. In so doing, they drew on colonial discourse, framing the infrastructure of radio broadcasting as a "public good" to which all should have access. Inclusion in the infrastructure of broadcasting, they argued, was a key metric of recognition within the shifting terrain of the Kenya colony. What emerged was a technopolitical network geared toward forwarding "parochial" interests, to be sure, but this process was driven from below.[5]

Broadcasting "Development": Radio and Fashioning
"Intermediate" Subjects

By the 1940s, and across the colonial world, colonial governments hoped to use policies of "social welfare" to offset popular forms of discontent.[6] As it pertained to broadcasting, administrative plans to use radio as a tool of social welfare were guided by emergent conversations regarding the danger of "the masses." Indeed, pursuing "development" through "mass education" over the airwaves was imagined as a prophylaxis against the specter of disorder. Administrators were particularly anxious about the social and political implications of what they referred to as ongoing "social changes." These changes, they argued, had generated discontent which could be scaled up, leading to "social upheaval"—a potential for rupture which they attributed to the emergence of "mass consciousness."[7] Newly mediated messages, heavy with developmentalist promises for the future, administrators hoped, would be capable of shoring up the center of the anxious colonial state.

It is no surprise, then, that the pedagogic function of radio was matched by its purported disciplinary capacity. As the Chief Native Commissioner (CNC) wrote: "African broadcasting" offered "tremendous opportunities . . . [for] controlling and regulating . . . [Africans'] social, economic, and political advancement."[8] The colonial state appreciated that people's social and political worlds were set against an expanded scale in the years following the war. This was threatening, administrators claimed, because it led to a misdirection in blame with people ascribing "responsibility . . . [to] groups or individuals, especially those in authority."[9] The goal was not to repress these reactions, but to guide and mold them.

Radio listening was to "broaden the outlook" of audiences, turning "the African into a self-governing and responsible citizen."[10] Citizenship proper was not, of course, on the table in this period. Instead, the Advisory Committee on Education in the Colonies imagined an intermediary status. As the report noted:

> The realization of citizenship begins in a small unit where common loyalty and common interests are expressed in daily activities and mutual service. Within these small units people hold tenaciously to what may appear as a narrow sectionalism, operating behind barriers which divide them from their fellows. Contact with the modern world is breaking down these barriers and expanding the social and political horizons of the people. But the conception of common citizenship is vastly wider. It may be that intermediate behind the small sectional units and all-embracing citizenship we shall have to envisage some form of 'national units.' It is going to be a delicate task to enlarge the narrower loyalties until they have reality and meaning in a wider circle.[11]

Educative radio programming was to offer listeners a framework of belonging that was "intermediate" after "the small sectional units . . . [but before] . . . all-embracing citizenship."[12] The Committee was not altogether wrong in its assessment of the historical conjuncture it was facing. The politics of intermediacy were indeed "delicate." What it seems to have missed was that the meaning of "narrower loyalties" had gained their definitional and affective strength precisely against this "wider circle."

The Committee was not totally naïve, though. Its members expected that efforts to use "mass education" to enact intermediate subjects could engender other, potentially crosscutting, projects of affinity-building. Language occupied a central space in these discussions. Broadcasting, the Committee argued, would undoubtedly "have a profound influence on the development of . . . language,"[13] but it conceded that efforts at linguistic standardization would be routinely thwarted by what it referred to as "political factors and parochial attitudes."[14] Despite these problems, for administrators, it was self-evident that first Kiswahili, and later English, would become the *lingua franca* of the Kenya colony, dominating both education and radio broadcasting. This would be an incremental process that would require building infrastructural attachments, as nothing could "make people listen in to programmes which do not appeal to them."[15]

Listeners, as this suggests, could not be presumed to exist *a priori*. They were not a pre-existing pool, but rather a potential that had to be cultivated to be tapped. And yet, administrators remained undaunted. Responding to critics' claims that the expansion of broadcasting was premature—seeing as Africans had not demonstrated an interest in the medium—G. S. S. Hutchinson, Manager of Information Services, Kenya, was indignant. "This is not," he retorted, "a valid argument . . . The existence of broadcasting facilities creates the demand for them," not the other way around.[16] Hutchinson was not altogether wrong in his assessment. But questions of attachment-building remained: How was a desire for listening to be cultivated? What forms of listening were most desirable? What kind of broadcasting network would give complete coverage? Which receivers were best placed to accomplish these goals? Neither these questions, nor their answers, were ever simply ideological, but were guided and constrained by the affordances and limitations of broadcasting technologies, the political futures they appeared to portend and, importantly, their cost.

Materializing an Infrastructure

For the first time, and coming on the heels of the Colonial Development and Welfare Acts of 1940 and 1945, financial aid from "Imperial sources" was made available to "promote the growth of broadcasting in the Colonies."[17] This marked an ideological and material about-face as administrators increasingly understood development as entailing more than simply investments designed to increase revenues—this being in keeping with the logic of colonial "self-sufficiency." Instead, emphasis was placed on provisioning "social services" for the African majority. Colonial Secretary Creech Jones was emphatic on the importance of this development, writing in a 1949 circular that there was an "urgent need" for the "wide development of Colonial broadcasting."[18] The availability of these monies was contingent on their being used to forward a vision of radios as "instruments of social and educational advancement." The infrastructure, administrators argued, was uniquely poised to act as a technology of state-directed development.[19] All seemed to agree on this point. To this end, £1 million was made available for colonial broadcasting schemes. To be divided among Britain's colonies, this was, in truth, a paltry amount, which saw Kenya receive £27,130.[20]

Nevertheless, with the availability of funding and the contracted monopoly of Cable and Wireless Ltd. (C&W) set to expire in 1956, the colonial government, for the first time, took seriously the possibility of constructing a state-owned broadcasting network. And so, in 1954, the Kenya Broadcasting Commission—the first of its kind in the colony—began its work.[21] It was clear to this body that despite the availability of funds, the material affordances of broadcast, in tandem with continued fiscal constraint, would have to shape the network to come.

When the technologists of the Commission published their report, they speculated about the utility of two wavelengths: medium and short. Medium waves were the cheaper option, but, traveling closer to the earth's surface, and contacting the ionosphere closer to their point of origin, they serviced a more constrained geography than shortwaves. Moreover, the capacity of these waves to propagate through the air was contingent on atmospheric conditions. And wavelengths in "tropical Africa," experts remarked, were burdened by higher noise levels, which presented an acute problem, reducing "the effective range of medium wavelengths to a small fraction of their range in temperate regions."[22] The propagation of wavelengths in the "tropics" was also subject to seasonal variation, with radio noise intensity reaching the maximum in March and April, and October and November—"the periods of the 'Long' and 'Short' rains respectively."[23]

Medium waves were not only subject to meteorological and atmospheric interventions, so too were they subject to topographic variation. The Great Rift Valley, running some 7,000 miles from northern Syria to central Mozambique, was one such topographical intervention. The terrain of this great rift varies, ranging from salt flats some 152 meters below sea level to a series of towering mountains, some of which reach 17,000 ft.[24] Just north of Nairobi, the site of C&W's main transmitter in these years, the Rift reaches its deepest point. The presence of this ancient Rift presented problems for medium waves, which travel along the "surface of the earth," making issues of topography of paramount importance, as the frequency, while able to surmount "minor hills up to 1000 ft or so . . . [were] stopped by mountains." But it was not simply vertical interventions that had to be considered but also the "electrical conductivity of the ground over which the wave must travel," for moving over dry, "sandy soil reception becomes poorer . . . while over marsh or water (which has high electrical conductivity) reception remains good to much greater distances."[25]

Shortwave seemed to circumvent some of these problems. Traveling at a higher frequency than medium waves, and bouncing off the ionosphere at a greater distance from their sites of transmission, shortwaves have a unique spatial reach. But shortwaves remained vulnerable to atmospheric changes more common in the tropics. As the Commission noted: "The useful energy from a shortwave transmitting aerial travels skywards until it meets reflecting layers, some 200 miles up, which return it to the earth for reception." But reflecting layers varied "daily, seasonably, and over an 11-year cycle," occasionally disappearing "altogether during periods of electrical disturbance in the atmosphere."[26] In addition, shortwave frequencies were overcrowded and subject to competition. And while different geographies could, theoretically, share a single short wavelength for use at different times, Africa was in the "unfortunate position of being in competition with Europe."[27] But modes of transmission could not be considered without considering modes of reception. Not only were shortwave receivers costlier than medium wave, but they were "more elaborate" and "less easy for the unskilled user to tune."[28] Shortwave technology may have been more expansive, in sum, but it was also more complicated and more expensive.

The same affordances of shortwave that made it the most optimal mode of broadcasting for communities living at a distance from Nairobi, concurrently made shortwave technologies objects of concern.[29] Indeed, the capacity of shortwaves to bounce off the ionosphere and thus travel great distances brought to the fore questions of the forms of social and political affinity that "listening in" on shortwave receivers might enable. As the 1954 Report of the Kenya Broadcasting Commission noted, shortwave transmissions could be subject to interference from the more powerful shortwave stations used by "other countries." Despite the fact that there was some international control governing the allocation of transmission frequencies, it was not "completely effective." This was not mere prognostication. As the Commission reported, foreign transmissions were being received in "increasing numbers . . . especially in Africa."[30] Evidently, wavelengths could not be constrained by the same logic of territorialized control that animated colonial imaginaries. And so discussions regarding topography and airwaves, ideal transmitters and receivers, were not merely technical questions, but issues which verged on "higher politics."[31]

The Problems of "the Market"

When it first secured its monopoly, and absent metropolitan investments, C&W's presence in the region had seemed like a boon to the state, allowing it to offset the costs and risks of the venture, all the while guaranteeing the state access to C&W's network. C&W, though, was only contracted to offer

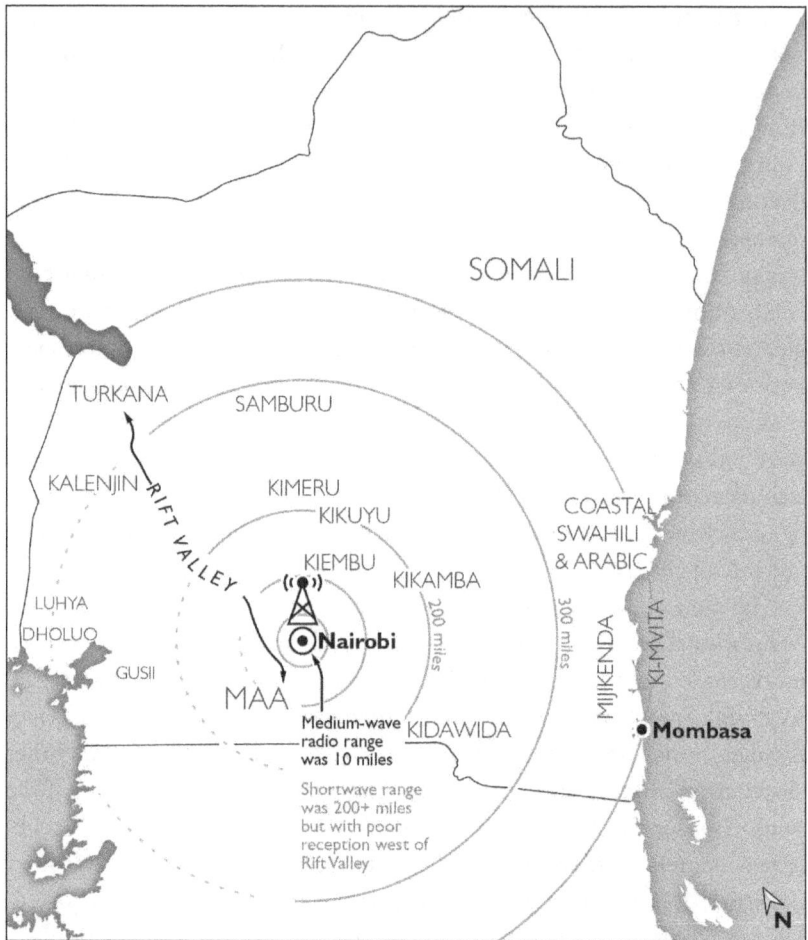

Map of the limitations and reach of the Kenya colonial broadcasting network, with major linguistic groups' geographic placements

services to the remunerative white settler and South Asian populations during its tenure.[32] This racially circumscribed mandate was viewed with increasing anxiety by administrators. The racial composition of targeted listeners notwithstanding, C&W's reach had grown. In 1931, the number of license holders in Kenya was just over 300. By 1959, this number had risen to 30,000.[33] As for African communities, the initial contract granted the government the right to use C&W equipment and studios for "public purposes" at a rate of KSH 20 per half hour.[34] As we will see, by the 1950s, these relations of infrastructural dependency appeared, to many, to be a dangerous liability. Having examined the terms of the agreement between C&W and the government, J. B. Miller remarked disparagingly that it was "clearly a makeshift arrangement."[35] Philip Mitchell was in agreement, arguing that the "monopoly of Cable & Wireless" was "ridiculous" and should have been "terminated."[36] The state was in a bind, however. C&W's position was firm; it was protected under contract.[37]

Insofar as the government was dependent on C&W's administrative ambitions regarding broadcasting were subordinated to the market-driven interests of the company. Indeed, while C&W had been deemed a "public utility company," it was operating first and foremost as a "commercial concern," and as such had little interest in investing "capital with no prospect of securing revenue."[38] And this profit-driven model was built into the network itself, translating into incomplete coverage across the Kenya colony, with some areas enjoying "fairly good reception, whereas many others had extremely poor reception or none at all."[39] Administrators routinely lamented that C&W's monopoly over broadcasting had impeded the state's ability to extend broadcasting to African communities. C&W had little patience for such complaints. When the government appealed to C&W, requesting that it jettison licensing fees on receiving sets for Africans, C&W refused to compromise, arguing that the fees, though well beyond the reach of many would-be African listeners, were its main source of profit.[40] Licensing fees were not the only issue. C&W's infrastructures were also beholden to its commercial ventures, which prevented the government from using C&W's transmitters for broadcasting to Africans at "optimum listening hours."[41] Constraints in time would prove to be a problem in the years to come.

It was not simply in material terms that this infrastructural attachment was a liability, but in ideological terms as well.[42] Administrators and metropolitan radio technologists routinely complained that C&W's commercial interests threatened to undermine radio's purported status as a "public good."[43] The British Broadcasting Corporation (BBC) lamented that C&W refused to rebroadcast BBC programs that it did not consider to be of "general interest."[44] According to the BBC, this was based on a "narrow conception of the responsibilities" of the corporation, charging C&W with routinely disregarding responsibilities Parliament had placed on the BBC.[45] Among these was the mandate to broadcast items deemed important to the "Commonwealth," even if they did not "comprise the most popular programmes."[46]

Commercial arrangements with C&W were not the only ones with which the Kenyan government and the BBC had to reckon. While metropolitan monies had been made available for the extension of broadcasting, the Colonial Office (CO) was not willing to take on the full financial burden of developing a robust radio network in its colonial holdings. As one administrator wrote, in choosing between the "broadcasting and receiving sides," the CO had decided to concentrate "upon the former," leaving the issue of receiving technologies to "the market."[47]

The incongruous forms of affinity that shortwave technologies seemed to hold out took on particular urgency due to the Treasury's refusal to make monies available for the design and construction of receiving sets. It was left to administrators to communicate these technopolitical concerns to designers and manufacturers as they discussed the parameters of appropriate reception facilities for African listeners.[48] While some argued that "pre-set tuning" might offer a solution, ensuring that "whomsoever possesses such a receiver cannot be 'got at' by propaganda from other parts of the world," others argued for centrally controlled broadcasting in the form of communal listening sites or wired broadcasting.[49] The latter had the advantage of being a closed system. This would be a boon for "broadcast programmes . . . brought from the studio to the listener by wire, instead of by ordinary wireless transmission, [eliminated the problems of] interference, fading, and distortion."[50] But most importantly, wired broadcasting ensured there was "no risk of interference from unwanted programmes broadcast on a nearby wave band."[51] "Central Control" was in "complete charge of every programme at every stage."[52] These arguments turned on the proposition that a simple technological fix could constrain the social and political futures that colonized communities could access over the airwaves. This infrastructural attachment, administrators hoped, could delimit people's access to the increasingly competitive world of radio broadcasting. These options had the added advantage of circumventing the "problem" of users whose "inquisitive nature" led them "to pull [apart] the pieces of most kinds of European equipment that comes within . . . reach."[53]

But Kenya's human geography, itself the product of colonial labor policies, mitigated against the use of wired broadcasting, as well as medium wave. Kenya's African population, administrators noted, tended to live in less concentrated settlements in the countryside, making the proportion of the population that could be economically served by either medium wave or radio rediffusion "much smaller than in Europe or many other countries."[54] Finally, there was the problem of infrastructural neglect, which left few rural areas electrified by the 1950s. These areas, it was noted, were likely to remain dark for "some considerable time."[55] This would have to guide the design of receiving sets, as any radio put in circulation could not depend on a direct energy supply, and would thus have to be battery operated.[56]

The CO thus labored to pressure commercial manufacturers into designing sets specifically suited to "the tropics." By and large, these efforts failed, confirming that radio manufacturers were not "progressive, imaginative, or inclined to take risks."[57] Market-making operated differently in the colonies than in the metropole. In the metropole, the design and manufacture of receivers could be "left to the ordinary process of demand and supply."[58] In the colonies, by contrast, both supply and demand were mere prospects.

The issues were threefold. First, receivers produced for African listeners had to be within the reach of an African "clerk class" who, administrators argued, could not afford sets exceeding £5.[59] Second, all technical components would have to be tropicalized to "resist the inroads of insects, dust, and damp," if they would be able to operate in the region.[60] But most importantly, manufacturers had to build a prototype for a "market which seems to him potential rather than actual."[61] The trick would be to get manufacturers "to realize that such a poor population could and would buy radio sets." This required cultivating a pool of listeners as a market, developmentalist subjects as consumers.

These arguments for market-making were not framed as anathema to radio's purported function as a "public service." Manufacturers needed only "see the vast potential market for the poor man's radio *and* the part which radio could play in the development of the colonies."[62] If this market of consumers could be cultivated, it would open a global market for "British industry."[63]

But administrators were cautious about the degree of marketization that was desirable in their colonial holdings. While creating a class of African consumers for manufactured receivers was amenable to administrators, after consideration, they outright rejected the possibility of initiating broadcasting to colonial subjects on a commercial basis.[64] Administrators were particularly concerned because, they claimed, listeners looked to radio broadcasting as "a Government service," presenting a "very real danger that any products advertised would be considered as having official blessing, and Government would tend to be blamed for any shortcomings found in them."[65] The prerogatives of commercial broadcasting were simply out of step with the prerogatives of what had "*come to be known as* 'public service' broadcasting," the Commission contended.[66] Commercial broadcasting conceived of only one type of listener: the consumer. Public service broadcasting, by contrast, was concerned with various interest groups—themselves internally differentiated—who, together, made up "the community of the nation."[67] After much discussion, the Commission agreed that broadcasting in Kenya "should be controlled by a single, independent public corporation," with limited "'spot' advertising." As we will see, the government's insistence that radio broadcasting was a "public service" would become the foundation on which people would demand that the government provision communities with broadcasts in vernacular languages, granting them access to the infrastructure not as homogenous listeners but as linguistically particular constituents.

In a moment of evident frustration, one commercial firm queried the administration, asking was "there something rather odd and exceptional in the idea of conducting a service to the public on a profitable basis?" Here he hit the nail on the head. From the earliest days of colonial occupation through to the 1930s, private capital was widely held to be the engine of infrastructural expansion, but in this period—and in the context of an emergent social welfarist model of "community development"—administrators felt that there was indeed something "odd" about the prospect of leaving the "public good" to the whims of the market.

These difficulties notwithstanding, listenership was on the rise in these years. In 1951, the government claimed that broadcasts were reaching 1.5 million listeners per year out of an estimated population of over 6 million.[68] By the mid-1950s, the Saucepan Special had entered the market. This shortwave receiver, designed by Harry Franklin, an information officer posted in Northern Rhodesia, was specifically designed to withstand the conditions in the "tropics:" "clumsy handling by . . . set owner[s], the hazards of long transportation . . . extremes of temperature and humidity, the ravages of white ants, borers, etc."[69] Tropicalizing this technology, however, required more than merely modifying its technical components. It had to be "tropicalised" in "cultural" terms as well. This was the reason for the set's color: blue. Franklin, based on his experience in Northern Rhodesia, contended that "one tribe or another [were] superstitious about every other colour."[70]

The set was sold for £5, and it was reported that Africans were buying up the "quantity of cheap wireless sets on the market" in "large numbers," with 1953 estimates putting this number at 5,000.[71] Evidently, the "pump had been primed" and people were "listening in."[72]

Infrastructural Neglect

The state's sudden investment in the field of broadcasting and its commitment to "social welfare," while framed as an ethical project of "development," was in truth a response to the specter of "disorder." The Kenya colony had been wracked by a series of labor actions in the 1940s, culminating in a general strike in 1947, which transcended the parochial imaginary of the colonial state as laborers spread the strike from Mombasa to Nyanza. This movement was not only capacious in spatial terms. Labor organizers drew on globally

The "Saucepan Special" shortwave receiver designed by Harry Franklin that extended the reach of broadcasts in colonial Kenya

circulating discourses as they made bids for inclusion in regimes of labor that proffered protections to metropolitan workers.[73] In so doing, colonial communities mobilized and consolidated affinities that transcended the geographic boundaries of the colony, in the process demonstrating the limits of colonial conceptions of the appropriate boundaries of social and political affinity in its colonial holdings. The 1947 strike had, indeed, been the initial prompt for the Commission, invigorating conversations about the need to expand the colony's broadcasting network on a territorial basis. But dockworkers and railwaymen were not alone as they worked to enact alternative forms of social and political affinity.

In Mombasa, anxieties about the emergence and fortification of extraterritorial networks were rife in these years. In the early 1950s, a Mombasa District Annual Report noted that the "direction" of the "Arab community" was "becoming established on an East African rather than on a territorial basis."[74] People were drawing institutional, political, and social connections between Zanzibar and Mombasa, with the "East African Arab Congress" organizing meetings in Zanzibar in 1953.[75] While "Arabs" on the coast were not, themselves, internally homogeneous—there being a "serious rift in the Arab community"—many sought closer relations with Zanzibar and argued with increasing urgency for the sovereignty of the Sultan over the coastal region.[76] But years of infrastructural neglect ensured that gaining control over people's informational worlds would prove to be difficult. This issue was not limited to the coast.

Following the war, the administration had actively scaled back broadcasting. Reliant as it was on renting C&W transmitters, broadcasts to Africans trebled around nine hours per week. By 1951, broadcasts directed to Africans were limited to Kiswahili and Kikuyu, with the latter receiving a mere thirty minutes per week—this, in part, based on expectations that first Kiswahili, and then English, would come to supersede vernaculars as the language of communications in the Kenya colony.[77]

A decline in programming was matched by government directives that actively dismantled the network that had been operative during the war. In 1950, to the consternation of many in the western part of the colony, administrators and inhabitants alike, Nairobi removed government sets from the Welfare Centers dotting Nyanza's landscape.[78] By the end of the year, the thirty receivers that had been dedicated to Africans had been withdrawn, leaving all remaining sets in the hands of South Asians and white settlers.[79] It seems likely that the white settler lobby, always allergic to the government funding services for Africans it deemed frivolous, was behind this shift. For many, though, the rationale behind this decision was inscrutable. Returns from 1948 indicated that Africans in increasing numbers were listening in at "Welfare Centres, Information Rooms, and Schools," leading one administrator to remark on the successful work that had been done to "popularize vernacular broadcast programmes."[80] And yet, by the end of the year, vernacular-language broadcasting had been jettisoned.[81] Pecuniary constraints had led the state to dismantle the infrastructural attachments it had forged with listening communities during the war. Listeners and local administrators were incensed, and memories of government provisioning would shape the protests to come.

Broadcasting to the coast received somewhat more airtime, particularly following the general strike of 1947, which saw the launch of Sauti ya Mombasa—to be renamed, Sauti ya Mvita, or "Voice of the Coast"—a Kiswahili-language program operating in a coastal variant, Ki-Mvita.[82] In 1948, the government expanded airtime from a mere fifteen minutes to forty minutes, and "audiences had increased very considerably."[83] By many accounts, Sauti ya Mvita was popular. But the state's fears were only momentarily assuaged by this rise in listenership. While extending control over inland communities in places like Nyanza would come to appear to be beset by the problem of too few shortwaves. On the coast, the problem was the inverse, these communities were subject to altogether too many signals.

At the forefront of administrative concerns, in this regard, were two other events that held out the promise of futures that threatened the integrity of the colonial state: the Mau Mau uprising and the Egyptian Revolution. Taken together, these events seemed to press against the boundaries of the Kenya colony, concurrently threatening to make Kenya's borders irrelevant. The political unit itself appeared to be at stake. For many, these events demanded the mobilization of a radio network not as an infrastructure of expansion, but of containment and counterinsurgency—the new network a means of constraining the imaginaries of Kenyan communities.

Broadcasting Affinity: Radio and the Development of Other "Intermediate" Networks

On July 23, 1952, the Egyptian Free Officers staged a coup, which led to the ouster of King Farouk, who was viewed by many as being a puppet of the British, and an ineffectual ruler besides. Within Egypt, Gamal Abdel Nasser's formal assumption of power entailed a violent authoritarian crackdown, seeing large numbers of communists and Islamists sent to concentration camps, and effectively subordinating the labor movement to the state. However, Nasser's rhetoric outside of Egypt was not one of containment, but offered a vision of political affinity that transcended the boundaries of the territory. As he wrote in the *Philosophy of the Revolution*, "If I were told that our place is limited by the political boundaries of our country I . . . do not agree . . . the era of isolation is now gone. Gone also are the days when barbed wires marked the frontiers separating and isolating countries, and every country must look beyond its frontiers to find out [from] where the currents that affected it spring."[84] As noted by James Brennan, for Nasser, Egypt was uniquely placed to foment these border-transcending attachments, situated as it was at the crossroads linking Africa to the Middle East.[85] What was required was an infrastructure of connection, one that enabled the materialization of this "united struggle."[86] Nasser imagined radio broadcasting as being just such an infrastructure. As Radio Cairo proclaimed, broadcasts directed to the Swahili Coast linked "the fighting people of Africa with the Arab peoples who are also fighting for freedom, peace and prosperity."[87]

To this end, in 1953, Nasser launched "Voice of the Arabs" or "Saut el Arab." This Arabic-language broadcast was beamed on shortwave, its messages traveling over the airwaves and finding homes at nodal sites across continents. As *The New York Times* anxiously reported, it could be "heard and understood by the restive peoples from Algeria to Uganda and from Iraq to the remote sheikhdoms at the southern tip of Arabia."[88] These messages readily reached the Swahili Coast, and immediately people were listening in.[89] Nasser hoped to use radio broadcasting as a means of geographically extending his political project. Generating attachments between people and Radio Cairo's network seemed to offer the revolutionary government an infrastructure of anti-imperial coordination.

The content of these broadcasts was diverse, ranging from critiques of American neo-imperialism, to support for the Mau Mau uprising, to condemnations of the apartheid state in South Africa.[90] Transcripts of the speeches of African nationalists, such as Kwame Nkrumah, were put up over the airwaves.[91] A new lexicon of critique was disseminated as listeners were referred to, often by Nasser himself, as "my brothers." Broadcasts spoke of a united future as seemingly discrete struggles were linked. Out of this would emerge a "new African personality" primed to enact an emergent concept: "African unity."[92] This effort to enact a new, non-territorialized vision of connection, did not turn on flattening the particularities of listeners, but worked to suture people's heterogeneous experiences together, putting them to work in the project of anti-imperial struggle. In geographic terms, this vision oriented people toward the north. In ideological terms, this vision oriented people toward an independent future, though one not framed by the borders of colonial states.

As new audiences tuned in, anxieties among administrators heightened. Reporting on a 1954 trip to the coast, the Governor of Kenya, Evelyn Baring, gave a lengthy description of the state of broadcasting in the region. "The broadcasts which have been made [in Kenya] are widely listened to and by all accounts are a great success in Mombasa. But at present in all the North Coast and in most of the South Coast it is impossible to pick them up."[93] It was estimated that on average 20,000 people were listening in, no paltry number, but it was estimated that there were a prospective 100,000 listeners. The 80,000 being missed could potentially be brought into the listening fold if only the signal were stronger.[94] And these material constraints had dire political ramifications. This technical lumpiness had created space for other, deeply troubling, forms of listening as broadcasts emanating from New Delhi and Cairo were "easily heard."[95]

The Governor not only overestimated the popularity of government broadcasts, but he underestimated the popularity of those coming from Cairo. Other observers offered a corrective. Not only were the broadcasts emanating from transmitters sited at various elsewheres "easily heard," but people on the coast were "avidly" seeking them out as they used their shortwave receivers to tune in to Saut el Arab, with a smaller number listening to Arabic broadcasts from Omdurman and All India Radio.[96]

An audience assembled for a broadcast program in the Coast Province of Kenya, African Affairs Department, Annual Report, 1952, KNA

As administrators debated prospective solutions to what they anxiously referred to as the "problem" on the coast, they were forced to confront the limited reach of Nairobi's incomplete network. Broadcasts from Nairobi were "transmitted on [C&W's] two low power transmitters, one on the medium wavelength, giving a regional service within 10 miles of Nairobi, and the other on the shortwaves, giving fairly adequate coverage in most parts of the Colony." But, the report continued, "reception from Nairobi is poor in Nyanza Province and the Coastal districts."[97]

The limited reach of these transmissions was not the only problem. From the outset, information experts had struggled to develop messages for broadcast that would appeal to Muslim communities living on the coast. And often, they failed. People on the coast argued that BBC's programming was "too classical and high brow." The music was also not liked, alienating would-be listeners who would have preferred to hear broadcasts of readings from the Koran.[98] Criticisms of both the BBC's Arabic- and Nairobi's Swahili-language programming persisted through the 1950s.[99] In 1954, "Arab" women living on the coast complained about the programming directed at them, arguing that "such broadcast talks conflict[ed] with their customs."[100]

African Knowledge-Workers and Experts

The infrastructural weakness of the Kenya Information Office (KIO), in tandem with a history of underinvestment in the infrastructures of broadcast, required a makeshift solution. On the coast, this took the form of Sauti ya Mvita. With a limited budget, they turned away from central planning toward community development, relying on "volunteers" who were given a free hand in developing programming, on the agreement that broadcasters would not introduce "politics" over the airwaves. These conditions ensured that the directions this station took were largely guided by prerogatives of the African information workers who peopled its one shabby office, which was overseen by the Municipal African Affairs Officer.[101]

These knowledge-workers were largely pursuing their own vocations when they decided to devote their time and energies to Sauti ya Mvita. As Sheikh Hyder Kindy—a Swahili man born on the coast, and a career civil servant and information officer—recalled: "Volunteers were called for and a few enthusiastic Muslims came forward to offer their services." Their aspirations in this were both "intermediate" and global. As Kindy recounted, those who volunteered felt "that this would give them an opportunity to propagate their religion through this powerful medium." Administrators had hoped to use Kiswahili as the language of intermediacy—but a secular Kiswahili, not one bisected by religious affiliations.[102] Suffice it to say that the incomplete networking of the state not only created space for other forms of listening, but also required that it mobilize the skills of African knowledge-workers and experts.

Acting as infrastructural prosthetics, these men used their position within this lumpy network to pursue their own projects of affinity-building, and the work of these men was eclectic. As Kindy recalled, the team extracted news bulletins from the *Mombasa Times* from whence they were translated into Kiswahili, taped music from the BBC, and, later, Radio Cairo.[103] The initial half hour program was divided into an "opening with the Zanzibar National Anthem, a Koran reading of about five minutes and then a news bulletin, after which there would be some music—mostly Arabic and Swahili songs."[104] To capture both Kiswahili- and Arabic-speaking Muslim populations, Sauti ya Mvita began broadcasting in both languages, and administrators were pleased with the results, arguing it was "proving increasingly effective in diverting attention from Arabic broadcasts from Cairo and other foreign stations."[105]

But the politics of the coast were complicated in these years. As Kindy recalled, while his aspiration had been to use Sauti ya Mvita to "propagate" Islam in eastern Africa, he was in an awkward position, rejected by black Africans for being "too

Coast Province Department of Information office, Mombasa, Kenya, circa 1950

Muslim," and by Arabs for being "too African."[106] The stakes of these politics would heighten by the late 1950s, when Muslims on the coast began calling for *Mwambao*, or coastal autonomy, against the perceived threat of "up country Africans."[107] Indeed, if Kiswahili was imagined to be the language of intermediacy framed by the borders of the colony, on the Coast, as we shall see, Kiswahili mapped on to a kind of identity politics framed against a different set of geographic and religious coordinates, politicizing Kiswahili broadcasting in critical ways.

While Sauti ya Mvita had gained popularity in these years, largely due to the African knowledge-workers attached to the station, the state's continued reliance on C&W's transmitters ensured that technics remained a problem. As Kindy recalled, "There was no doubt that *Sauti ya Mvita* programmes were extremely popular wherever they were heard throughout the province. It was therefore unfortunate that the transmitters then placed at our disposal by Cable & Wireless Ltd. had insufficient power to permit our broadcasts being heard over a wider area of the Coast."[108] In 1955, and despite efforts to improve the strength of the transmitting station, Mombasa's signal was routinely swamped by other, and often hostile, stations.

The technopolitical anxieties of the state were not lost on leaders on the coast. Powerful sheiks and liwalis actively leveraged administrative fears, arguing that if Nairobi wanted to regain control over people's imagined futures, the government would need to cater to Arabic-speaking populations. To do so, the state would need to install a more powerful transmitter in Mombasa.[109] This demand for infrastructural inclusion would ultimately be realized in 1956, when a transmitter was finally installed at the coast. But here, again, the state found itself playing catch-up.[110]

By 1954, Radio Cairo had expanded its programming, launching Sauti ya Cairo, a Kiswahili-language broadcast designed to reach Kiswahili speakers in Kenya, Tanganyika, Uganda, and the Belgian Congo.[111] These programs responded to the demands of Muslim listeners. Initially thirty minutes long, daily news, political commentary, and cultural programs followed readings from the Koran. By 1955, the program was forty-five minutes in length, which was increased to an hour in 1958 and, finally, an hour and a half by 1961.[112]

The Suez Crisis of 1956 raised the stakes within this competitive radio sphere. The British, pursuing a strategy of informational isolation, deployed force, bombing Radio Cairo's transmitters within days of their invasion of Egypt.[113] Initially, this appeared to have "worked." But within a few days, "repairs" had evidently been made, and the signal was "heard again at the Coast." People resumed listening in.[114] Efforts to "lure" listeners away from Radio Cairo "and its various attractions" had failed.[115]

By 1957, the Voice of Free Africa, by many accounts the most virulent in its critiques of empire, joined the queue.[116] Administrators and foreign observers were alarmed by the vitriolic messages crossing into British airspace. As the *New York Times* wrote: "Cairo Radio, burdening all wave-lengths from dawn until after midnight in almost every language of the area, wheedles and cajoles, browbeats and foments the restive and unhappy peoples in every corner of the area."[117] Even those not actively listening to the broadcasts were subject to their influence, as broadcast messages "were spread rapidly and elaborated through coffee-house gossip."[118] While the spread of this information "by the tongues of listeners" was derided as "rumor" in the "colonial lexicon," many of these "agitators" were drawing on global infrastructural attachments enabled by the introduction of shortwave broadcasting.[119]

Radio Cairo's strategy did not turn on flattening the particularities of colonized peoples' struggles. Instead, its broadcasts invited listeners to situate their grievances against a larger frame of reference. Of this, administrators were well aware. "Considerations in determining action [regarding the future of broadcasting] are the political asmosphere [*sic*] in the Sudan, Somalia, Zanzibar and

Egypt ... [there is a] need to counteract this influence as soon as possible," wrote the Chief Secretary in 1955.[120]

Cordoning off nodes of discontent within the empire was a core concern of the British in these years.[121] It was a strategy that, they hoped, could operate with the same logic within the space of the Kenya colony as it would across the empire. Radio Cairo threatened this strategy. By 1958, Radio Cairo was broadcasting in over twenty languages, including Somali and Amharic.[122] But Radio Cairo's efforts to firmly tether its programming to the lives of prospective audiences were not limited to considerations of language. Radio Cairo expended a good deal of energy generating programming that would come across as firmly embedded in the lives and concerns of the communities listening in on their shortwave receivers. While perhaps hyperbolic, a sense of the labor of this knowledge-work was offered by Ray Vicker of the *Wall Street Journal*:

> Each program is preceded by long studies. Included are studies of the customs of each people to whom a broadcast will be beamed. Attention is paid to the time they sleep and awake, the time they start and end work and the average time-distance between work and home. This helps decide the time the program will be beamed. Other studies delve into the political position of the country, its economy, the type of government and major problems occupying the attention of the people . . . Each program is presented by a native of the country to which the message is beamed, assuring that all idiom and methods of presentation will be authentic to listeners.[123]

Mobilizing African knowledge-workers from the communities it addressed, Radio Cairo worked to embed its programming in the "warp and woof" of everyday life.[124]

Listeners were united, then, not necessarily by the simultaneity of their listening, nor by the homogeneity of the programming, nor by their geographic contiguity, but by programming that both addressed the particularities of their struggles with those located across the colonial world. By holding their experiences in the same analytic frame as those of colonized communities located further afield, Radio Cairo sought to impress on people that their myriad grievances were the function of a singular system. The message of unity in difference was what appeared most threatening to colonial powers.

Evidently, it was true that the airwaves knew "no frontiers" and radio was indeed capable of inviting people to participate in imaginaries that were both larger and smaller than the boundaries of the colony.[125] And this was precisely the fear. On the coast, administrators were anxious that a form of unity based on a shared commitment to Islam could scale up, leading to the formation of broad-based sentiment as listeners fortified attachments among their struggles. It was thus "politically very important that . . . [Kenyan programs] reach Moslems throughout the Coast area."[126] While broadcasts from Nairobi were audible in Mombasa, they could not be heard beyond a few miles of the island. This was not, evidently, a problem for shortwave broadcasts emanating from India and Egypt.

The Technopolitics of Broadcasting the Future

The constellation of issues brought together by Radio Cairo's broadcasts mirrored its technics. The strength of its signals, broadcast from at least eleven shortwave transmitters in the Nile Delta area by 1958, in tandem with the "truly remarkable increase" in shortwave receivers amongst African communities, lent Radio Cairo an aura of invincibility.[127] And Nasser was not wrong in attesting to the unique geographic position Egypt held in fomenting new attachments in the region. As noted by the *New York Times*, Radio Cairo had "become the most far-reaching voice to be heard between Longitude 20 and Longitude 60 East and from Latitude 40 North to below the Equator."[128]

While this was a politics of connection that recognized popular desire for linguistic and cultural differentiation, the central message was clear. As Osgood Caruthers of the *New York Times* wrote in 1958, "One theme never varies . . . that is that the foreigners—'white dogs,' imperialists, bloodsuckers and colonialist oppressors . . . who have dominated most of the area for more than a century, must either leave voluntarily or be driven out."[129] The technological reach of these stations, for many administrators, seemed both to precede and to be generative of newly thinkable political futures; the reach of the medium portending the message.

This is not to say that listeners were single-minded audiences. People were savvy in their listening habits, preferring to piece together information provided by myriad different media houses, rather than accept without question the messages of a given one. As noted by an administrator in 1956 following a trip to the coast where he visited "various coffee houses," people were listening to a "tremendous number of programmes from the Arab world . . . [as well as] the BBC."[130] This was not, in other words, passive listening, and these were not passive listeners. Theirs was an open, though discriminating, informational world.

While administrators framed the problem of Radio Cairo as being at once political *and* technological, the only acceptable potential solutions were framed in explicitly technopolitical terms. What was needed was the mobilization of new technological networks that could materially foreclose the political languages and potential futures thinkable to listeners. "You are well aware," wrote the Provincial Commissioner on the coast, "of the reasons for improving the local broadcasting along the Coast . . . Increasing numbers are listening to Radio Cairo and Radio India . . . a substantial improvement to this service [is] . . . urgently necessary for political reasons."[131] Initially, the government proposed a technopolitics of time, arguing that they need only increase the hours being broadcast over C&W's transmitters from forty-seven hours per week to more than eighty hours.[132] But this ambition was beset by problems. Not only were the airwaves crowded, so too was the clock.

Historically, Kenya's Kiswahili broadcasts to the coast ended at 7:30 p.m.[133] Egypt, wise to the African Broadcasting Services' schedule, ran programs that began at 7:00 p.m., continuing until 9:00 p.m.[134] Extending the Kiswahili broadcasts to 9:00 p.m., one administrator noted, was essential, and had been "made doubly urgent by the opening up of Swahili broadcasts from Saut el Arab, the Cairo station . . . [which contained extremely] undesirable material."[135] But even this seemingly simple fix was not feasible. Again, the government's relationship with C&W was an impediment. The company, responding to government requests to extend broadcasting times, indicated that it was unable to lease the government its equipment in the evening hours. These transmitters, the company explained, were occupied at night when they were used to transmit to Australia. This was a commercial agreement, and it was protected under contract.[136]

If the government was to succeed in political terms, it would have to strike out on its own in technological and fiscal terms. And as the administration would learn, simply extending the hours of broadcast would not suffice. It was contending with other, more sophisticated, technical networks which transmitted "on a wave length so close to that used by ourselves at the Coast that . . . [our] transmissions are frequently swamped."[137] Not only did the signals from Egypt appear to be more local due to the clarity of their reception, of course, but the messages themselves seemed to be more firmly embedded in the local milieu to which they were directed.

By all accounts, an expanded material network was the only solution to the expansion in the imaginative horizons that Radio Cairo offered to audiences on the coast. What was needed was a new transmitter of containment, one that would broadcast on a medium wavelength, receiving signals parallel to, but distinct from, the shortwaves coming from Cairo. In moving from discussions of expanding the hours of broadcast to discussions of expanding the infrastructural reach of the state's broadcasting network, the colonial administration's strategy shifted from a technopolitics of time to a technopolitics of space.

As early as 1955, Kenyan administrators had argued that "the need for a Regional Transmitter at Mombasa . . . [was] almost an Emergency matter."[138] But the labor and expertise required to get the new station up and running was considerable, especially given the problem of finding an appropriate site for the new transmitter, which required determining not only the quality of the soil, but reckoning with concerns over atmospheric "noise." The staccato rhythm of infrastructural expansion was out of step with the temporality within which new relations of affinity were being forged. As one frustrated administrator wrote: "This is an extremely urgent matter, unfortunately it is also a highly technical one. We shall have to be very careful to avoid being bogged down in a morass of megacycles and kilowatts."[139]

Vernacular Broadcasting and the Politics of Infrastructural Inclusion

If Radio Cairo offered listeners the possibility of generating new social and political attachments over the airwaves, putatively domestic dynamics, too, encouraged people to engage in the work of consolidating and fortifying new attachments. In 1952, the government declared a state of emergency in response to what government officials referred to as Mau Mau (see chapter 3).[140] Men and women fighting government forces and "loyalists" from their strongholds in the forests of Central Kenya and the Rift Valley firmly rooted their critiques within a Kikuyu epistemology of "civic virtue." While the language of Mau Mau was Kikuyu, the government feared that the movement could transcend ethnic boundaries, spreading to other ethnolinguistic communities.[141] Policy decisions in these years complicated these politics of identification and connection.

In 1954, the colonial administration began implementing the recommendations of the Swynnerton Plan. Designed to increase agricultural outputs, the plan inaugurated land consolidation and the formalization of tenure. As old borders were redrawn, there was downward pressure on people to firmly root themselves within the logic of a territorialized and ethnicized narrative of belonging.[142] This was matched by other transformations, most notably the 1954 Lyttleton Constitution, which, among other things, created room for African and Asian participation within the government. National political parties were banned, but political aspirants were invited to create parties bounded by locality. While mitigating against large-scale political mobilizations was one of the goals of the colonial government, the disproportionately low levels of African representation was bound to cause conflict.[143] Taken together, these politics seemed to demand, as a form of protection, that people firmly tether themselves to ostensible homelands defined in ethnic and linguistic terms. The colonial state would seek to capitalize on and direct this emergent and competitive realm of differentiation through broadcasting.

If, in the late 1940s and early 1950s, the administration hoped to use Kiswahili broadcasts to generate "intermediate" subjects, listeners had their own ideas about what networks of affinity they hoped to access through the radio. In the mid-1950s, one administrator reported that "the most interesting feature of the period has been the clear indication that African listeners are no longer content with radio material which merely fills air-time. They have become increasingly critical in their selection of programmes and throughout have shown that they wish to help, by constructive suggestions, what they regard as *their own* radio station."[144] African listeners were, indeed, discerning and were increasingly making their critiques and desires known.

In response to this renewed interest, the African Broadcasting Service (ABS) launched the "African Broadcasting Club" in 1954, and a listenership magazine called "Habari za Radio." Groups were to organize themselves by "neighborhood," the goal being to establish an "intimate feeling of belonging to and being identified with the programmes broadcast" by ABS. In keeping with the politics of intermediacy that the government hoped to cultivate, "Habari za Radio" was a predominantly Kiswahili-language monthly. The KIO used the magazine to get a handle on listeners' responses to government broadcasting. But it was also a space where listeners were invited to express their tastes and desires.[145] It was taken for granted that capital-P politics would not be permissible, just as politics were not permissible over the airwaves. And yet, what the administration at times seemed to view as the apolitical domain of language was, for listeners, a deeply political domain of future-building.

Letter writers were encouraged to see themselves as individuals involved in a collective project. This fortified a vision of radio broadcasting as a "public good." As the magazine cajoled: "Remember this work of RADIO, is yours . . . if you hear anything broadcasted by A.B.S [that] is true, then, write short letters."[146] More verbose commentaries would be ignored, letter writers were warned. There were limits on space.[147] By the late 1950s, the magazine claimed to have a membership trebling around 750—a small group—but that it had a reading public of 5,000.[148] People actively responded to "Habari za Radio's" solicitations for feedback, ABS claiming that it received thousands of letters per week from listeners. It had to be acknowledged, though, that the reach of the magazine, like the reach of the paltry broadcasting network itself, was incomplete, with some areas conspicuously absent on the ABS's Club map.[149]

Letter writers seemed to take seriously the claim that radio was theirs, and that it was a public service provided by the government. Over the years, the government had actively shored up this vision by provisioning free listening facilities—rediffusion services and public address points—policies that were designed to "'prime the pump' and build up a listening habit."[150] By 1954, it was clear that an interest in radio had been cultivated. "The market," administrators argued, could be left to do the rest. "With the present popularity of the programmes and the quantity of cheap wireless sets on the market which Africans are buying in large numbers, this object [of priming the pump] can be said to have been achieved. The Department will provide no further free listening."[151] People were incensed by the contraction in the infrastructure of broadcasting in these years, which, to many, appeared to constitute a form of infrastructural exclusion.

As popular experiences of the technopolitics of the state were increasingly shaped less by technological intervention than by infrastructural and technological absence, people made their discontent known. Within the pages of "Habari za Radio," listeners argued that the government should subsidize their access to receivers, seeing as it was a public service. As Suleiman Abdullah wrote: "I am an A.B.S. club member, but I don't know what to do because I don't have my own radio and my parents are poor [so] they can't get one . . . [I'm] asking if the A.B.S. club officials would agree to buy me a radio then I shall be paying little by little every month."[152] The African Broadcasting Services' response to these demands was tepid. It was simply no longer its responsibility to provision sets for its listeners.

The government had changed the terms on which radio could be experienced as a "public good." But it was based on this vision of the infrastructure, and the services that it proffered as "their radio," that people made claims on the state. Historical infrastructural attachments ensured this would be the case. Nowhere was this clearer than in discussions surrounding the language of broadcast. It is evident that access to the infrastructure of broadcasting was, for these letter writers, critical to their apprehension of whether the colonial state was including them and their communities in "this work of radio."[153] And here, listeners would be able to leverage their demands.

Letter writers offered both critiques and praise of government-directed broadcasts supposedly orchestrated in their name. People living in Kenya's Central Province complained that only Kikuyu and Kiswahili speakers were accounted for in the programming. As Angelo Daudi of Meru wrote, "The problem arising is, our listeners don't understand well Kiswahili and Kikuyu. So, I would like to ask if it is possible to get like 15 minutes of programming in [the] Meru language." The response was predictable. "In Kenya, there are 82 tribes," should the administration try to cater for each? "Or shall we ask everyone to learn Kiswahili and English so that we can be able to communicate together happily?" Kiswahili was to be the language of intermediacy.[154] But here, many administrators seemed not to have understood the politics of language in light of political and administrative reforms.

These demands were about more than simply having one's language represented. As one Kimeru-speaking letter writer wrote, "Mr. Murungi mixes Kimeru with Kikuyu during his broadcasts and as a result of this, everything gets mixed to such an extent that most of the listeners have lost interest in the programme. We are tired of his poor Kimeru broadcasts." This linguistic hybrid was not wanted by listeners, who demanded "clean Kimeru."[155] For many letter writers, the stakes of such debates were not insignificant. By the mid-1950s, and with independence a vague outline on the horizon, minority communities feared being swamped by majorities. For some, in Central Province, these groups feared the domination of Kikuyu-speakers. For others, along the coast, the question was whether the region would become an "Arab-Muslim" zone, or a Christian "African" one.

And people leveraged the fears of the colonial state in working to get it to accede to their demands. As Ole Aomo wrote: "I am a Maasai man, who lives in Narok, and I would like to say that because the Mau Mau have already entered my area, Maasai programs should be introduced . . . [so] those who have already joined the Mau Mau . . . may be told the truth . . . [I] am sure this will help."[156] If on the coast, the threat of Radio Cairo offered listeners one point of leverage, inland Mau Mau offered another. And people mobilized the threat of the movement and its capacity to spread along the inscrutable lines of the "bush telegraph" in their bid for government recognition of their linguistic particularities.[157]

By the 1950s, evidently, "parochial pride" was not something to be unraveled. Instead, it had been reformatted as a—if not the—most desirable form of social and political identification. Vernacular programming—administrators hoped, and cultural patriots would join them—could shore up notions of community bounded in place and by language. It would be these modes of identification, not the more elastic modes historically available to people, that would take precedence. In the mid-1950s, one administrator reported that: "Nyanza Province were very keen on programmes in their vernacular languages" and that the circumstances demanded that the "claims of Nyanza Province should receive . . . consideration alongside those from the Coast."[158] And so, in 1953, Dholuo broadcasts were restarted for fifteen minutes per day. Programs in Luhya, Gusii, Kinandi, and Kikamba broadcast over communal sets followed.[159]

In contrast to Radio Cairo's strategy of forging connections across difference, the colonial state's strategy was to deploy difference as a means of undermining emergent attachments. By 1953, these efforts were fortified as lorries arrived heavy with receiving units for communal listening. Plans to create a market for receivers were reignited, and information officers recommended that wireless sets be given to loyal chiefs to ensure that the technology became "known" and desired "in the African areas." This, it was hoped, would lead to "increased sales to individuals."[160]

If "Habari za Radio" offered one platform for listeners to engage in comparative work as they measured their infrastructural inclusion against that of others, these issues also came to the fore in more expressly political arenas. Despite the strategic widening of Kenya's political sphere pushed through under the Lyttleton Constitution, African representatives were not invited to broadcast their opinions over the airwaves, and this caused no small measure of upset.[161] African representatives—Nyanza-born Tom Mboya, Member from Nairobi among them—were incensed.[162] Tellingly, Mboya argued that broadcasting was a "public service," the public paying the costs of licensing fees, and so Elected Members, as members of the public, were "as much entitled as the Government to put forward . . . views on various aspects of Government policy and programmes."[163] While C&W had obstructed the state's capacity to provision radio to African communities as a free "public service," requiring Africans to pay licensing fees, for Mboya, this offered a point of leverage. African radio owners had invested in the broadcasting infrastructure of the territory, whether it was privately owned or not, and so, they seemed to argue, it was partially theirs.

Mboya was joined by others, such as Ronald Ngala. Born in 1922, in Gotani in Kenya's coastal region, Ngala attended Alliance High School and later Makerere, where he received his teaching certification, becoming a teacher and, later, a headmaster on the coast.[164] In 1957, Ngala was elected to the Legislative Council as the coast's representative. Following the promulgation of the Lyttleton Constitution, which formalized Ngala's status as a politician, the politics of the coast amped up, making it a highly contested space of competitive forms of differentiation. Swahili-Muslims and Arabs on the coast feared being swamped by "up country" Africans. Coastal Africans, for their part, feared domination both by coastal Muslims and by larger African ethnolinguistic communities "up country," leading to what one administrator referred to as the increased "political awareness of local Africans."[165]

Ngala, like many other leaders of the period, was an "ethnic patriot."[166] And like other ethnic patriots, he set to work constructing a timeless history of "his tribe," the Giriama, titled *Nchi na Desturi za Wagiryama* ("The Land and Tradition of the Giriama"), which, importantly, he composed in Kiswahili. On entry into the Legislative Council, Ngala began advocating for the particularities of coastal African communities. In 1957, the District Annual Report noted,

"March saw the first African elections to be held in this country when Mr. RG Ngala was elected as member for the Coast Constituency. Since the elections there has been a marked increase in political activity, particularly among Africans, who are treated to a weekly diet of nationalist politics at meetings held in Tononoka."[167] Ngala's politics, administrators reported, were "full of nationalist and racial slogans."[168]

Ngala was quick to link the need for the autonomy of African communities on the coast to radio broadcasting. He argued that ethnolinguistic groups on the coast needed to be represented in government-provisioned radio broadcasting. Why, he asked, did Sauti ya Mvita fail to live up to its name and create broadcasts that would appeal to all on the coast?[169] His demands did not stop there. During a debate on information estimates undertaken during a legislative council session, Ngala argued that Sauti ya Mvita should dedicate more airtime to African listeners, in particular by broadcasting in Giriama and other coastal vernaculars.[170]

To these requests, the Chief Secretary responded that the "provisioning of programmes for Africans was in hand, but that broadcasting in vernaculars would not be possible because of the limited funds available."[171] Here, the administration missed the point. Comprehension was not the issue. Kiswahili-language broadcasting on the coast appeared to these listeners to portend a future wherein the region would be dominated by the interests of Muslims, both "Arab" and "Swahili." Kiswahili, for these listeners, was not the language of intermediacy but the language of partisanship. For these critics, vernacular broadcasting might enable them to leverage their "parochialism" into political representation on the national stage.

In 1960, Ngala joined representatives of other minority constituencies to found the Kenya African Democratic Union (KADU), with himself as the leader. KADU, in keeping with Ngala's radio politics, argued that independent Kenya should be a federated union, this being the only way to prevent Kenya's minorities from being swamped by the dominant ethnolinguistic groups. By the late 1950s, the balanced provisioning of services had become one register by which communities interpreted their position in relation to the colonial state. The stakes of these relations were high, with a future independent state coming into view as an increasingly robust outline on the horizon. But Ngala's seemingly parochial politics in these years did not stop him from traveling to London with Nyanza-born Tom Mboya to protest colonial efforts to use the new constitution to bifurcate a nation-in-becoming.[172] For many people at this time, politics was a game of hedging. Absent a secure future, it was left to these leaders to play the game of both the specific (language/"tribe") and the general (nation/"African").

And demands begot other demands. In the mid-1950s, Somali-speaking listeners argued that the state should provide Somali broadcasts for communities living in the Northern Frontier Province. While in the past the claims of this historically marginalized population could have been ignored, by the end of 1957, it was clear that Radio Cairo's Somali-language broadcasts, heavy with "vicious propaganda," were being "listened to by the Somalis." And so the price for ignoring claims for inclusion via the recognition of difference appeared to be impossible, making it of the "greatest importance" that a "good Kenya service be provided."[173]

The specter of Mau Mau certainly haunted the colonial imaginary when administrators discussed the future of the colony and the role of broadcasting in shoring up a beleaguered center, but so too did the foreign broadcasts bleeding into Kenyan airspace. Broadcasting in Nyeri, it was noted, was geared toward "rehabilitation," whereas broadcasting on the coast and Nyanza was "preventative." The airwaves, heavy with colonial visions of the future, touching down in communities to the east and to the west could be subtler. In these regions, "many a pill," one administrator argued, could be "slipped in without their realizing it."[174] The state was savvy that people in this twilight moment were responding to conflicting demands and the uncertainty of the present. And so it settled on conceding to these demands, hoping that the politics of infrastructural inclusion premised on popular demands for the recognition of linguistic difference in the domain of broadcasting could be beneficially divisive, allowing the state to retain its fragile hold on a territorial unit: the Kenya colony.

Conclusion

In 1956, the state's contract with C&W finally expired. The following year, and in the wake of much anxiety and hand-wringing over the siting of new stations in both Nyanza and on the coast, new transmitters had been installed in both regions to address what was banally referred to as the "problem" of each.

In coming to this solution, the state self-consciously deployed technics to effect political ends, hoping the network could shore up the borders of the Kenya colony in this moment of flux. As noted by then-Director of Information, Watkins Pritchford, "The whole object of installing a medium wave transmitter is to try and ensure that listeners do not need to turn to the short-wave band at all. If we can keep them listening consistently on the medium wave band we can ensure that our programmes are not interfered with by outside stations and by avoiding the necessity for the listener to use short wave band at all we may be able to wean him away from the powerful shortwave broadcasts from Cairo, Omdurman and All India Radio."[175] While the media sphere remained saturated with threats, Cairo was the "principal antagonist" in this "radio war."[176]

The animus for infrastructural expansion, however, was a response to home-grown demands for infrastructural inclusion premised on government claims that radio broadcasting was a "public good." As infrastructural networks were variously expanded and rolled back, people engaged in comparative work as they calculated their relative access vis-a-vis an increasingly hard set of borders that separated themselves from other ethnolinguistic communities. It so followed that people did not make claims on the state as homogenous listeners, but as linguistically particular constituents.

"Tropicalising" technologies was a colonial trope premised on a vision of incommensurate "Otherness." Listeners reformatted this vision when they argued that the other public of which they were a part was not, in fact, the "masses," but rather more discrete categories that demanded accommodation. Theirs was a politics that linked the recognition of difference to equal airtime, that linked political inclusion to access to infrastructures. The state had laid the groundwork for this, and these demands were ultimately heard loud and clear. The recognition of linguistic difference as a metric of inclusion was on the table, so far as broadcasting was concerned. These demands mitigated against the linguistic telos that would have had first Kiswahili, and then English, as the lingua franca of the Kenya colony.

In acceding to these demands, the state sought to reinforce the parochial and ethnolinguistically bounded focus of African political activity by mobilizing a new infrastructural network. As A. M. Dean of the Department of Information wrote: "Present and possible future political trends in the Colony argue strongly in favour of Government owned Broadcast Service in all languages."[177] This was not simply inclusion, but an inclusion that turned on the capacity to capitalize on—and to make antagonistic—difference. As one coastal administrator wrote: "What we need to do . . . is to make both African and Arab feel that they were being well catered for, and even for each to feel that they are getting 'a better deal' than the other community."[178]

5

The Politics of Divisibility

Safaricom and the Remaking of
the Corporate Nation-State

One morning in September 2014, I hopped on a *matatu* heading northeast away from Nairobi's Central Business District. I alighted by a lime green pedestrian bridge connecting East and West Nairobi, bypassing Nairobi's long-awaited Thika Superhighway below. Crossing the bridge, I confronted a large billboard marking my destination. Familiar lime green letters marched neatly across the face of this sign, and the many others peppering the landscape of the sports grounds: Safaricom Stadium Kasarani. This name was new. A rebranding that came on the heels of a large investment into the facilities made by Safaricom—at the time of writing, the region's largest corporation, and dominant telecom and financial services provider.[1]

The proprietary claim implicit in the presence of the new name frames how visitors interpret, and are interpolated by, the grounds of the sports complex. The stadium, historically a key symbol of the nation, now hails Kenyans with an altogether different brand of belonging, that of corporate citizenship. In a few discrete places, the old name remains in smaller, red font: Moi International Sports Centre. In this space of cultural and sporting spectacle, Safaricom has displaced another giant, Kenya's dictatorial second president, Daniel arap Moi.[2] Reflecting on the meaning of this displacement, I joined my fellow travelers, many dressed in their Sunday best, as we approached the hulking stadium. The event: Safaricom's Annual Shareholders Meeting.

A friend had facilitated my access to the meeting. His sister was a shareholder, but, he told me, she was so disappointed in the performance of her shares—and the resulting dividends following the massive oversubscription during the Initial Public Offering (IPO) six years earlier—that she had no interest in attending the Annual General Meeting (AGM). This enabled me to participate in her staid, a deal secured by the scrawl of a signature and the inscription of a national ID number on a form downloadable from Safaricom's website.

But a few years earlier, access to the internet was limited to "cybers" and the homes of Kenya's elite. By 2014, smartphones proliferated across the Kenyan landscape. These material transformations have been accompanied by the emergence of a new set of imaginaries, both of which are largely attributable to Safaricom and the "miraculous" rise of mobile telephony. Indeed, techno-enthusiasts worldwide describe Kenya as a "hub of technological innovation," the continent's "Silicon Savannah."[3] Within Kenya, people at all tiers of society traffic in this rhetoric, which has moved outside of the circumscribed field of technological innovation and is used to index a whole range of projects—from an individual's aspirations for the future, to a national shorthand for how "development" is thought.[4] The promises of "the digital" also shaped the aspirations of would-be shareholders who viewed buying into the digital communications company as a means by which they could secure their fiscal futures—personal enrichment a means of *becoming* digital.

As I made my way across the complex, I became increasingly perplexed by the confusing semiotics of the space. Every few feet, one was confronted by signage. Safaricom's telltale italicized lime green font nested amongst the dark green, red, and black of the Kenyan flag. Most striking within this palette was how the green of the national flag, the green symbolizing the struggle to regain control over lands lost to white settlers—itself a contested symbol of the ethno-politicized struggle that precipitated independence—was juxtaposed against a green of a lighter hue, what Kenyans refer to as "Safaricom green." Here, the signs of the corporation and signs of the nation at once seemed to jockey for preeminence, while concurrently existing in uneasy cohabitation—an awkward conviviality splashed across walls, banners, and flags.

Approaching the stadium, I was handed a small box containing a sandwich, a can of soda, and a Safaricom-branded T-shirt. This welcome, I later learned, was a vast improvement from the year before. In 2013, the company had provided no material offering in exchange for attendance. As shareholders were quick to point out, Safaricom had not even provided transport to the stadium. People had been incensed by this deviation from the script that guides interactions between patrons and clients. In registering their critique, Kenyans called the corporation to account according to enduring expectations regarding the proper operation of politics in Kenya, a register wherein politicians routinely pay (and are expected to pay) people to turn out for rallies and political meetings. Shows of political loyalty are rarely freely given in Kenya.

But shareholders are not clients as conventionally understood. Their investment in the relationship binding themselves to Safaricom has been secured by shillings, their shares "freely" bought on the Nairobi Securities Exchange (NSE). Shillings, rather than promised votes, are what ostensibly tethers Kenyan shareholders to their corporate patron. Meeting attendees were acutely aware of this transmutation in conventional relations of dependence.[5] They appealed to the company both in their capacity as client-citizens and in their capacity as corporate shareholders, a mix that reveals how the company is reconfiguring not only the meaning of citizenship, but the nature and location of the attachments binding Kenya's publics to both the state and the corporation.[6]

My sense of the complicated place the company occupies in Kenyan social, political, and material life heightened as the day progressed. As I walked up the steps of the stadium, a brass band started playing. "*Wimbo wa taifa*" (national anthem), the woman next to me whispered to her companion as they stopped, bowing their heads. As the last notes faded, we proceeded into the belly of the building. Safaricom's management was seated on a raised dais in the center of the stadium. Shareholders sat in bleachers gazing down at this international cadre of men and women who were flanked by flags and bunting boasting the colors of the Kenyan national flag, the secondary green of Safaricom conspicuously included. The meeting combined practices that typify shareholder meetings the world over with the model of a Kenyan *baraza*, or political meeting, where equally normative codes of behavior dictate the enactment of attachments binding clients to their political patrons.[7]

The semiotic doubling of Safaricom's green jockeying for space with the green of the Kenyan flag at the 2014 AGM was not seamless, gesturing toward an ambivalence regarding the place of the corporation in everyday life. This is not altogether surprising. Everyday Kenyans of various stripes find themselves and their lives attached to the corporation. Accessing money, provisioning for a family member or friend, gaining access to small loans, making a call, sending an SMS, gathering provisions following calamity, or watching a football match—all these activities are routed through Safaricom's infrastructure. The multiplicity of its domains of operation renders Safaricom—both as a material network and as a discursive and visual presence—pervasive. Its ubiquity and reach through the sinews of everyday life is impressive, rivaling the reach of the state in ways both visible and invisible. As Safaricom's penetration into the intimate lives of its customers expands, it generates new infrastructural attachments, what industry insiders refer to as "stickiness."

Infrastructural attachments also characterize the relationship between the Kenyan state and the corporation. Today, the state is not only a major shareholder of the firm, but Safaricom is the largest single taxpayer. The state thus has a "double-dipping" interest in the firm: first, as a recipient of dividends, and second, to shore up the revenue regime of the government—the firm, put simply, acts as an infrastructural prosthetic aiding in the reproduction of the state.[8] While Safaricom's embeddedness in this Kenyan milieu is not accidental, it can be a liability for both the firm and the Kenyan state, each of which works assiduously to walk this line between proximity and distance.

This chapter seeks to understand this dialectic of attachment and detachment—both in material and symbolic terms—by exploring how Safaricom, the region's largest company and a multinational corporation, in its intimate imbrication with the Kenyan state has come to produce state-like effects.[9] It begins by considering the colonial origins of telephony—a racially exclusive network that mapped onto the hierarchies that structured the Kenya colony more broadly. The postcolonial state worked to undo this historical inequity when it established the Kenya Post and Telecommunications Corporation (KPTC) as an arm of the state, a set of attachments that promised infrastructural connectivity for the citizenry of the newly independent nation.[10] Quickly, though, it became clear that the discretionary logic that guided infrastructural provisioning in the colonial period under conditions of fiscal restraint remained in place, resulting in the growth of a network that was lumpy in its reach, and exclusive in its coverage. This history sets the stage for Safaricom's birth in the early 2000s, and deepens our understanding of the disputes that it engenders in the present.

Safaricom emerged at the interstices of the changing proclivities of international governing bodies, transformations in national policies, and global shifts in telecommunication technologies. In Kenya, as across much of the world, the 1980s marked an ideological about-face. A reemergent faith in the capacity of "the market" to remedy state "inefficiencies" set the stage for the privatization of state assets. This "wisdom" gained purchase as the World Bank worked to reframe "public goods." Reports such as "Infrastructures for Development" argued that infrastructures were not "public goods," as hopeful development thinkers from the 1940s through the 1960s had held (see chapters 3 and 4), but ought to be treated like any other commodity: as objects of consumption best regulated by market mechanisms.[11] The watchword of this period was "transparency." It was only by disembedding the corporation from the state that "innovation" and the "efficient" provision of services could be ensured and the "public good" protected. Safaricom was born of these transformations. An erstwhile state-held entity, Safaricom began its path to privatization in the late 1990s, culminating with the company going "public" in 2007–8 when it was listed on the NSE.

In this regard, and in broad strokes, the history of Safaricom resembles stories of privatization in the context of neoliberal reforms evident elsewhere. However, we lose much by assuming the uniformity of policies operating under the banner of "neoliberalism"—privatization among these—and their effects.[12] In what follows, I explore privatization as an incremental and protracted process that turned on the production of divisibility—the discursive, epistemic, and material processes that transformed a series of entities that "were formerly thought of as either organic or classificatory 'wholes'"—an infrastructure, a corporation, a national asset—into objects of calculation subject to division.[13] As new calculative spaces were carved out by Kenyan lawmakers, investment bankers, and corporate officials, the boundaries demarcating the borders among these entities, and the relations among them, were redrawn.

But privatization was not simply about numbers and calculation. Kenyan observers routinely debated the qualitative implications of these transformations, drawing on historically resonant ideas regarding the appropriate relations among infrastructures, governance, and political belonging as they debated the possible outcomes of the wholesale adoption of this market rationality. In doing so, they raised fundamental questions about how the production of national assets as divisible commodities would generate new forms of detachment that would unbundle the material tethers linking Kenyans to the state, simultaneously undermining Kenyans' rights over the infrastructure.

Producing the infrastructure—a "public good"—as a divisible entity concurrently brought to the fore uncertainty regarding who or what was responsible for "development." As this history of the present will demonstrate, the onus for "development" has incrementally moved from being framed as the responsibility of the state, as thin as that promise had been, to being placed on the shoulders of Kenyan citizens, who have been reconstituted as a market of investors and innovators in the process (for the latter, see chapter 6). Far from entailing a simple withdrawal of the state, the production of divisibility has entailed generating new infrastructural attachments by reconfiguring the boundaries between public and private, state and corporation, citizens and shareholders. New and discomfiting forms of both attachment and detachment—encapsulated by the reemergence of the corporate-state and the emergence of the shareholder-citizen—have been produced in these shifting frontiers.

As this suggests, creating new domains of calculation had effects in spaces seemingly external to "the market," rendering another conceptual series—"the public," "citizenship," "development"—thinkable as partible entities subject to divisibility and new forms of distribution. The category of "the public," in particular, was pressed into the service of multiple projects, as actors variously framed "the public" as "indigenous" Kenyans, Kenyan citizens, ethnicized political subjects, would-be shareholders, and sources of capital. These processes of marketization through the production of divisibility have reconfigured conceptions of obligation, as citizen-shareholders, who have been reformatted as sources of value, call on Safaricom to act *like the state* from which it has been incrementally unbundled, rooting their claims not simply in a language calculation but in a language of ethics.

The Colonial Origins of the Postcolonial State

While Safaricom's rise as a corporate-state is recent, this institutional dispensation is not novel. Indeed, the history of telephony in the region has been characterized by continual efforts to draw and redraw the boundaries between the corporation and the state. Following the collapse of the Imperial British East Africa Company in the late nineteenth century, the colonial state contracted the Eastern & South African Telegraph Company (later the Eastern Telegraph Group, which became Cable and Wireless Ltd., see chapters 3 and 4) to unroll a telephone network.[14] Despite the corporate provenance of this network, the state saw in this infrastructural attachment a means of consolidating its tenuous sovereign authority in the region. Inaugurated in 1888, the network was mainly used for administrative purposes. This all changed in 1908, when the service was extended to private users, namely white settlers. By 1933, the communications of Kenya, Uganda, and Tanganyika were unified under the authority of the Postmaster General.[15] While slow to expand the network, by 1961, two years before Kenya's independence, the number of phones had increased from 12,000 in 1946 to nearly 48,000.[16]

If, for much of the colonial period, it was taken for granted that telecommunication services would be provided for by commercial firms, and would service a relatively wealthy, white, and small proportion of the population, this was not the case following independence in 1963. By this point, it was widely accepted that utilities such as telephony were for the "public good." These were to be "natural monopolies," administered by the state and shored up by the resources of the people.[17]

Across much of sub-Saharan Africa, though, newly independent states had to develop political ideologies that reflected their limited revenue regimes. In Tanzania, this new spirit was captured by the concept of "African Socialism." In Kenya, this ideology was captured by the country's new national motto, *harambee*. As Prime Minister, later President, Jomo Kenyatta proclaimed in a speech on the eve of Kenya's independence, "As we make merry at this time, remember this: we are relaxing before the toil that is to come. We must work harder to fight our enemies—ignorance, sickness and poverty. I therefore give you the call: HARAMBEE! Let us all work hard together for our country, Kenya."[18] *Harambee*, Kiswahili for "all pull together," would become the rallying cry of Kenyatta's Kenya African National Union (KANU) government, providing the roadmap for how "development" was going to be achieved by the independent state, and offering a thinly veiled warning to those advocating for regional autonomy. Emblazoned on Kenya's coat of arms, *harambee* served as a reminder to Kenyans that conditions of austerity would be the normal state of affairs in independent Kenya.

Harambee was not simply a means of managing fiscal constraint. As the Kenyan bourgeoisie took the reins of power, they reacted much as the colonial state had in its years of most forcefully defending white settlers. Seizure and distribution would both proceed unevenly, often to the benefit of co-ethnics.[19] In the popular imagination, Kikuyu communities in general were privileged by this arrangement. Closer to the truth, it was predominantly the Kikuyu elite who were the beneficiaries.

These dynamics would structure the fate of telephony in Kenya, but the early independence period was one that appeared rife with possibility for evenly distributed and government-operated infrastructural networks. In 1967, Kenya, Uganda, and Tanzania formed the East African Community (EAC), with communication services operating under the banner of the East Africa Posts and Telecommunications Corporation (EAPTC). This liberation from metropolitan corporate dependence was largely a fiction, with EAPTC continuing to rely on the British corporation, Cable and Wireless Ltd. (C&W), which held a 40 percent stake in the firm.[20] Nevertheless, network growth continued apace. By the end of 1969, there were 350 telephone exchanges operating across the region. By the end of 1970s, teledensity reached 0.73 per 100 people. This distribution was lumpy at best, with 99 percent of lines servicing urban communities. The late 1970s saw the dissolution of the EAC and the dismantling of the EAPTC. In 1977, the Kenya Post and Telecommunications Corporation was established as a "traditional monopoly," the government having purchased C&W's stake.[21] The state could now proceed with unrolling its own telecommunications network to service the citizenry—prospective infrastructural attachments that would tie Kenyans and their futures to the independent nation.

This aspiration proved to be short-lived. Those in government used the KPTC as a vehicle to funnel resources and provide employment in exchange for political support.[22] As a government report noted in 1979, parastatals had become "personalised institutions" that reflected the ethnicized favoritism of Presidents Jomo Kenyatta and Daniel arap Moi, who used the firm as a "conduit for rents and rewards."[23] These processes accelerated in the late 1980s, with political pluralism and liberalization on the horizon.

By the 1990s, it was clear that this bargain had produced questionable results. Out of the 5,050,000 households in Kenya, less than 5 percent had a landline.[24] Those lines that did exist remained overwhelmingly concentrated in Kenya's cities.[25] Upkeep was difficult; rainy seasons waterlogged facilities, shutting down services for extended periods of time.[26] The firm's cash flow was negative; accounting practices were imperfectly executed. The Corporation was in dire straits. Rather than its profits entering into the state's revenue stream, the government was now forced to transfer money to the KPTC. The firm's role in shoring up the state had faltered. The infrastructural attachments binding the KPTC to both the state and the citizenry continued to fray. For critics in government, in order to "reverse" these "trend[s]," the government needed to divest "some of its investments to Kenyan investors."[27] As we will see, this would not entail the retreat of the state but rather its reconfiguration, as the levers of accumulation (both licit and illicit) were redistributed across new locales.

The Production of Divisibility and the Birth of the Corporate-State

Safaricom's complicated relationship with the Kenyan state, then, extends back to its founding moment. And, like many origin stories, this one proceeded through a series of perceived betrayals. While reforms operating under the aegis of the World Bank and the IMF had begun in the 1980s, by 1996, enforced restructuring and privatization of telecommunications was specifically laid out as a condition of aid.[28] But it was not until the mid-2000s that the state began breaking up the last of its nationally held assets. This process was one of fits and starts. In 1998, a new Information and Communication Technology (ICT) policy was enacted, which split the KPTC into three discrete entities: the Postal Corporation of Kenya, the Communications Commission of Kenya (CCK), and what was to become Kenya's monopoly telephone service provider, Telkom Kenya (TKL).[29] The division of the KPTC legally authorized the production of divisibility, a process of "unbundling" that rendered portions of the asset available to private, foreign investors.[30] The policy, however, did not go uncontested, and privatization remained a "matter of concern," as many debated the implications of this new legal architecture.[31]

In establishing TKL as a parastatal, the government built on historical forms of boundary-crossing, mobilizing the state to deepen the penetration of private, foreign capital. The colonial government had facilitated the establishment of a number of such hybrid bodies. Following independence, however, the state conceived of parastatals less as vehicles for foreign capital, than as a means of enticing Kenyan investors who, it speculated, would eventually purchase government-held shares and engage in a kind of patriotic capitalism.[32] These bodies took myriad forms, ranging from industrial firms to agricultural corporations, from the Kenyan Film Corporation to TKL. By the time of the 2007–08 IPO, the dreams of the early independence period would be realized. However, rather than more closely binding Kenyans to the state through their participation as shareholders, the IPO would see Kenyans more closely bound to a corporation that had come to look more and more like the state.[33]

Proponents of the disaggregation of state assets argued that the move would increase both "transparency" and "efficiency." As David Mwiraria, then-Minister of Finance, argued in Parliament in 2004, the rationale for privatization was clear: "In the past, incidents of mismanagement in public enterprises have imposed heavy financial costs on both the economy and the Kenyan taxpayers. As a result, the public has paid for services which were either not rendered or when rendered, were of a low quality. In general, this sub-sector has been largely under-performing." Spotty service provisioning was not the only issue. "In some instances," he continued, "these enterprises hired excess labour while engaging in inappropriate procurement procedures. The result has been overpriced purchases of assets, real estates or outright theft through purchases of [the] wrong equipment."[34] Public enterprises, he charged, were being used illicitly to secure private gain. Rather than provisioning services and enabling the development of robust infrastructures that bound people to place, those managing state enterprises stood accused of unraveling the material networks that enabled connectivity, separating the materiality of infrastructures from their purported function: the provision of services necessary to secure the "public good." These actions placed an undue burden on the Kenyan population, as government-appointed managers sucked capital out of state coffers. Informally, these predatory actions socialized state enterprises' debts by transferring most of their "[debt] burden to the Exchequer."[35] These practices were particularly vexing in the context of telecoms, for, as Moses Akaranga put it, communication infrastructures were critical to "the development of any place."[36] The division of the KPTC, and the subsequent creation of TKL were designed to protect both the Kenyan public—here conceived of as citizen-taxpayers—and the state from the perceived predations of government officials. Only through these reforms would national development be possible.

In private, observers at the American Embassy tended to agree. However, they were quick to argue that as a remedial step designed to ensure transparency, the bifurcation of the KPTC had been a failure. So, too, did these observers question TKL's integrity as a semi-autonomous parastatal: "To no one's surprise, the CCK's latest annual report notes TKL's utter failure to translate its monopoly status into improved services [in landline provisioning] and infrastructure . . . Nearly all Kenyans can recount anecdotes which together paint a picture of an inefficient, badly managed company widely seen as more a platform for political patronage and corruption than a genuine service provider."[37] These observers claimed that government foot-dragging had beleaguered the path to liberalization, as contracts to private corporations were routinely thwarted in what observers perceived to be efforts to protect ethnic "cabals."[38] Challenging a long-held ideology of distribution that links political patronage to (albeit unequal) service provisioning, these critics positioned patronage as an anathema to the "public good." The message was clear: the politics of patronage were an obstruction to infrastructural maintenance and expansion, both of which had implications for service delivery.

There was some truth to this. The extension of Kenya's landline network continued to proceed slowly and unevenly. By 2002, Kenya had 327,000 landline connections servicing a population of over 31 million, ranking relatively wealthy Kenya 23rd on the continent.[39] Services, parliamentarians claimed, were mainly nested in Nairobi and its immediate hinterlands, "leaving most of the country unwired in the information age."[40] Kenya's entry into the "information age" would be heralded by the incremental privatization of Kenya's communication infrastructure, a process that was widely held to be the only way to bring together what was needed: the holy trinity of foreign expertise, technology, and capital. For proponents of privatization, the selling off of state assets would not reproduce asymmetric relations of power between putative centers and peripheries. The goal, wrote then-Minister of Finance Chris Okemo in 2001, was to cultivate new, "strategic partnerships." This was especially important "when . . . dealing with enterprises that involve technology and management expertise."[41]

The value of these relationships was not to be framed in terms of dollar (or shilling) amount alone, but crucially turned on a qualitative assessment of the emergent relationships born out of the selling of state assets. "What was needed," Okemo argued, was "somebody who would bring value to the corporation not just by acquiring assets but . . . [by bringing] technology and management skills that will enhance the value of that enterprise; that will lead to efficiency of services." His framing was cautious, however. Care must be taken, he stated, to protect "Kenya's interests first . . . all other interests are secondary."[42] It was amidst such debates that the Privatisation Act of 1999 had been passed by Parliament. Oddly, then, privatization was tentatively devised as a means of both protecting "the public" and securing the "public good."

Safaricom made its first appearance in the days bracketing the division of the KPTC, though the specifics of its origins remain disputed. According to Safaricom, the company was created in 2000, having gained access to TKL's subscriber base and network—in other words, its "wealth in people," and its wealth in things.[43] Soon thereafter, Vodafone Kenya Limited acquired a 40 percent interest in Safaricom by contributing $20 million in cash. TKL and British-owned Vodafone contributed $33 million and $22 million, respectively, as their contributions for the portion of the license fee required to operate cellular services—though TKL's contribution came in the form of its network and subscriber roll.[44]

However, according to the Public Investments Committee on the Accounts of State Corporations, established in 2007 to investigate Safaricom's origins, Safaricom was registered as a wholly-owned subsidiary of the KPTC on April 3, 1997. In January 1999, the newly formed parastatal, TKL, transferred a 30 percent stake of Safaricom to Vodafone Kenya Limited (VKL), followed by another 10 percent in October of that year.[45] This latter 10 percent, while seemingly unknown to the public in 1999, would become important in years to come. Divergent timelines notwithstanding, both narratives held that by the time of the IPO, Vodafone held 40 percent equity while TKL held 60 percent, the latter portion of which was subsequently taken over by the government, along with a range of debts, in December 2007. Privatization, evidently, did little to prevent the socialization of corporate debt. While these timelines would be hotly contested in the months leading up to the 2007 IPO, what united these founding narratives were the numbers. Indeed, in the series of debates and intrigues that would assemble around the company, the numbers themselves—60/40—emerged as the objects over which people argued.

Foreign proponents of liberalization were unconcerned with the specifics of Safaricom's origin story. For their part, the growth of Safaricom confirmed a re-emergent orthodoxy: development was best effected when driven by global capital and expertise operating under the conditions of a liberalized economy. As noted by a US Foreign Service Officer:

> TKL's failure as a fixed line service provider stands in stark contrast to the roaring growth of Kenya's mobile phone industry—one of the country's rare economic success stories. Mobile services were commercially launched in 1993, but the market didn't start to take off until 1997, when SafariCom, [sic] a wholly-owned subsidiary of TKL, was established. In

May 2000, TKL sold 40% of Safaricom to Vodafone UK, which assumed management of the company . . . the market took off. Mobile network connections surpassed the number of fixed lines in 2001 and by 2004, total subscribers exceeded 2.8 million. Networks have been rolled out quickly, and close to 60% of the population now has cellular signal coverage. This rapid growth in mobile is both a cause and a consequence of TKL's failure in the fixed line market.[46]

These numbers, and the calculative rationalities underwriting them, worked to depoliticize liberalization. They offered "proof" that privatization had "succeeded," enabling both transparency and smooth service delivery. In a sentiment that was coming to the fore amongst policy-makers, foreign lenders, and industry leaders alike, inefficient government agencies needed to be replaced by responsible and efficient corporate management structures and models.[47] If a new assemblage of managerial techniques and structures of ownership was the solution to state "failure," new technologies were to be its arsenal, allowing Kenya to "leapfrog" over landline infrastructures altogether and launch the country into a mobile future.[48]

Kenya's entry into the mobile market in this period was *not*, however, unique. The launch of mobile phone networks in Kenya was part and parcel of a global shift toward mobile telephony. In 1995, there were approximately 91 million mobile phone users across the globe.[49] By 2006, this number had risen to 3.3 billion subscribers.[50] Kenya was a part of this trend. By 2006, 26.9 percent of the adult population were mobile subscribers.[51] While Kenya's rates of adoption were reasonable under the leadership of Vodafone, it was not until these new mobile networks were reformed and made "peculiarly Kenyan" (see chapter 6) that the country embarked on the path toward becoming Africa's "Silicon Savannah."

Global trends notwithstanding, for US observers—unconcerned as they were with the material and symbolic implications of privatization—market growth in Kenya was framed as the natural consequence of the company being reformed by the injection of private foreign capital and expertise, both of which accompanied the distribution of Safaricom's ownership structure across new locales. It was these investments to which they attributed the generation of new markets in mobile telephony, itself unquestioningly positioned as an index of national development. But this was to be a wireless, rather than wired, development—a development secured through the privatization of national assets and infrastructures, rather than through the extension and maintenance of state-owned infrastructures. This would serve an integrative function, binding diverse parts of the country into a single network.[52] These infrastructures were not a means of generating attachments between people and the state. Instead, Kenyans were becoming tethered to an interesting hybrid, Safaricom, by a series of os and 1s.

Privatizing "the Commons" and the Problem of Infrastructure

These transformations were not lost on the Kenyan public for whom the fact that Kenya's telecommunications sector, under the auspices of a parastatal, TKL, had been ceded to foreign interests caused much consternation. The numbers themselves—40 percent being held by British-owned multinational, Vodafone, and 60 percent being held by TKL—were central to these debates. Under the terms of the Privatisation Act of 1999, no more than 40 percent of a state asset could be foreign-held.[53] This regulation indexed an attempt to fix in place and render in numerical form the nature of the new relationships being established as state assets were privatized. And yet, these numbers had little to say about the nature of the new relationships being enacted as Kenya's telecommunications sector was rendered divisible. Specifically, critics voiced uncertainties regarding how the incremental divisibility of state assets would upset the status of infrastructural access as an attribution of Kenyanness.[54] These issues came to a head in 2004 as a bill authorizing the further disaggregation of parastatals, TKL among them, was being debated in Parliament.

For some, allowing foreigners a stake in state infrastructures would parlay into a kind of neo-imperial relationship. For these critics, the enforced privatization of state enterprises was in the interest of foreign investors alone; a cynical means of creating new markets for metropolitan capital, their own markets having long since dried up. For others, the question of Kenya's sovereignty was front and center. The colonial history of infrastructure service provisioning hovered just below the surface in these debates. As C. Kilonzo stated in the National Assembly: "Privatisation is a new form of colonization. It cannot be anything else. Very soon, if we sell all the parastatals then all the managing directors will be European."[55] Though all seemed to agree that services would be improved, not everyone was convinced that the benefits of these improvements were worth the costs.

For these opponents of privatization, the question turned on the status of the asset in question. For many, state assets were "public resources" that should be protected and retained as whole, indivisible parts of the national patrimony.[56] As Boaz Kipchumba argued in April 2004, Kenyans had a naturally endowed right to these networks which were "owned by members of the public."[57] For some, this was a matter of history. As Kerow Billows argued: "There is no way we can sell Telkom Kenya; a company owned by Kenyans, in which they have invested their time, efforts and resources for 40 years."[58]

Others moved away from history, tapping into a grammar of rights and ownership rooted in a deeper genealogy of belonging, firmly rooting TKL and its infrastructures within the logic of the patria. It is "our heritage," one parliamentarian emphatically stated.[59] These critics argued that Kenyans' proprietary rights over telecommunication infrastructures were rooted in a logic of primordialism. Kilonzo, for example, argued that while he, too, supported the new privatization bill, it was essential that the majority of shares "should be sold to Kenyans, and to indigenous Kenyans for that matter. If we are going to pass and support this Bill, it has to be on condition that it is only going to benefit Kenyans, and it must be indigenous Kenyans."[60] In invoking indigeneity rather than citizenship as the metric of rightful ownership, Kilonzo firmly framed Kenya's telecommunications sector as one of the "resources of this country." They were the legitimate purview of the legitimate holders of the land, defined not as citizens but as firstcomers.[61] Like conflicts over the continent's subsoils, this debate turned on a logic of autochthony.[62] On this framing, the resource ought to remain indivisible, a commons over which indigenes held a naturally endowed right.

These issues were likewise hotly debated in the Kenyan blogosphere, where some argued that the privatization of state assets was a form of enclosure suited to the digital age. As one blogger wrote, "If you are awake, you should have noticed something very strange has been happening since [the] 1980's. From this period, we have witnessed [the] massive sale of basic industries, or *natural monopolies* by states. For instance . . . Kenya Post and Telecommunications Corporation (KPTC) . . . As usual, the priests will yell, *efficiency*. They said the same as they enclosed common lands in England and Scotland which was then extended to other continents in [the] form of colonialism. *We are in the midst of the greatest enclosure of commons ever witnessed by mankind.*"[63]

For this blogger, state assets were a commons that the government should retain outside of market relations. Kenyans had naturally endowed rights to these services and networks, rights that were undermined as these entities were commodified and rendered divisible. For this blogger, sixteenth-century England offered the appropriate analog for the forms of expropriation entailed by the privatization of Kenya's "natural monopolies." The "priests" of this new era were international governing bodies and their allies in the Kenyan government. The dogged pursuit of privatization was so single-minded that it had the air of religious zeal. Numbers for this observer were evidently not secular signs of accounting.

But policymakers were in a fix that pit the material affordances and requirements of infrastructures against an ideology that linked infrastructural connectivity to the rights and endowments of belonging. As the Assistant Minister of Co-operative Development stated, while parastatals had been "founded using the taxpayers' money . . . parastatals were expected to offer services and they were supposed to be of help to Kenyans at large. I am saying this because I realize that when we speak about privatization, we are mainly talking about the improvement of the infrastructure."[64] Critics of privatization were in a double bind. To privatize would be to remove publicly held assets from the hands of Kenyans who were their naturally endowed owners. Dividing these assets, they seemed to argue, fundamentally altered their status *vis* Kenyanness. But allowing the company to remain in the hands of the state as a "custodian of . . . enterprises" would prevent "the improvement of the infrastructure" and thus fail "to be of help to Kenyans at large."[65]

Anxieties over the status of Kenyanness in relation to infrastructural provisioning precluded a simple economistic reading of value. As Peter Kenneth argued during the same discussion, "On the issue of valuation . . . we need to be careful. We need to have the correct machinery and procurement procedures in place. This is normally a very difficult subject in privatization. *I am aware that not all of us can agree on certain values which are pegged on various assets of a company*."[66] Establishing consensus was beset by other problems. Not only did they need to mobilize calculative techniques that could establish the monetary value of existing assets, but also mechanisms that could also forecast how these assets might generate profits in an uncertain future. The problem, as Kenneth saw it, could only be expressed in the past conditional. Drawing on history, he argued that there were bound to be disputes as to what "the real value . . . could have been." Establishing future value turned on mobilizing techniques of speculative accounting, but the accuracy of the resulting values in both quantitative and qualitative terms was far from settled. Indeed, lurking in the background was uncertainty regarding how the divisibility of state assets might reconfigure the terms of Kenyanness. Perhaps not all parastatals ought to be subject to divisibility. Some, as Kenneth suggested, ought to be retained as a whole by the government "on behalf of its people."[67]

For opponents, then, the concern turned on the relationship between materiality and meaning. Was the improvement of services worth what some viewed as a ceding of state sovereignty? Did this new relationship not simply re-rehearse colonial dynamics? And, perhaps most importantly, what were the implications for belonging? The "value" of foreign investments, evidently, did not simply inhere in the amount of foreign capital and expertise brought in to run what were formerly state-held assets. Quantities in these discussions were meaningless without a close reading of the mechanisms and relationships that modified their qualities.[68]

Privatization in the form of the IPO seemed, to some, to remedy this impasse. State assets would be rendered divisible in the form of shares, but the selling of those shares would be undertaken in a bid to protect the patria, to ensure that Kenya's natural "resource" returned to its natural owners: Kenyans. This debate reframed the terms of earlier discussions over privatization. Public assets would, indeed, return to Kenyans, but not as autochthons or firstcomers, nor as undifferentiated Kenyan citizens, but as individual shareholder-citizens.

"The Question Which Vexed the Committee's Mind Was: 'Who Is Mobitelea Ventures?'"

One evening in 2015, a colleague, Kevin, and I were out for *nyama choma*, a Kenyan specialty, with a friend, Simiyu, and some of his colleagues, all of whom work at Finance Against Poverty (FAP).[69] FAP's research is concerned with understanding the patterns and networks of the poor so as to better include them in emerging financial markets. This mandate, which goes by the shorthand M4P (Markets for the Poor), rests on an increasingly prevalent postulate: The private sector, rather than governments or NGOs, is best placed to ensure development. According to this developmentalist wisdom, the poor ought not be viewed as people in need. Instead, this rationale holds that the poor should be viewed as both entrepreneurs and a potential market to be tapped through the marketization of everyday life.

M4P is not limited to FAP. This ethos was formalized following the publication of C. K. Prahalad's 2006 book, *The Fortune at the Bottom of the Pyramid: Eradicating Poverty through Profits*. In this oft-cited text, Prahalad, both a management professor and a consultant, argued that "we stop thinking of the poor as victims or as a burden and start recognizing them as resilient and creative entrepreneurs and value-conscious consumers." The bottom of the pyramid, he continued, was not simply a base of consumers, but was a "source of innovations," innovations that could be scaled up and marketized (see chapter 6).[70] This near-orthodoxy is propounded by Safaricom, Simiyu's former employer, as well. Indeed, Safaricom routinely attributes its success to the company's investment in "knowing their customers."

As the night wore on, talk turned to Mobitelea, a "shadowy" company that emerged as a shareholder of Safaricom in the period leading up to the IPO. Despite much talk of the company, which was registered in Guernsey, an island in the English Channel, very little was known regarding how Mobitelea gained shares in Safaricom, nor did anyone seem to know—or was willing to name—who was behind it. The moniker, Mobitelea, had puzzled Kevin and me. We asked the group if any among them knew the story of Mobitelea's origins. Jim, an innovations leader at FAP, laughed. Standing up and walking away from the group as his cellphone rang, he turned to the table and shouted: "Everyone knows! Mobitelea: Moi, Biwott, Telecoms!" It was so simple, so elegant, so obvious. While observers in the global north had argued that privatization was a means of eliminating "corruption," Jim's exegesis of the "shadowy" firm's name pulled back their carefully crafted curtain.

Gideon Moi, son of the second President, Daniel arap Moi, and Nicholas Biwott, a powerful political operative under Moi, had leveraged their power to carve out a piece of this great liberalized cake. Privatization had, evidently, done little to put an end to Kenya's powerful political elite's mobilization of state infrastructures for personal enrichment—it simply institutionalized these dynamics. Safaricom may have displaced the legacy of Daniel arap Moi at the newly renamed Safaricom Stadium, but Moi—and the politics of patronage and statecraft for which he stood—had, evidently, remained embedded in Safaricom, the "transparent" imprimatur of Vodafone notwithstanding.

These awkward infrastructural attachments returned to public consciousness during the second phase of privatization, which took the form of an IPO. Proponents of the 2007 IPO celebrated this period of expanded divisibility. Despite actively courting international investors, they argued that the IPO was a means of democratizing the ownership of the corporation, by placing it in the hands of a distributed public of Kenyan shareholders. This was accompanied by a reconfiguration in how development was being imagined. No longer linked to the state and the partisan ethno-politics that characterized its operations, development was instead increasingly viewed as being the proper purview of Kenya's multiple citizen-publics. Within this newly distributed developmentalist landscape, Kenyans were reconfigured as sources of capital, a market of small investors whose private hoards could be mobilized, enabling both development and profit generation through the marketization of everyday life.

By 2007, Safaricom was largely unrecognizable as the progeny of TKL. Vodafone-appointed South African CEO Michael Joseph's leadership impressed techno-enthusiasts globally. From a company with only 200,000 registered lines, Safaricom had assembled a subscriber base of over 8.1 million, fast becoming the largest corporation in the region.[71] Much of this success is attributed to the company's launch of its widely lauded value-added service M-Pesa, which enables mobile-to-mobile money transfers. Building on extant social practices of those at the "bottom of the pyramid"—namely the routine movement of remittances from Kenya's cities to the countryside—M-Pesa promised to de-materialize the movement of money, enabling Kenya's poor and middle classes, long-excluded from the formal banking sector, to "send money home." The wild success of M-Pesa shaped what is, today, Safaricom's guiding business model: the everyday tactics and strategies of the poor could form the basis for developing new services and locating new frontiers of accumulation (see chapter 6 for a full discussion).

This para-ethnographic work is at the center of Safaricom's commitment to "knowing its customers," and it has proven to be a wildly profitable business model. Indeed, the generation of new markets that happened in these years demanded the expansion of infrastructures, as the increase in M-Pesa users put pressure on the existing bandwidth, leading to overcrowding on the 2G spectrum. And so the company purchased the country's first Third Generation (3G) license from the industry regulator, the Communications Commission of Kenya, for $25 million in the same year.[72]

M-Pesa has generated widespread international attention, widely lauded for "including" populations heretofore "excluded" from financial markets—for "banking the unbanked." The language of inclusion, from the beginning, situated M-Pesa in an altruistic register, positioning it as having democratized development by following the lead of the poor, rather than imposing top-down, centralized development initiatives. This notional democratization of development was tellingly synergistic with the rationale underwriting the IPO, which was similarly framed by its proponents as a means of democratizing de-velopment by distributing the company's ownership, in the process returning to Kenyans what was rightly theirs: Safaricom, a now divisible part of the na-tional patrimony.

In the months leading up to the IPO, however, critics began querying the legal status of the mechanisms by which Safaricom had been rendered divisible. Under the agreement struck between TKL as 60 percent shareholder of Safari-com and Vodafone as 40 percent shareholder, the Government of Kenya would transfer TKL's 60 percent to the Minister of Finance. Twenty-five percent of these shares would be sold to the public on the NSE. Under the terms of this agreement, the majority of shares would be held by a combination of the Gov-ernment of Kenya and the Kenyan populace. But the agreement allowed for a future wherein Vodafone would become majority shareholder.[73]

This move would have shifted the geographic center of ownership beyond Kenya's borders. On the face of things, however, Vodafone seemed more concerned that the company's top leadership, not the majority of its shares, be the purview of various global elsewheres. Indeed, rumors held that Vodafone had been willing to forego this majority stake in Safaricom if the government of Kenya agreed to cede its right as majority shareholder to appoint the company's CEO.[74] This clause in the agreement, which, by 2014, had led to the appointment of one South African and one Guyanese-British CEO, was popularly held to have been a means by which the company sought to distance itself from the Kenyan state, both with regard to the protracted nature of Kenyan ethno-politics, and the politics of patronage that are widely held to be the animus for "corruption."[75]

Revelations that came on the heels of the announced IPO called into question Safaricom's desired separation between corporation and state. The Public Investments Committee tasked with examining Safaricom's origins focused squarely on the discomfiting attachments connecting the company to the state. Specifically, critics honed in on the numbers, the status of the exchanges, and how they had been valued. The Public Investments Committee outlined the confusion. Safaricom had lacked the liquidity to purchase the licenses requisite for its operations, leading Vodafone and TKL to outlay the necessary monies in the form of an interest-free loan.[76] Why, the Committee asked, were the monies exchanged framed as a loan rather than equity? At issue was the nature of creditors' oversight, rates of repayment, and uncertainty regarding "the background, nature and implication of the said loan."[77]

The character of the exchange was beset by other irregularities. As the Committee wrote, the deals under question had not generated a paper trail, the "absence of a written agreement between the parties . . . [rendering] the Committee unable to confirm the term[s] of the said loan."[78] For members of the Committee, the lack of documentation, and the discursive authority it bore, smelled of the bad old days of the politics of patronage. Not only had privatization failed to deliver on transparency, but uncertainty regarding how these proprietary numbers were to be evaluated made it difficult to discern who (or what) was really driving the company.

The Committee revealed, moreover, that the 1999 law, which mandated that 60 percent of the shares remain in Kenya, had replaced an earlier policy. Initiated in January 1997, the Telecommunications and Postal Sector Policy Statement stipulated that any new investor could hold no more than 30, rather than 40, percent equity in the company. According to the second origin story, the company had been established in 1999, meaning the original 1997 regulations were in place at its inception, rendering the altered proportion held by Vodafone illegal. If this second origin narrative was true, how, when, and why—Kenyan parliamentarians queried—did it come to pass that this policy was changed? More than timelines were in the crosshairs. According to the parliamentary committee, "By the time of its inception . . . Telkom Kenya, on behalf of the Government of Kenya, owned 70 per cent of Safaricom. The remaining 30 per cent was owned by Vodafone Kenya Limited on behalf of Vodafone PLC."[79] The "vague whole" embedded in the publicized 60/40 split was disrupted as Kenyans learned of the prior agreement, which had held Kenyan ownership stable at 70 percent.[80]

The real uncertainty, however, honed in on how the numbers had been valued. Specifically, the Committee revealed that it was not clear that TKL had received *any* capital in exchange for the 10 percent equity that it had shuttled to Vodafone Kenya. As the Committee report recounted: "Mr Sammy Kirui, Chief Executive, Telkom (K) Ltd informed the Committee that, while he is not sure what necessitated the reduction of TKL shares in Safaricom from 70% to 60%, available records are not explicit [regarding] whether TKL was paid for the reduction."[81] By uncoupling 10 percent from the original 70 percent, the network was revalued and transformed into a gift to be freely given—a cessation of both sovereignty and a selling-out of citizens. By giving a proportion of Kenya's natural resources away, the government had disrupted the terms of its fiduciary relationship with Kenyan citizens, an agreement wherein it held a proportion of state assets, but only "in trust" for the Kenyan public.[82] Requests for information from the relevant parties only deepened suspicions, with the Committee revealing that Vodafone had only "verbally" requested that its stake be increased from 30 to 40 percent, a request the board of directors would approve "subject to . . . *obtaining an official request in writing* from Vodafone."[83] This documentation was not forthcoming.

Furthermore, the "Committee found out that contrary to what is in the public domain, the company has more than two shareholders. The third shareholder is a shadow firm known as Mobitelea Ventures Limited . . . whose directors are also obscure."[84] Not only was this contrary to the 1997 stipulations governing the privatization of state assets, but it also contravened the shareholders' agreement, which mandated that all shareholders be notified of a potential sale, and that the identity of new shareholders be made known.

When pressed on this, Vodafone executives—Michael Joseph, ostensibly chosen as CEO because his status as South African inured him from imbrication in Kenya's complex politics of patronage, included—refused to disclose the status of Mobitelea, though no one denied that a deal had been struck. As a letter written by Gavin Darby, the Chief Executive of Vodafone Africa stated: "[I]n 2002, Mobitelea Ventures . . . was offered an opportunity to acquire 25%

of VKL by Vodafone Group. *This was in return for the advisory role it played as a local partner of Vodafone Group on local business practices and protocol challenges associated with investing in Kenya.*" Safaricom representatives here tried to assuage fears. This new investor, though its identity remained obscure, was not "foreign" but was a "local partner." And so, Darby seemed to suggest, the 5 percent of the company held by Mobitelea did not shift Safaricom's center of gravity outside of Kenya. To the contrary, the deal had been struck in a bid to embed the company more firmly in Kenya's social fabric. The deal was struck, in other words, to thicken infrastructural attachments. This gift, while seemingly freely given, entailed troubling political intimacies. Parliamentarians were indigent, demanding to know how these shares were "given to the shadowy company for free," arguing that they were "public funds!"[85]

Vodafone representatives tried to assuage fears about the estrangement of public property and trust that had attended the movement of the shares. "Cash" had, indeed, exchanged hands, they claimed in an about-face.[86] Sammy Kirui, CEO of TKL, argued that the reduction in shares may have been necessitated by the "need to raise funds for the expansion of Safaricom's network."[87] Contrary to what advocates of privatization under the banner of "transparency" had been saying for years, here Kirui seemed to argue the politics of patronage were *not* an obstruction to infrastructural expansion—they were its condition of possibility. It was not that corporate actuarial logics had come to determine the operations of the state, but rather that the corporation mobilized numeracy to obscure the fact that it was, like state-managers before it, emulating tactics of Kenyan statecraft.

Seeking to neutralize critiques, Vodafone officials indicated that there was nothing unusual in this; it was standard corporate practice. "When Vodafone makes investments in new territories," Darby told the Committee, "it is not uncommon that it works alongside a partner, who typically gives advice on *local* business practices and protocols and various challenges associated with investing in a *new market.*"[88] If, for Kirui, the deal with Mobitelea had provided value in the form of capital needed for "local" infrastructural expansion, Vodafone executives argued that the expansion of the company and its infrastructures turned on mobilizing "local" knowledge about "practices" and "protocols," and that the value of this outstripped any irregularities that had enabled the movement of shares among the contracted parties.

These debates brought into relief an emergent developmentalist regime. "Local knowledge," mediated by foreign expertise and capital, was necessary for the generation of new infrastructures. This coupling of the infrastructural and the "para-ethnographic" has since become the modus operandi of Safaricom, a distributed vision of developmental agency that attaches ever-larger swaths of the Kenyan population to the company as sources of knowledge and sites of innovation.[89]

These responses did not satisfy critics in Parliament: "It is appalling that Vodafone Plc a Company registered in the United Kingdom involved itself in underhand activities of [a] corrupt nature aimed at depriving Kenyan citizens of billions of shillings through m/s. Mobitelea Ventures, despite the fact that the U.K. has been in the forefront of campaigning against corruption in developing countries."[90] These critics rejected Vodafone's labors to depoliticize the numbers. This algebra of intimacy was hypocrisy at its worst. Under the guise of a foreign multinational corporation, the networks of patronage had, evidently, not disappeared. Instead, like infrastructural expansion, they were outsourced and distributed amongst the island of Guernsey, the Nairobi offices of Safaricom, the State House, and the metropolitan headquarters of Vodafone. What in the past had been criticized as illicit networks of patronage by foreign observers, Vodafone rebranded conventional market research. The attribution of corruption, evidently, mapped onto the racialized geography of the postcolonial world order.

On the heels of these troubling revelations, critics called for an immediate cessation of the IPO, which would have mobilized popular holdings to capitalize the firm, in the process transferring the individuated wealth of middle-class investors to the government of Kenya and to the "shadowy" firm, Mobitelea. These critics challenged the government of Mwai Kibaki to make good on its promises to end corruption. "If it was transparent as it claims to be," it should have no problem getting to the bottom of this "murky" deal, they argued.[91] These irregularities needed to be resolved and "all those disclosures . . . made" before the IPO was allowed to go forward.[92] This was not merely a matter of political wrangling. Many Kenyans firmly located the IPO within a longer genealogy of state corruption.

The context of these revelations, however, coming as they did on the eve of the IPO, offered Kenyans the opportunity to weigh in. As one blogger wrote, "If you are against corruption then it is a good idea . . . to stop corruption dead in its' [sic] tracks and what better place to start than to cause the Safaricom IPO to fall flat on its' [sic] face."[93] By withholding their hard-earned shillings and boycotting the IPO, Kenyans could actively work to remedy the misdeeds of the past. This argument, though in a slightly different register, marked an extension of the logic justifying privatization that had animated parliamentary debates since the early 2000s. It was the very expansion of private ownership through the divisibility of state assets that seemed to offer a means of protecting the "public good."

If critics of the IPO centered on past irregularities, its supporters directed attention to how the IPO would usher in new futures, not least by ensuring the most "efficient" division of labor was enacted as Safaricom was redistributed amongst the contracted parties. This reconfiguration of infrastructural authority, they argued, was the surest path to achieving national development. As former President Uhuru Kenyatta argued when acting as Minister of Finance in the aftermath of the IPO, "The desired objectives of the privatization of State corporations . . . [was] not a question of just selling assets but [was a means] . . . to improve infrastructure, [the] delivery of services, [and to] reduce the demand on Government resources . . . [thereby] allowing business to do what it does best and the Government to regulate."[94] By transferring the responsibility for infrastructural maintenance to the private sector in the form of semi-privately owned utilities, problems of breakdown and infrastructural reach would be mitigated. The state, for its part, would be strengthened in regulatory terms as it was scaled back in infrastructural terms. Both the "public good" and the state would be revivified and extended. This, then, was not a retreat of the state but a recalibration of the relationship between private and public, both within the state and, as we shall see, in the public imaginary.

Indeed, Safaricom and sympathetic voices in the government framed the IPO as an opportunity for Kenyans to repossess what was naturally theirs: Safaricom—an infrastructure, a network, a future, now imagined as constituting a divisible piece of the national patrimony. As the Minister of Finance, Amos Kimunya, was quick to remind his audience: "It is not Safaricom that is issuing shares. It is the Kenyan Government that is divesting part . . . of the 60 per cent that we own on behalf of the Kenyan people, 25 per cent is what *we are saying that we do not have to own on behalf of Kenyans when they can own it directly*."[95]

This was framed as a democratization of Safaricom's ownership structure, but it had political entailments that fundamentally brought to bear questions of responsibilities and rights—namely, those framed through the categories of "citizenship," "development," and "patriotism." As Kimunya stated, the IPO would enable "the Government . . . to . . . expand the opportunities available to Kenyans of all walks of life, so that they can participate in their own wealth creation, while at the same time, helping to build a new and better Kenya."[96] The national patrimony would be democratized, its ownership distributed across the Kenyan populace. People would be able to develop themselves as they developed the nation.

This notional democratization of state assets was a reframing of earlier discussions, which positioned the telecommunications infrastructures as a commons. The commodification of the asset had made it divisible and subject to new distributions. In this new framing, it was not Kenyans writ large that had a right to these resources, but those with the means to buy back what was rightfully theirs in the form of shares. With this came a reconceptualization of development, which was now imagined as being the responsibility of Kenyans, not only in their capacity as taxpaying citizens, but in their capacity as a market of investors, a shareholder nation of patriots. This emergent theory, that brought together citizen-shareholders with the corporate-state, incorporated older ideas of citizenship and statecraft, but suffused them into entirely new ideas of "public accounting," wherein citizens were reformatted as "categories of contributors" whose hard-earned shillings would shore up both the corporation and the state as people freely entered into new contractual relationships governed by the marketplace.[97]

The 2007–2008 Election Year and the Making of a Nation of Shareholder-Citizens

If supporters of the IPO argued that it would conjure into being a new public of citizen-investors, the timing of the planned IPO articulated with concerns regarding existing Kenyan publics. The IPO was set to take place in late 2007, just prior to the much-anticipated presidential election, which pit Raila Odinga of the Orange Democratic Movement (ODM) against incumbent Mwai Kibaki of the Party of National Unity (PNU). For many, the de-ethnicized and "transparent" politics promised by Kibaki in 2002 had not been borne out. Key government positions continued to be occupied by members of the Kikuyu old guard.[98] The Kroll Report, the contents of which circulated in the blogosphere, confirmed that business during Kibaki's tenure had proceeded much as usual, as powerful political figures continued engaging in a range of irregular (though by no means uncommon) practices which "fleeced" the Kenyan public.[99]

Led by Odinga, opponents voiced concerns that monies made from the sale would be used as a "campaign war chest," leading ODM to petition to have the IPO halted through the courts. ODM's legal arguments, while deemed "frivolous" by the court, delayed the staging of the IPO.[100] While the actual voting proceeded without incident, delays in releasing the results, widespread accusations of election rigging on both sides, and government- and opposition-sponsored aggression, led to a cycle of violence and retribution through the first months of 2008. This violence led to the deaths of over 1,100 Kenyans and the displacement of an estimated 500,000.[101]

People claimed that new communication technologies were largely to blame for the violence, citing the virulent SMSs flying over the cellphone networks crisscrossing the country. If, to date, Safaricom had labored to present itself as at once embedded and disembedded in the Kenyan state, as at once intimate with and estranged from its complicated politics of patronage, here it tried to weigh in as an ostensibly nonpartisan arbiter between ODM and PNU. Apparently at the behest of then-Minister of Internal Security John Michuki, Safaricom enrolled its network, sending the following message to its subscribers: "The Ministry of Internal Security urges you to please desist from sending or forwarding any SMS that may cause public unrest. This may lead to your prosecution. From: SAFARICOM." In appealing to its clients not as customers but as Kenyan citizens, Safaricom tried to leverage the attachments generated by its networks in a bid to refashion sentiments away from the interethnic conflict and toward national unity.

While Safaricom directed its message to all subscribers, some interpreted this as further evidence that Safaricom was in the pocket of the PNU government, revealing the difficult line the company walks between attachment and detachment. As one blogger wrote, "One fails to understand how Safaricom, owned partly by Vodafone PLC and the Kenyan public through the Government of Kenya, can allow its network to be misused for political gain at this critical time while it failed to do absolutely nothing when fake propaganda was circulating within its network prior to the general elections. Perhaps the profits were too sweet to turn down, huh? The latest short text message is in fact tantamount to Safaricom turning into a state broadcaster and I dare add, [is] a direct infringement on the privacy of its subscribers."[102] This digital observer rejected Safaricom's efforts to distance itself from the Kenyan state. Any indication that Safaricom was working with the Kibaki government undermined its ostensible separation from Kenyan politics, conjuring up memories of Moi-era Kenya, during which the media largely operated as a mouthpiece for the government. Safaricom's mobilization of its networks for political purposes not only indicated its attachment to Kenyan politics, but demonstrated that it was willing to operate *like the state*.

By April 2008, the violence had come to an end with a negotiated agreement between the two sides. For foreign observers, Safaricom's ability to proceed with the IPO was "as much a symbol of political reconciliation as of economic common sense." The power-sharing agreement between Kibaki and Odinga, with Odinga joining the government as Prime Minister, "set the stage for political stability and reform." Further, "in this new context, [Minister of Finance,] Kimunya was able to have a private lunch with Odinga, an erstwhile bitter political enemy, just prior to his IPO announcement. In a well-choreographed piece of political goodwill, Odinga expressed his support for the IPO, and Kimunya later commented that Odinga also expressed a personal interest in participating in the IPO as an investor."[103] For this observer, in staging this about-face, Odinga tacitly confirmed that the divisibility of state assets through their commodification opened up new possibilities for Kenyans, not as a bifurcated body of rights-bearing citizens, but as a class of citizen-shareholders. The disinterested whims of the market, so the argument went, could supersede Kenya's protracted ethno-politics.

Irrespective of irregularities, Odinga argued, the point remained that Kenyans were the rightful heirs of the company. Given this, he continued, it was critical that "real efforts should be made to make these purchases more accessible . . . [to] Kenyans *in all the regions* and *in all walks of life*." In a thinly veiled accusation, Odinga here suggested that ethno-regional patterns of infrastructural stratification were alive and well. But timing is everything. And so, he concluded: "Let me encourage as many Kenyans as possible to fully participate in this offering."[104] Both opponents and supporters of Odinga were flummoxed by this about-face. Indeed, on the Kenyan blogosphere, lively debates were staged between ODM and PNU supporters, between Kenyans living abroad and those within the country. As one blogger succinctly wrote: "We're torn."[105]

Some supporters argued that Odinga had done all he could, and that Kenyans had to respect the decision of the court when it rejected his appeal to halt the IPO.[106] Others were incensed. Having so staunchly opposed the IPO, Odinga's newfound support for Safaricom's expanded divisibility demanded explanation. Some argued that by supporting the IPO, Odinga had, in effect, sanctioned the Kibaki government's successful rigging of the election. As one blogger wrote: "The rigging of the 2007 elections was ALL ABOUT the Safaricom IPO."[107] For these critics, the decision of the court was meaningless. As another wrote: "ODM opposed [the] safaricom IPO on principle and they OWE Kenyans a credible explanation" for this "change of heart."[108]

In the run-up to the election, Odinga had promised an end to the "corruption" with which the Kibaki administration had come to be associated. His change of heart called this commitment into question. This was particularly galling because Odinga had cast his bid for the presidency on the basis of a logic of class that, in Kenya, articulates with popular perceptions that Kenya's postcolonial past has been characterized by ethno-regional patterns of (infrastructural) stratification. His support for the IPO seemed to confirm that, at the end of the day, he, like Kibaki, was simply another *samaki wa kubwa* (big fish)—that his class-based allegiances had overcome his capacity to act for a bloc of voters in a context where class and ethnicity overlap in ways that are as commonsense as they are misleading. As one Kenyan living in the United States wrote,

"When we told you that there was nothing fundamentally different between PNU and ODM you called us all sorts of names. We told you over and over again it was all about eating, eating, eating and more eating and that ODM was simply a vehicle to get to the Kitchen where the goat was being roasted and the Ugali was being cooked. Now if you knock at the Statehouse, Raila himself will open the door wearing a Chef's uniform complete with top hat."[109] This metaphor would not have been lost on Kenyan readers for whom eating is a sign of social and material largesse, which, if not properly distributed, quickly tips into accusations of antisocial forms of wealth management.[110]

If roasted goat was the metaphor used to describe how the IPO would enrich Kenya's political elite, then *sukuma wiki* (push the week), indicating the role that the green roughage plays in sustaining people in periods of material want, was the foodstuff used to describe the negligible value Kenyans-cum-would-be-shareholders were poised to gain. "For the price of Sukuma Wiki . . . you can own a piece of the country's biggest company, for the price of a match-box, you can own a piece of a company that is currently being investigated by the British Serious Fraud Office in London. For the price, less than *matatu* fare from Kangemi to Safaricom headquarters on Waiyaki way, you can own a piece in [the] . . . company."[111] As these debates roared on in the comment sections of the blogosphere and in bars across the country, Kenya's politicians worked to shore up an inclusive vision of the IPO. It was an opportunity, they argued, to put an end to ethno-politics by inviting everyone, irrespective of region, irrespective of class, irrespective of ethnicity, to take part in the patria. *Sukuma wiki*, in their hands, was reframed as the "family silver." As Kenyatta reminded potential Kenyan investors in the aftermath of the IPO, "We . . . [were] hoping . . . we . . . [could] broaden the base of these enterprises and get Kenyans, through the Stock [E]xchange, to own shares and to participate in the family silver. . . . These companies now have a broad shareholding. . . . Kenyans from every corner . . . are participating in the Stock Exchange as owners and receiving dividends from these entities."[112] Shareholder-citizenship, Kenyatta argued, offered a solution to the problems of partisanship and ethno-politics that had led to the election violence. If Kenya was plagued by "political tribalism," then the "family silver"— mediated by the market and under the ostensible bipartisan lime green of Safaricom, itself managed by a British multinational—was leveling.[113]

Corporate branding in the months leading up to the IPO shared this framing. The slogan of the campaign, "Safaricom IPO. The future, let's celebrate it together," offered a forward-looking vision which was unencumbered by violence and the divisive politics of ethnicity. Here, Kenyans of all stripes, "Kenyans from every corner," were invited to buy into the lime green of the digital patria, unburdened by the divisive politics of ethnicity that shaped the election dynamics. This was, following Dinah Rajak, a branding that positioned the IPO as an opportunity to engage in a kind of "patriotic capitalism."[114]

Not all Kenyans were convinced by this message. As one blogger wrote of what was coming to be called "SafariCON: The Bitter Option" (a play on the company's slogan, "Safaricom: The Better Option"), the "Safaricom IPO has ... gone political and it follows that it will go tribal—if it hasn't already. In the end, the Safaricom shareholding register will likely read more like a provincial rather than a national roll call. . . . There are those . . . who will disregard the need for transparency and a clean fair market and [who will] go ahead to buy the shares based solely on where the leaders backing the IPO were born."[115]

If divisibility had been framed as a means of democratizing Safaricom's ownership structure by allowing Kenyans "from every corner" to access what was rightfully theirs as heirs of the country's resources, bloggers such as this one pointed to the problems of this vision. In reframing Kenya's telecommunications sector not as a commons but as a divisible and commodifiable asset, the government had not only untethered one of the cords binding Kenyans to the central state but, concurrently, had created the conditions where *some* Kenyans, those defined by "tribe" and region, would have preferential access to the national patria due to their connections with Kenya's governing elite.

Whether because people wanted a piece of the undifferentiated "family silver," or as a defensive move to ensure that Kikuyus did not constitute the shareholding majority, Kenyans turned out *en masse* to buy shares when Safaricom finally went public on June 9, 2008. These would-be shareholders were hoping for a windfall, the IPO a speculative gamble that might secure their financial and imaginative futures. The IPO was the largest to have been staged in the region, and the company, controversy notwithstanding, was firmly rooted within the logic of the patria. With this came an ethics of expectation actively cultivated by lending institutions that promised "huge returns instantly."[116]

Aspirations bristling, thousands of small investors trekked to the NSE to witness Kibaki start the day's trading.[117] The minimum buy-in was 2,000 shares and cost KSH 10,000 (approximately $100), not an insignificant sum for many. The excitement generated by the IPO led to the company being oversubscribed by a shocking 532 percent, raising over $800 million for the government.[118]

The public's disappointment was swift and palpable. Refunds following the oversubscription were issued slowly. Excitement over the IPO continued to wane as people discovered that returns on their investments would be small. As one Safaricom employee and shareholder explained, the period following the IPO was one of great frustration. Kenyans were well aware that the company was the largest and most profitable in the region, and so the IPO produced dissonance as shareholders sought to reconcile the "real value of Safaricom and the value of shares." This left people feeling like Safaricom was not treating Kenyans "with the dignity we deserve[d]."[119]

But, equally significant, the IPO signaled a transformation in how development was being conceptualized both by the corporation and within the apparatus of the state. Government officials began mobilizing Kenyans' massive investment in the IPO as evidence of the "inherent fund raising potential among ordinary Kenyans in rural and urban areas."[120] This involved the redistribution of developmental agency. "Why," one parliamentarian asked, "must you look for money from the Treasury, when you can get it from Kenyans? So long as you float attractive . . . bonds, people will buy them just the same way they did with Safaricom."[121] Here, Kenyans were being asked to subsidize development not through the contribution of taxes as undifferentiated citizens, but by buying into the capitalization of increasingly privatized assets and infrastructures as investors. Then-Minister of Finance Oburu Odinga, agreed. The oversubscription of Safaricom had demonstrated "that Kenyans have a lot of resources that can be tapped for development purposes." From a vision of infrastructures as a natural resource, the naturally endowed purview of the albeit shifting category of "Kenyans," this new rationality saw Kenyans not as having an inalienable right to infrastructures-cum-resources but as themselves the bearers of resources. From autochthons to citizens, Kenyans in this most recent iteration were imagined as citizens-cum-shareholders-cum-investors—a new market, a market of patriots. And so we return to where we began: Safaricom's 2014 Annual Shareholder Meeting.

Material Ethics in Safaricom's Kenya

Much had changed in the years between the 2008 IPO and the 2014 AGM. Safaricom's share of the market had massively expanded, with the company having secured over 21 million subscribers out of a population of 44 million. Its total revenue had soared to $1.4 billion, a large proportion of which was comprised of fees associated with its breakout service, M-Pesa.[122] As Safaricom expanded from being a mere telecommunications firm to a financial service giant, Kenyans became ever more tightly attached to the corporation and its infrastructures.

Safaricom's foray into financial services has been matched by boundary-crossings of other kinds. In recent years, Safaricom inaugurated myriad initiatives under the banner of corporate social responsibility (CSR). By 2019, the

Safaricom Foundation occupied a place of pride on the company's website, boasting the efficiency with which it has unrolled infrastructures and programs normatively framed as being the responsibility of modern states: education, health services, clean water, and disaster relief. At that time, Safaricom's website reminded visitors that over a million people had benefited from the foundation's disaster relief efforts; 830,000 Kenyans had received specialized health care; 680,000 children were now learning in newly constructed classrooms; 232,000 people had been "economically empowered" through community-based income generation projects; and 172,000 people had gained access to clean water. The page concluded: "The total contribution of Safaricom Foundation towards Kenya's development so far is Kshs 1.8 billion. As it turns ten, it continues to work countrywide to improve the quality of life of Kenyans."[123] Not merely corporate branding, many Kenyans expect Safaricom to step in to provide services normatively associated with the state. As one Safaricom employee explained, when something terrible happens, people ask: "What is Safaricom doing [to help]?" However, there are definite limits to this state-like behavior. Safaricom is in the business of "transforming" the lives of the Kenyan people (generically construed), customers (a more limited category of persons), and shareholders (who are also part owners of the entity). And it is precisely this tripartite vision that subjects Safaricom to forms of claims-making by Kenya's multiple publics.

Safaricom actively works to modulate the registers of attachment and detachment that characterize its relationship with the Kenyan nation and the Kenyan state. Retaining a monopoly over the contours of these boundaries requires constant work and vigilant labor to police. Routinely, though, Safaricom has to contend with moments of breakdown as it seeks to control how its state-like status is mobilized by Kenya's publics. These moments of breakdown were brought into full relief at the 2014 AGM staged at the newly rebranded "Safaricom Stadium."

When Safaricom's shareholders spoke at the AGM, they began their comments in a manner that is telling of the seemingly contradictory space Safaricom occupies in the public imaginary. Half of the questioners began by following the globally circulating script of conventional practice at an AGM. These speakers began by naming the nature of their relationship with the corporation: "My name is James Ochieng. I am a shareholder." The other half localized this globally circulating script, prefacing their comments with language evocative of a *baraza*. These speakers began: "Asante Sana, Bwana Chairman." Holding these two modes of appeal in a single analytic frame allows us to see how these questioners sought to root their claims on the company in two contiguous registers. First, in their capacity as shareholder-citizens, and second, in their capacity as citizen-clients. Like the twinning of "Safaricom green" with that of the national flag, these modes of address were not isomorphic, with speakers switching between these two registers over the course of their questions.

Many questioners began by congratulating the company for reforming its corporate practice in a manner that conformed to the expectations of the politics of patronage. "Mr. Chairman, we have been entertained today. We have had transport, we have had some lunch, we have had some giveaway. This meeting is a big improvement in the history of Safaricom," this speaker continued, to which the crowd responded with enthusiastic applause. Critical to this enthusiasm was the sense that the company had really "listened" to its shareholders, just as it really "listens" to the needs of customers.

Offering "some giveaway" was a concession. In past years, Safaricom publicly laid out the costs of hosting such an event in an effort to explain why it would no longer offer shareholders gifts. This was an effort to deflect accusations that it was acting as a poor patron. The corporation's arguments for the wisdom of fiscal restraint had, evidently, fallen by the wayside in the face of the voluminous complaints of its shareholders. As another shareholder remarked, the corporation's board was a "listening board [which] understands that last year we were complaining. . . . [Today,] at least you have given us a gift, you are a listening board, and you have heard." In conforming to these demands, Safaricom demonstrated its willingness to occupy the conceptual space of a political patron, offering material gifts to thank loyal shareholders.

If praise for Safaricom's hospitality was premised on the politics of the gift, then in other instances, attendees called the company to account specifically in their capacity as shareholders. In these instances, shareholders contrasted their paltry dividends against the massive profits and operating costs of the company: "Mr. Chairman, as you have said in your report . . . the company has done very well. . . . I want to remind you [that] these shareholders have been holding this company when it was not showing any improvement . . . Mr. Chairman, while you make very good business, there is a point of heavy finance costs. Can this be explained because it's . . . very heavy. It's consuming our dividends." In statements such as these, the company was asked to account for the poor returns being offered to shareholders—constituent components of the company; indeed, its part-owners. Speakers such as this one demanded that company spending be reformed, and profits maximized, in order to ensure the greatest returns to shareholders. In so doing, they followed the economistic logic that supposedly typifies an AGM.[124]

The line between shareholder and citizen-client became more blurred as people honed in on their frustration with the company. Interestingly, the moments when the tensions engendered by the company's state-like effects were most visible was in discussions of the company's projects operating under the banner of CSR, which were raised by the majority of the questioners. "Mr. Chairman," one began, "I am pleased with your speech . . . and . . . that you are actually transforming the lives of Kenyans to be better but we . . . shareholders . . . I want to ask the other shareholders, are our lives better today than . . . when we bought these shares?" Is it not the responsibility of the company, he continued, to focus equally on the shareholders, to "transform their lives" as "individuals" *and* "shareholders?" "Mr. Chairman," he concluded, "as I give you a credit . . . remember also I'm subtracting because our lives are suffering." This algebra of praise and critique mapped onto the twin subject positions questioners occupied as, on the one hand, shareholders, and on the other hand, Kenyan-nationals. One questioner was quite explicit in this doubling. "[Please] be kind," he began. "Go back to yourself, and support your people, Kenyans, and think about us, shareholders." Here the line between the two subject positions reached the apex of indistinction. The company, ostensibly a "Kenyan" corporation, had failed its shareholders precisely in its capacity as a state-like patron. Its failure to provision for the well-being of Kenyans as client-citizens in this capacity was scandalous. "Don't you feel ashamed?" the speaker concluded, to a round of rousing applause.[125]

Conclusion

In July 2008, satirical columnist Sunny Bindra published a column in Kenya's largest newspaper, *The Daily Nation*. The tone of the piece, titled: "Music, Lights, Freebies: Welcome to the Kenyan AGM," was ungenerous:

> The amazing success of the Safaricom IPO confirms that we are on our way to becoming a shareholder democracy, does it not? Hundreds of thousands of new shareholders have been brought into the bosom of capitalism, and are basking in the promise of the new wealth that will follow - yes? Anyone who thinks we are en route to a mass-capitalist society needs to visit the Annual General Meeting of any listed company. . . . [Most shareholders] haven't the faintest clue what they are there to do, or indeed what 'their' company does. They are simply like children at the circus, waiting for the show to begin. And a show it is. Many companies now organise dancing and singing troupes to perform at AGMs, thereby

accepting that there is no serious business to be conducted there. They give out freebie packs—to much pandemonium. I have often heard shareholders (and even financial journalists) rating companies according to the quality of the free lunch provided.[126]

Bindra's condescension notwithstanding, the journalist was not wrong in noting the peculiarities that characterize Kenyans' modes of address at Safaricom's AGMs. But attendees' appeal to the company as both client-citizens and shareholders would be misunderstood if simply framed as shareholders' naive misunderstanding of the nature of the relationship binding themselves to Safaricom. This doubling of the shareholder and client at the AGM was born out of a set of overlapping histories that rendered Safaricom's relationship both to the state and to Kenya's multiple publics ambiguous, a precarious relationship shaped in no small measure by a corporate strategy that tacks between embedding and disembedding, between attachment and detachment.

As we will recall, the early debates surrounding privatization had positioned state infrastructures as a commons to which all should have rights of access. While much has changed in the intervening years, a residue of this thinking persists into the present. This line of thinking certainly formed the backdrop of the IPO, as Safaricom and sympathetic voices in the government framed the public offering as a means for Kenyans to buy back what was rightfully theirs. When shareholders called Safaricom to account in this double register, then, they drew on this deeper genealogy of rights to access, which were historically premised on their status as Kenyans, not necessarily as undifferentiated citizens, but as citizen-clients. This logic they overlaid with a logic of obligation premised on a new set of relations brought into being as they moved into the position of shareholders. To turn to this deeper history, what first appears contradictory is revealed as being a historically rooted claim to access shored up by the postcolonial state; one located in a shifting reading of infrastructures—telecoms included—as a part of the national patria. Safaricom has capitalized on the shifting set of attachments binding its infrastructures to the state and the nation, even as it works to police these boundaries. Understanding shareholders' modes of address requires addressing these longer histories of the company in Kenya and locating the dialectics of attachment and detachment that have become the company's central business strategy, a strategy that is by turns both a source of profit and a potential liability.

These tensions map onto the company's own self-representation, which acknowledges it is responsible to a tiered public. As then-CEO Bob Collymore stated in closing the 2014 AGM, "Our most singular collective aspiration in

Safaricom is transforming lives. We not only transform lives through corporate responsibility initiatives, but also in our products and services. The Safaricom team of over 4000 employees do so, and . . . [do so] knowing full well that [we are] . . . only relevant if we transform the lives of our shareholders, our customers . . . and the communities in which we live. Asante sana." While, in abstract terms, Safaricom is committed to "improving the lives" of an undifferentiated category of Kenyans, the very terms of its engagement truncate this possibility. As noted by Collymore, the company's commitment is to a divisible Kenyan public: "shareholders," "customers," and "communities."

Shareholders at the AGM, too, were well aware of the limits of Safaricom's mandate. Indeed, these limitations framed the critique of the final speaker when he invoked the language of shame. Here, he called the company to account for its performance not in a register of economics, but in a register of ethics. Following Laura Bear and Nayanika Mathur, this speaker sought to register his critique in a language that called into question the morality of a contemporary moment where the "public good" has been rendered a calculable object, generative of "limited social contracts and precarious intimacies."[127] In so doing, this speaker drew on an older language of expectations that situated the nature of Kenyans' fiduciary relationships with Safaricom as, at its base, guided by an ethics of patronage that, issues of "transparency" notwithstanding, he tried to leverage as a mode of economic and political critique.

Understanding the corporate-state entanglements that have seen the interweaving of the fates of the government and the firm, I have argued in this chapter, is crucial to explaining the "peculiarity" of the 2014 AGM. In the chapter to follow, I explore how the firm's infrastructures themselves have been reformed and made "peculiarly" Kenyan. By directing attention to a more prosaic set of infrastructural attachments, I show how the labor of Kenyan knowledge-workers and everyday users has been quietly subsumed by the firm, underwriting both the development of new services and frontiers of accumulation. It is to these stories that the final chapter turns.

6

Safaricom's Austere Labor Regime

The Expropriation and Subsumption of Affective Work

In 2021, the Central Organization of Trade Unions (COTU) (Kenya) lambasted Safaricom's new CEO Peter Ndegwa, branding him the "most dangerous CEO the company has ever had when it comes to protecting workers."[1] By downsizing and forcing existing employees to reapply for their jobs, the General Secretary of COTU argued that Ndegwa was causing "anguish, despair and depression among its [6,000] employees in his mission to maximize profits." This was particularly galling because, not only was he "the first Kenyan Safaricom CEO," but because it is Safaricom's "workers who build the reputation and capacity of . . . [the] company," enabling the firm "to realize profits." As is the case in instances of corporate restructuring the world over, Safaricom's management stood to gain from squeezing more labor power out of fewer workers, benefiting shareholders and managers in the process. However, what made this particularly offensive to the General Secretary was not reducible to a critique of standard corporate practice geared toward profit maximization, but rather that Ndegwa, a Kenyan, had pursued this strategy. This critique would not have been lost on Ndegwa, who formerly worked for East African Breweries, another brand that is widely celebrated as homegrown in the region. If Kenyanness sat at the center of the General Secretary's critique of Safaricom's CEO, it also seemed to inform his understanding of how the company generates profit from the labor and contributions of "workers" whose national identity uniquely prepared them to aid in building the "reputation" of the corporation, ensuring that its operations and services shore up the "peculiarly" Kenyan identity of the firm. But this pool of 6,000 workers comprises but a minuscule proportion of Kenyans whose livelihoods are contingent on their attachments to the corporation.

A month earlier, Safaricom had faced down another scathing critique of its labor practices—this time from Donald Kipkorir, a prominent Advocate of the

High Court of Kenya and longtime critic and sometimes competitor of the firm. Following a ruling in the United Kingdom that saw the Supreme Court and House of Lords determine that Uber drivers are employees of the firm rather than simply users of the platform, Kipkorir tried to make the same case for the approximately 160,000 people that run the M-Pesa kiosks that are ubiquitous across Kenya's urban and rural landscapes—"Every village," he noted, "has an M-Pesa shop."[2]

M-Pesa is a mobile-to-mobile money transfer service that allows Kenyans to send and receive small amounts of digital currency that they can later withdraw from M-Pesa agents as cash. Like Uber, Kipkorir argued, Safaricom's ownership of the proprietary rights of the M-Pesa platform not only allows it to seize the value circulating through its networks, but those who download the app also have no say regarding its terms. This is as true of M-Pesa users as it is of agents who, he argued, also cannot negotiate the "certain agreed commission or share of profits" they pay to Safaricom. Given this dispensation—namely, the firm's monopoly over the infrastructure of exchange—those running M-Pesa kiosks should be "recognized as Safaricom employees." His opinions on jurisprudence were informed by specifically imperial histories. As he wrote, there are few "judges in Kenya or even in Singapore, Canada or South Africa who will demur with decisions of [the] UK Supreme Court and House of Lords. How can a child dissent from his mother?"

What is remarkable about Kipkorir's claims has less to do with the debate over whether those attached to the platform are employees or users, an argument that has been unfolding internationally, than with the implicit and prescient observation that M-Pesa users—those sending and receiving money—and M-Pesa agents—those moving money in and out of the system—have found themselves in a shared predicament. Safaricom's monopoly over the infrastructures of exchange that underwrite M-Pesa enables it to seek rents both from customers and, he seemed to imply, from M-Pesa agents.[3] However, it is not simply the firm's capacity to seize value in the form of rents that ensures the company's dominance in the region, but its capacity to seize information from the millions of user (and agent) transactions in the form of digital traces that Safaricom's data analysts and product developers read for predictive and profitable insights. This data-driven para-ethnographic work both guides the generation of new services by tracking the activities of those at the "bottom of the pyramid," and generates credit scores that enable Safaricom to offer small loans to Kenya's largely "unbanked" population.[4] While the corporate-state entanglements that have facilitated these peculiar attachments were not named by Kipkorir, one might imagine that he tacitly acknowledged something of the complexity of the situation in his closing remarks. Citing William Shakespeare's *Julius Caesar*, he concluded: "Mischief, thou art afoot. Take thou what course thou wilt."

In the last chapter, I explored the forms of attachment that have seen the interweaving of the fates of Safaricom and the state in Kenya. This chapter tracks how these corporate-state entanglements have enabled the daily knowledge, work, and strategies of M-Pesa agents and users to become firmly attached to the infrastructures of the corporation. Drawing on interviews and ethnographic fieldwork, I explore the daily affective and material work of the women and men who work as M-Pesa agents for Safaricom. On one reading, the commission-based fee structure, long working hours, and lack of protections suggest that M-Pesa agents are best understood as exploited labor. However, I argue that the uncompensated affective and social labor of these people working at sites across the country constitutes creative forms of work that enable infrastructural maintenance and expansion. While essentially free in Safaricom's accounts, this work critically underwrites the firm's success. Put simply, while their labor is formally exploited, the broader forms of work required to build and maintain the social and material networks on which Safaricom depends are expropriated. Following Nancy Fraser, this work is "treated as costless in capital's accounts, [and] it is expropriated without compensation or replenishment and implicitly assumed to be infinite."[5]

This dynamic extends beyond Safaricom's labor regime, capturing the majority of Kenya's adult population. Safaricom understands users' interactions with its platforms as an asset for the company, which, following Bill Maurer, capitalizes not only on fees but on "accessing and leveraging vast troves of transactional data," rendering users "everyday designers . . . [of] mobile money."[6] These data are used to develop what Kate Meagher has described as "frugal innovation[s]" as designers draw on "locally embedded knowledge and networks" to create new markets.[7] Indeed, as everyday relations are commuted into service use and move from agents' kiosks to Safaricom's headquarters, they are refined into data that form the basis of new services and, ultimately, the consolidation of new markets and frontiers of accumulation through the appropriation of "marginal gains."[8] These infrastructural attachments have rendered the firm and its services "peculiarly" Kenyan, a dispensation that is critical to understanding Safaricom's capacity to generate massive profits, which totaled over $417 million in 2022.[9] To unravel these processes, we begin with the life and work of one M-Pesa agent, Peninah. It is to her story that this chapter now turns.

Selfie of Peninah in her M-Pesa kiosk, Zimmerman estate, in the suburbs of Nairobi, Kenya, 2014

Affective Work and the "Human ATM"

"The customers, they know if you're stressed, it's not good," Peninah told me. "If you're not nice with them, they know. They have to come here and find you happy; otherwise it's bad for business."[10] At the time, Peninah worked as an M-Pesa agent for Safaricom, managing a small green kiosk in Zimmerman, one of the many lower-income neighborhoods that today flank Nairobi's periphery. As an M-Pesa agent, Peninah's primary role is moving money in and out of the system, accepting deposits and processing withdrawals through the window of her shop. Through this opening, she shouts out greetings and exchanges gossip with passersby, neighbors, friends, and fellow "entrepreneurs"—a category favored by both Kenyans and developmentalist thinkers these days.

In 2014, when Peninah allowed me to join her in her shop, she was in a hard place. She struggled with her in-laws—she had no rights as far as they were concerned. While she had birthed and was raising their first grandson, his father, Maina, had not gone to visit her parents, nor negotiated the terms of their union—the important first stages in initiating *mahari* (bridewealth payment). Absent this, her family could not intervene in contexts where her partner or his family were treating Peninah unkindly. These issues were never far from her mind.

Living quarters were tight. Her apartment, which she shared with her son and his father, was next to that of her would-be in-laws, who also routinely took care of her son. Her shop was just outside of the gate and to the left of the apartment complex in the neighborhood, and she was constantly subject to the surveillance of Maina's mother. Peninah had distant hopes of traveling to the United Arab Emirates, a site to which many young Kenyans travel in the hopes of pursuing more desirable futures. But this would entail a series of sacrifices. Not least, she would have to travel without her son. While this aspiration hovered in the background of conversation, in the day-to-day, her temporal horizons were more constrained. Her commission, which in 2014 hovered around KSH 10,000 per month, barely covered her expenses, which included the KSH 6,000 per month it cost her to rent the kiosk. She often paid a portion of the rent for the apartment—though technically this was Maina's responsibility—and contributed money for school fees. All of this on top of buying groceries, prepaid electricity tokens, and clothing for her son, Kamaish. Money was tight.

From Peninah's perspective, her ability to establish and maintain intimate relationships with her customers required that she calibrate her moods. This, too, is work. On many occasions in this period, I would find Peninah sitting behind the grating of her shop looking tired and unhappy. The "stress" she was under in her personal life, she explained on numerous occasions, was acute, and it was a problem for business.[11] Following Arlie Hochschild, this affective work requires Peninah to "induce or suppress feeling in order to sustain the outward countenance that produces the proper state of mind in others."[12] While never acknowledged in Safaricom's accounts, this work "calls for a coordination of mind and feeling."[13] Peninah was, evidently, engaged in a careful act of self-curation, which she viewed as being critical to extending the social relationships essential for the smooth operations of this marketplace.[14] While this affective work does not "simply mirror the commodity logic" embodying "distinctive . . . grammars," it is nevertheless crucial to making Safaricom's network operate.[15] Despite the centrality of affect in Peninah's own narration of her work, this labor is linguistically, conceptually, and materially devalued by industry insiders, who refer to M-Pesa agents using the nomenclature: "Human ATMs."

This designation is revealing. In much of the global north, the Automated Teller Machine (ATM) was designed to ease banking transactions, enabling customers to deposit, withdraw, and transfer funds without making recourse to the cumbersome, costly in capital's accounts, and often unavailable human teller.[16] While not occurring on the factory floor, this was, in any case, a classic story of automation: the actions of the human worker were delegated to the machine, displacing the human worker in the process. This transformation, though, relied on the presence of an efficient electricity grid, coordinated computing, and the proliferation of brick-and-mortar banks. The Human ATM, then, names a peculiar, historically specific hybrid.

In Kenya, the ATM could not simply replace the human teller. The limited reach of financial and other infrastructures precluded such a possibility. In this austere infrastructural geography, rather than the machine replacing the human, humans are taking the place of never fully networked infrastructures. And yet, far from impeding "innovation," celebratory narratives of M-Pesa specifically cite Kenya's historically poor connectivity, limited infrastructural reach, and lax regulatory structures in accounting for the success of the mobile-to-mobile money platform. M-Pesa is, on this framing, an ur-example of technological "leapfrogging." Such narratives of technological leapfrogging elide the reliance of contemporary capitalism on older formations of precarious, often feminized and racialized reserves of labor power from which surplus is expropriated, and whose progenitors can be easily discarded in the interest of profitability.

Rescripting Kenya's Austere Infrastructural Landscape

After decades of lamenting the "failure" of development in sub-Saharan Africa, the discourse that emerged around M-Pesa was an about-face. Narratives of the incapacity of the state to provide for its citizens gave way to celebrations of the merits of this austere infrastructural geography. The success of mobile telephony was at the center of this emergent discourse. It was precisely the absence of landline infrastructures and the "inefficiency" of state monopolies over telephony that paved the way for the rapid takeoff of mobile phone usage under the management of Safaricom's parent company, UK-based Vodafone. While infrastructural lumpiness inhibited fixed-line use, they claimed, it enabled unprecedented connectivity in mobile phone markets.[17] The state's "inefficiencies" had been replaced by the "efficiency" of private, foreign corporations. Focusing on the technical and managerial issues sidelined the political implications of these reconfigurations.

The timing of this rescripting of state "failure" was important. By the mid-1990s, the relationship among service provisioning, infrastructures, and governance had been renovated by international financial institutions (IFIs), such as the World Bank and the IMF. No longer were infrastructures framed as "natural monopolies" best built and maintained by the state. Instead, the private sector was best poised to foster development. These ideological shifts entailed governmental interventions that established "seemingly depoliticized arenas within which private actors could pursue their 'economic' interests, free from overt 'political' interference."[18] Safaricom, as we shall see, benefited from these ideological reorientations.

These transformations heralded a redistribution in developmental authority. As elsewhere, the 1980s and 1990s saw a proliferation of NGO and private-sector initiatives across Kenya. The future-oriented language of state developmentalism, and its attendant promises of "modernity," fell into disuse as a more restrained vision of what was possible emerged.[19] "Development" morphed into "poverty alleviation." "Big push" investments were replaced with frugal, parsimonious, and often privately funded interventions, which operated according to the profit sensibilities of market actors, rather than the accounting standards of deficit and debt-financed states. The decline of narratives of state-led development led to the rise of the "micro"—microcredit, microenterprise, microfinance.[20] As the role of the state was reconfigured and "micro-logical" interventions became the norm, so too was the place of the poor undergoing renovation in development discourse, which valorized the unplanned, spontaneous activities of the poor—people associated with the so-called informal sector.[21] The market was seen not only as an efficient means of distributing resources but also as a technique for the reformation of subjects. Conditions of austerity would "unleash the entrepreneurial capacities of the poor." Newly responsibilized "entrepreneurs" would not simply "make do" but were "new sources of economic growth" whose patterned behavior could guide the accumulative logic of corporate firms.[22] With the decline of formal wage employment, M-Pesa agents—savvy and self-identifying "entrepreneurs" willing to work long hours to secure their commission—fit squarely within this reconfigured discursive and institutional framework.

If international financial institutions were in the process of redefining the relationships among citizenship, infrastructures, and governance, Safaricom was undertaking boundary-work of another kind. Safaricom's early "success" was in the field of telecommunications, *not* finance. It was not until 2007, with the development of M-Pesa, that Safaricom entered the financial services sector. Safaricom's ability to enter into the financial sector, too, was discursively and materially linked to historical dynamics: in this case, a thin regulatory structure had created a permissive "ecosystem," one within which M-Pesa—a hybrid that relies on telecommunication infrastructures to underwrite financial services—could thrive. Techno-enthusiasts and IFIs without fail cite

Kenya's lax regulatory structure as an important precondition for M-Pesa's success. Kenya's regulators, by offering the firm a "letter of no objection," had wisely "allowed the scheme to proceed without hindrance, and ensured development led regulation, unlike in many markets where development lags behind regulation."[23] Within four years of its launch, M-Pesa had enrolled over 50 percent of Kenya's adult population.[24] This was not, however, a story of state withdrawal.

Safaricom began its life as a state-held entity, incrementally privatizing under pressure from international governing bodies (see chapter 5). And this storied history is crucial to understanding the firm's success.

Safaricom's dominance in the field of finance emerged following fierce struggles between Safaricom and the banks. If these struggles heralded the reconfiguration of the regulatory arm of the state, they did not entail the weakening of the state's fiscal arm. As Keith Breckenridge argues, Safaricom's monopoly status was enabled by government regulators, due in part to the joint ownership of Safaricom between British-owned Vodafone and the Kenyan state, "which gave the state a double-dipping interest in the company's enormous profits: first as shareholder and second as tax collector."[25] Within three years of its launch, Safaricom had leveraged this lack of distinction between public and private, transforming its tacitly state-sanctioned monopoly into enormous profits, with M-Pesa enrolling 15 million users who generated KSH 7.5 billion for the firm and its shareholders in 2010.[26]

The flexibility of Kenya's regulatory structures also critically shaped the emergent labor regime of which Peninah is a part. M-Pesa's "winning formula," one Brookings Institute report reads, resulted from the acuity of "Kenya's regulators, who correctly identified these agents as intermediaries rather than providers of banking services."[27] On this framing, Safaricom's M-Pesa agents are decidedly *not* financial specialists, nor are they maintainers of the infrastructure, much less system builders. M-Pesa agents simply provide services "just like those provided by machines that exchange coins for bills . . . They perform the functions of an ATM that allows cash withdrawals and deposits."[28] M-Pesa agents, then, are conceived of as mere "intermediaries," infrastructural prosthetics, Human ATMs, whose labor—while important to mitigating the problems of the lumpy reach of Kenya's banking infrastructure, the unpredictability of energy supply, and the limited connectivity of computers—is unremarkable.[29]

Within a year of its launch, approximately 2,000 of these "intermediaries" were stationed at kiosks across the country.[30] In the beginning, this was good business. The small number of agents working in the field meant that the commission-based pay structure—neither salaried nor waged work—could sustain agents. But agents, while under the managerial control of the company, are not formally employees.[31] As a result, Safaricom's responsibilities to agents are minimal. Agents' precarious status is matched by their compensation, which is calculated on the basis of the number of transactions they manage over the course of a month. Thus, Safaricom established a business model that—pundits lauded—could self-expand, enrolling agents as the network of end-users and their transactions grew. With the increased number of agents and kiosks came an increase in competition. The pool of potential profits shrank accordingly. As Peninah put it, these days "you walk this road and you maybe see fifteen [kiosks]. Imagine if it was only me."[32] Kenya's austere infrastructural geography—its lack of electricity and landlines—and M-Pesa's emergent business model led to the proliferation of people engaged in this supposedly unremarkable prosthetic labor. As one observer wrote of M-Pesa, "New technologies" are "leapfrogging the ones that exist in developed economies," helping to "solve problems arising from weak institutional infrastructure."[33] Today there are over 160,000 agents working across the country.[34] Focusing on the technics allowed observers to elide the work embedded in this new labor regime. Indeed, the technologies, such as they are, would be of little use without the work of Human ATMs who mediate among M-Pesa's platform, Safaricom's offices, and customers.

Affective Work as Maintenance and System Building

M-Pesa, though, is by no means a purely technological infrastructure. It relies on the affective, social, and maintenance work of people like Peninah who are responsible for explaining the system and its fee structures to customers, reading text messages to the blind, and collecting the ID numbers of new customers. Safaricom's reliance on the mediating work of its agents is in part due to its business model. Those at the "bottom of the pyramid"—defined as a person taking home an average of $2.50 per day—often make money on the day, making a subscription-based model untenable. Instead, Safaricom makes money off of M-Pesa each and every time a transaction is made. Peninah manages this infrastructural interface, providing the critical link between Safaricom and its customers, their shillings, and, importantly, their habits. Indeed, while this work might appear banal, it is essential to making Safaricom's network "hang together."[35]

From behind the gratings of their kiosks, Human ATMs enable users to deposit, send, receive, and withdraw money. The sequence unfolds as follows. First, a customer locates an M-Pesa kiosk. Once there, they deposit money into their M-Pesa account. The agent takes the customer's physical currency and deducts the amount deposited by the customer from their agent account, then moves the electronic value, denominated in shillings, to their customer's account. Once this part of the sequence has been completed, a user can transfer that value from their account to the account of a receiver. Having gotten notification of the transfer, the receiver can go to an M-Pesa agent to withdraw the electronic value sent to them as cash. This leads the agent's total electronic float—the total value stored in their agent account and accessed from their phone—to increase. Fees are not fixed, but depend on the amount of money transferred. But these broad outlines tell us little about the actual operations of the work this system devolves to M-Pesa agents. It is to the empirical details of this work that we now turn.

The money that Peninah makes at the shop is the primary income for her household, which is composed of herself, her son, and his father, Maina. Peninah's days are long. At 6:00 a.m., she wakes up to prepare porridge for her son before waking him and getting him ready for "baby class." She opens the shop at around 8:00 a.m. and closes, with breaks for lunch and dinner, between 11:00 p.m. and 1:00 a.m., depending on the day. Peninah's work is repetitive and often "boring" in her rendering.[36] There are spells of quiet when few people come into the shop and the work of an M-Pesa agent can appear to be largely rote.

Peninah is not subject to the iconic "time discipline" of labor sociology. In contrast to the closely monitored working conditions of factories or offices, as well as the countervailing pressure of workers—such as protection of working hours and weekends—M-Pesa agents compete as isolated individuals. They are free, even encouraged, to work long hours, and there is little in the way of collective organizing among agents. Safaricom does monitor their behavior, through infrequent in-person visits and ongoing data monitoring, and it actively shapes their behavior through text message nudges and reminders. The thrust of the model, however, depends on the mass availability of M-Pesa agents, whose work is required so that fees can be collected from end-users whenever they have the need or desire to send money or convert digital value into Kenyan shillings. Indeed, like an ATM, one of the services Peninah offers is access to financial services beyond the hours of the 9–5 working day of the human teller that the ATM in the global north sought to replace, but with a difference: Peninah is not, after all, an ATM. Maintaining these long hours, while hard on her body, and hard on her person, is critical to maintaining consistency of services, the precondition for tethering *her* customers to *her* kiosk.

Despite her long hours at the kiosk, Peninah's access to money is uneven and periodic. On both the scale of the day, and the scale of the month, she is perpetually hedging her financial futures. Peninah is cautious about how she structures her thinking about the future, and she has developed a series of strategies to mitigate her limited liquidity. In the main, Peninah purchases food to prepare for meals just prior to making them. This makes sense. The flow of money in and out of her shop means that she can take a little for lunch knowing that it will be replenished before dinner. There is no refrigeration in her home, moreover, making bulk perishables impracticable. These decisions, though, come as much from a place of fiscal constraint and uncertainty as they come from a place of practicality. Unexpected expenses, such as in the instance of illness, and the perennial and vexing uncertainty regarding whether friends and family will contribute to her imagined future (in the form of bridewealth), or that of her son (in the form of school fees), upset the balance that Peninah seeks to maintain.

Hers is a position of risk and relative vulnerability in which Safaricom's payment structure plays no small part. As an agent, her commission at the end of the month is determined by the number of transactions that she completes in an approximate 30-day cycle. In 2014, customers were charged KSH 15 to transfer KSH 1,000 to another mobile device, and KSH 27 to withdraw the same amount. The transaction fees are divided between herself and Safaricom at the end of every month. However, the number of transactions she will manage over the course of a month is by no means guaranteed. And Peninah's ability to secure her commission maps onto the ebbs and flows of the movement of money within the neighborhood.

While sitting with Peninah in her shop one day, a number of customers streamed in wanting to make deposits: "*Sina* float" (I have no float), Peninah responded.[37] Peninah's float is the total electronic value stored in her agent account. Every time a deposit is made, her float goes down. As her float approaches zero, as it did on this day, she has to refuse people wanting to deposit, instead waiting until customers wanting to make withdrawals arrive, their movement of money out of the system increasing her float.

I asked Peninah why so many people were depositing on that day. She speculated that it was because banks had closed early, it being a Saturday. But this was not the only reason. She continued, noting that times had been tough in the neighborhood—people just didn't have money to spend. Even those making deposits, she said, were using their money to pay bills. It is only when money is moving through the system and generating interchange fees that it generates monetary value.

Other days, this precarious balance of deposits and withdrawals was achieved. One afternoon, I joined Peninah working at the shop. For nearly three hours, there was barely a moment of pause. At any given instant, there were two to four customers waiting to be served.[38] I doled out airtime credit while Peninah took care of people making deposits and withdrawals. She speculated that it was busy because it was Sunday, a day when families make small outings after attending morning service. But, she continued, it also mapped onto another financial rhythm. It was close to the end of the month. With people being paid "from the 25th," this meant "a lot of business." And to her good fortune on that day, people themselves were doing the work of equalizing, deposits and withdrawals happening in balanced numbers and amounts.

The uncertainty of Peninah's commission, in other words, is compounded by the uneven but semi-predictable financial lives of her customers. As Peninah is quick to explain, there are definite and observable rhythms that govern the rates at which people use the system. These rhythms are structured along different temporal scales: Certain times of day are busier than others, as are certain times of month, and certain times of year. For example, whether making a salary or making wages on the day, the end of the month is a tense time as people evaluate how much of a stretch it will be for them to make rent. Other times of acute yet predictable financial anxiety are the periods before the beginning of school terms, when people go to great lengths to ensure they can assemble "school fees." Peninah's life is also structured by these rhythms but in a reduplicated way, as her livelihood is also contingent on the fact that these rhythms structure the lives of her customers and thus their use of her services. As Peninah succinctly put it, her livelihood "depends on what the customers are doing."[39]

Peninah's work, in other words, is that of both sociologist and ethnographer. She knows the neighborhood and its rhythms. She, better than anyone, knows when people have been paid, as well as the temporality of money sent, money stored, and money withdrawn. In part, this is because the rhythms of her customers' economic lives dictate her own. But her knowledge of "her customers" extends beyond the general patterns of financial life in the neighborhood. Many of these people are not strangers but intimates with whom she has developed relationships over the course of many months of work. Indeed, she routinely indicated the importance of moving away from generalizations to the particulars of each individual person. As she put it to me: It's important to know them, to understand them, as "every customer has their own problems."[40] For example, one of her customers cannot see very well so she reads the transaction texts to him.[41] Other loyal customers are not hard of sight but are illiterate, and it is left to Peninah to guide them through the service. Some know her son and buy him "smokies" (small sausages) being sold by another of "her customers" who has a small stand outside of her kiosk where he sells boiled eggs with *kachumbari* (salsa), and the small beef sausages that Kamaish, her son, likes so much. She knows the family of another customer and so, as I will discuss in greater detail below, is willing to bend some of Safaricom's regulations when carrying out their transactions. These relationships, despite their differences, are underwritten by a strong sense of trust crucial to generating the infrastructural attachments that Peninah both cultivates and maintains.

While these affective repertoires characterize many forms of feminized work, in the case of Human ATMs, they are matched by the daily labor of making the material infrastructure that grounds Safaricom's system work. Put simply, these intimacies, while crucial to securing Peninah's "customers," also enable her to enroll the broader social networks within the neighborhood that are essential to making Safaricom's business possible. Periodically when short of cash, for example, she heads down the road to Safi Roho, one of the many butchers in the neighborhood. An agreement produced out of routine, she regularly borrows KSH 10,000, which she uses as her liquid capital until the end of the day. Unlike in the context of the ATM, there is no large store of currency located behind Peninah's grate.[42] This infrastructure, such as it is, is nodal rather than formally networked, and it is reliant on relations officially outside of the purview of the system. Tapping into these wider networks is critical to maintaining Safaricom's infrastructure in this corner of Nairobi. Just "as poor people have to contend with fragmented physical infrastructure in all aspects of their lives," as Julia Elyachar notes, "they [also] have to invest more time in the maintenance of [the] infrastructure of communicative channels."[43] These arrangements, while officially outside Safaricom's model, are unofficially at its center. Both the back story and the front story of contemporary modes of accumulation, they are "arrangements . . . from which capital profits and on which it relies."[44]

Just as Peninah's affective prosthetic work stitches together the social relations necessary for M-Pesa's operations in Zimmerman to work, so too does her shop suture together Nairobi's infrastructural landscape. Indeed, Peninah's shop operates as an infrastructural node in a neighborhood characterized by the uneven delivery of services. While many in this area have access to electricity, blackouts are semi-regular, whether the result of the system going down or because people have not re-upped their prepaid tokens. Peninah's shop, by contrast, is constantly glowing at night, boasting the constant hum of *stima* (electricity), which Peninah pays for. Safaricom has, evidently, discharged the responsibility of infrastructure service-provisioning—required for its network of agents to undertake their work—to the agents themselves. But from Peninah's perspective, this is a business opportunity enabling her to offer another service: The charging of cell phones, for which she demands a flat fee regardless of the duration of the charge. For Safaricom, offloading the responsibility for generating infrastructural connectivity to agents allows the company to sidestep the problem (and costs) that characterizes life in Kenya's austere infrastructural geography, all the while encouraging the "entrepreneurial comportment" of its agents.[45]

Affective Work as Collusion

But these relations of trust, which seem to fall outside of market relations as conventionally understood, are also a potential liability to Safaricom. It is on the basis of these dynamics that Peninah and her customers routinely collude to evade some of Safaricom's more cumbersome regulations. The term "direct deposit" refers to one such collusion. Transferring value via M-Pesa incurs transaction fees, such that if you want your receiver to have access to KSH 1,000, you need to send KSH 1,000 plus KSH 15 for the transfer and another KSH 27 for the cost of the withdrawal, for a total of KSH 1,042. These fees are borne either in part by the sender (if they are willing to put part of the burden on the receiver), or in full by the sender (if they want to relieve the receiver of the burden of paying at the time of withdrawal). The latter is often the case with M-Pesa transfers moving from Nairobi to "up-country" Kenya, the location to which many Kenyans send their money to family, often elderly parents, living in the countryside. M-Pesa is, after all, a system premised on the promise of securing relations of kinship, allowing labor migrants of various stripes to "send money home."[46] If, however, the money is deposited directly into the account of the receiver, both the sender and the receiver avoid the secondary cost of withdrawal. This is expressly prohibited by Safaricom. As an agent, Peninah can facilitate direct deposits, but if she is caught, she risks being penalized.

Safaricom's "regime of perceptibility" is not, however, total.[47] For Safaricom to notice these transactions, they have to become visible to the system, which it achieves by virtue of the sensitivity of the digital network to the spatial coordinates of money moving in and out of the network. If, for example, a deposit is made to an account at Peninah's kiosk in Zimmerman and within minutes the money is withdrawn in Kerugoya, a town located some six hours away, the system will detect that a direct deposit has been made. But there are workarounds. If the receiver holds off a half-day before withdrawing the money, Safaricom's system will not be able to deduce with any measure of certainty that a direct deposit has been made. If this strategy of deferring the withdrawal is followed, as Peninah pointedly put it, "Safaricom can't see anything." Peninah deploys this strategy herself when she sends money to her mother and father who live "up-country." But she also does it for "her customers," those loyal regulars for whom she is both patron and client.

But there are other, more banal, evasions. Safaricom requires that its agents demand to see each customer's ID every time a transaction is undertaken. Peninah is uneven in how she applies this regulation. When I asked her about this, she told me that it was a matter of "knowing." "When I know you, I can withdraw for you without [seeing your] ID. If I don't know you, I can't trust you."[48]

M-Pesa was conceived of as a way to move money across distances, disembedding it from the everyday risks and social obligations that attend moving money by hand. The prestations of villagers that used to greet Nairobi's residents traveling "up-country" for a weekend or holiday have diminished as money can now be sent directly to recipients without traveling. This dematerialization of cash promises to—and does—satisfy the desire to send money across the country, unencumbered by social relations that are both a burden and a network of support. An early advertisement for M-Pesa captured this idea: Kenyan shillings flow from the phone of an urbane man to his parochial elders laboring in a field. But the abstraction this depicts—of a money infrastructure devoid of human mediation—is a fiction. The very functioning of M-Pesa is dependent on infrastructural attachments generated by those at the "bottom of the pyramid." These dynamics are not ancillary or necessarily subversive; rather, they are concurrently integral to Safaricom's profitability and occluded on Safaricom's balance sheets.

Affective Work and Value at the "Bottom of the Pyramid"

I was repeatedly told that Zimmerman, the neighborhood where Peninah lives and works, was but a "swamp" only ten years earlier. As Kenya's population has more than doubled in the past thirty-five years, peri-urban landscapes like Zimmerman have cropped up in many quarters around Nairobi's tiny central business district. Population growth has outpaced salaried job creation, leading to the growth of the proportion of the population who are outside the wage labor market.

Safaricom has benefited from these demographic shifts, as people flocked to become M-Pesa agents, the low commission-based pay structure notwithstanding. While in the early years, M-Pesa agents were forbidden from carrying out other forms of business from their kiosks, as the system grew these prohibitions became untenable. As Peninah told me in 2014: "About 5 years ago, they used to come by, and if they saw that you were selling anything else, like these padlocks and extensions, they would tell you to stop. They could even cancel your line. All you could sell would be maybe these things," she said, as she gestured to the back wall where chargers and phone cases hung. "These days," she continued, "that's not a problem."[49]

Peninah self-identifies as a good businesswoman, an entrepreneur. And her other business ventures have expanded her investment portfolio, offsetting the low commission she receives from Safaricom. The figure of the entrepreneur has a long and ethnicized history in Kenya, largely being associated with Kikuyu-speakers. Peninah, a woman from Central Kenya and someone who self-identifies as Kikuyu and as an entrepreneur, seems to affirm this narrative. But entrepreneurialism is also at the center of the lived realities of austere economies. Under conditions of constraint, the "'responsibilized' citizen comes to operate as a miniature firm, responding to incentives, rationally assessing risks, and prudently choosing from among different courses of action."[50] Peninah is not driven by the calculative practices of the marketplace alone, however. Instead, she has expanded her investments to manage the "turbulence" of her present, and in the hope of securing more desirable futures for herself and her son.[51] This economy of care does dovetail with the rhetoric of the entrepreneur, though, driving Peninah to expand her stock and the services she offers from her small, green kiosk.

Over the course of the time that I spent with Peninah (partly renting a room in "Maina's" apartment and partly visiting her shop in the afternoons), she significantly expanded her stock. These additions happened in a piecemeal way. One day, I arrived and saw that to her stock of phone cases, she had added batteries and chargers. On another day, she had added energy-efficient lightbulbs and the lines of wire required as Kenya's television network went digital. Some weeks later, I arrived at her shop to see that it had been adorned with thin, brightly colored ladies' belts and hair scrunchies, both of which were popular at the time. Within months, she had invested in a wooden structure. Hastily built at the cost of KSH 2,000 and erected in front of the shop, it housed plastic slippers and washing bins.[52] Peninah spends her days considering sartorial and technological trends, a lay market forecasting that draws on her observations from the kiosk, conversations with customers and friends, and her engagement with various media. She takes this work of cultural discernment seriously as a means of distinguishing her shop, attracting customers, and maintaining existing ones. In other words, it is not merely the rote fulfillment of mobile money transactions that make for a good M-Pesa agent, but the expansive care with which the kiosk is maintained and the work that agents invest in knowing their customers and attending to their needs and desires.

Peninah is a savvy businesswoman, to be sure. But her capacity as a shop owner is contingent on maintaining and extending social relationships that are conventionally held to be outside the rationality of the market.[53] This "entrepreneurial comportment," evident both in her expanding investment portfolio and in her strategies of collusion, motivations notwithstanding, are largely in line with the policy recommendations of bodies like the World Bank. Indeed, her activities are congruent with Safaricom's own business model, which operates in the register of altruistic capitalism. The goal is to "transform lives," by "unlocking" the "entrepreneurial" potential of Kenya's poor. This work of knowing her customers is critical to both her own future possibilities and those of the company and its expanding knowledge infrastructure, of which Peninah is a critical, though unremarked and woefully under-remunerated, component. Indeed, if the system's reliance on the Human ATM is occasionally a liability for the company, it is also at the center of Safaricom's ever-expanding knowledge apparatus. The company uses the data generated by Kenyans' use of M-Pesa as a baseline from which it develops new services. Agents play an important role in gathering these data.

Theoretically, every transaction that is undertaken at Peninah's shop is recorded both in her logbook and electronically. These data provide a map of information regarding person-to-person transfers, rates and values of withdrawals and deposits, and the locations of parties to a given transaction. All of these data are tethered to individuals. This was mandated by Kenyan regulators under the terms of a regulation called "Know Your Customer" (KYC), of all things. This regulation requires each M-Pesa account be linked to an individual, a connection secured when a new customer presents a valid ID on registration. All of this requires work. It is Peninah and others like her that collect and verify people's ID numbers. In this capacity, Peninah acts as a conduit between those at the "bottom of the pyramid" and Safaricom's offices. As this information moves, it is refined and subjected to techniques of assessment that enable its revaluation. In this capacity, she mediates between users and Safaricom's offices, where users' data are transformed into commercially useful knowledge.

Value at the "Bottom of the Pyramid": Infrastructures, Market-Making, and Reimagining the Commons

At Safaricom, we believe in investing in the future. Working hard towards the creation of technologies that positively impact the society and most importantly transform the lives of those around us is a core value that we hold dear. Milestones in the mobile money sector in the form of M-PESA . . . have put Safaricom on the global map. With more innovators and thinkers emerging daily, we endeavor to aid them in their road to discovery through interesting and encouraging platforms that enable them to reach their potential. This is because when they unlock their potential, they push us to ours and most importantly, the end result is solutions that better the lives of the human race.[54]

As Safaricom's website suggests, its goal is not simply to generate profit, but to create conditions that render visible and enable the expression of existing sources of potential profit that are simply lying in wait. One way that Safaricom seeks to "unlock" Kenyans' "potential" is through its commitment to "know its customers," who are also, importantly, potential innovators. This corporate mantra hinges on Safaricom's ability to mobilize and render valuable the "local." As we have seen, Safaricom invoked the value of the local in defending the irregular deal it had struck with Mobitelea (see chapter 5). Indeed, the clandestine movement of shares to Mobitelea in exchange for "local knowledge," a relationship the company framed as being the precondition for profit, ought to be seen as a precursor of what would become Safaricom's mantra for success—albeit in distributed form.

The origins of this, too, are storied. Sometime in the early 2000s, customers began complaining of network congestion on Friday evenings.[55] By this time, the company had nearly 700,000 subscribers. Though then-CEO Michael Joseph claims he never uttered the statement, popular memory holds that, in response to queries regarding network congestion, he placed part of the blame on the shoulders of Kenyans. "Kenyans have peculiar calling habits," he is rumored to have said. This led to a firestorm of critique. How dare this man, a foreign national no less, deign to make statements regarding what Kenyans "were like."[56]

While unbeknownst to him at the time, Joseph had hit on a key insight and guiding business strategy that is, today, cited as an explanation for the corporation's unparalleled success. Kenyans' "peculiarity" need not be a liability, Joseph learned, but could be transformed into an asset. Locating and understanding this peculiarity, making it legible, and capitalizing on the knowledge gained, was a means of more firmly attaching the company—foreign capital, technologies, expertise, management structures and all—in this particularly Kenyan milieu.

Operationalizing this insight as a business model required developing a sensitive and malleable knowledge apparatus. One that was capable of understanding, abstracting, and thus capitalizing on the "peculiarity" of a body of citizens-cum-users. For Joseph, the question became one of discerning what kinds of value might be located in the everyday practices and "peculiarities" of Kenyans. This brand of "ethical capitalism" does not imagine itself as being extractive.[57] Indeed, as seen in chapter 5, the logic of Markets for the Poor turns on a framing that does not hold the market in opposition to the social good.[58] Instead, proponents of this new rationality argue that marketizing the strategies of the poor concurrently generates profits and ameliorates the lives of those at the "bottom of the pyramid." This is a win-win narrative. This is a narrative of "financial inclusion" via "deepening" the reach of the "financial sector." It is to this deepening, and the data-analysis that sits at its center, to which this section now turns.

To get at this work, we have to travel with the data and move away from Zimmerman to Safaricom's headquarters. These imposing towers are located on Waiyaki Way in Nairobi's affluent Spring Valley neighborhood. In these skyscrapers, cloaked in opaque, reflective glass, Safaricom's recognized "expert" analysts engage in product development and refinement. These are sites of innovation, to be sure, but they are also "centers of calculation."[59] As these data move from sites across the Kenyan landscape, including Peninah's kiosk, to the offices of research, development, and design, Peninah's affective and social prosthetic work, as well as that of her customers, is commuted into new sources of value. In the process, the qualitative lives underwriting habits become the quantitative basis for the generation of new knowledge. But as these data are read for patterns, the social relations that enabled their unwitting collection are stripped of their interpersonal quality, revealing "special characteristics" only visible in the aggregate.[60] And yet, it is precisely in these unmarketized relations that analysts locate new forms of prospective monetary gain and on which new markets are assembled and from which profits are extracted.

James works as a financial analyst for Safaricom. When we met, he explained that M-Pesa had been a harbinger for how the company would come to approach the task of developing new products and services.[61] M-Pesa's beginnings, like that of Safaricom itself, were humble in origin. In 2005, two Vodafone executives, Nick Hughes and Susie Lonie, partnered with Faulu, a Kenya-based microcredit NGO. They began with a simple proposition: Could they develop a platform that would ease both the dispersal of these small loans and their repayment? When the service was launched, however, they noticed that pilot populations were putting the platform to uses unimagined by designers: They were using the platform to send small amounts of value to one another. A little tinkering and, voila, M-Pesa was born.[62] Put differently, people's reconfiguration of the system had transformed a banal loan repayment scheme into a "revolutionary" mobile-to-mobile money platform.

M-Pesa's peculiar trajectory led to an important innovation for Safaricom, which is today also its guiding business strategy: Users' patterns of use with one service could be used to guide the innovation of new services. Users, in this vision, have been reimagined as lay designers, albeit designers who are unaware of their role in generating new knowledge, and not cognizant that everyday life is being commuted into profit. This business strategy turns on locating sites of innovation on the cheap, the very cheap.[63] And yet, James insisted, this is a "collaborative way" of developing new services.[64] This form of knowledge-work is undertaken in an altruistic register. The goal is to "build communities" and "transform lives."[65]

Transforming lives through the generation of new services requires work, as data are read for patterns, patterns are read for potential value, and discovered value is used to develop new services. And "innovation" has begot "innovation." The team of M-Pesa observers tasked with locating and rendering visible this quiet and creeping value noticed, James recounted, the "queer behavior" of users. Over the course of many months, a new pattern became discernible to these analysts as they pored over the numbers that result from the over 10 million transfers people make per day. These users, they learned, were depositing more money into the system than the value they subsequently transferred. No longer simply a means of transferring value, users had reconfigured M-Pesa into a savings service.[66] And so, designers set to work constructing M-Shwari, a value-added service that operates as a platform for savings and provides a gateway for customers wanting to access small loans.[67]

Launched in conjunction with the Commercial Bank of Africa (CBA), Safaricom framed M-Shwari as a means of deepening the "financial inclusion" of the "unbanked" by developing unbiased but discriminating algorithms. In so doing, Safaricom capitalized on the popular perception that Kenya's financial sector is Kikuyu-dominated, with ethnicity largely determining access to loans. Safaricom's algorithm, by contrast—and despite the fact that the CBA is partially owned by the family of the former president, Uhuru Kenyatta—did not discriminate based on ethnic affiliation. This was a technological fix to a historical political problem that, from the perspective of many, sits at the center of contemporary inequalities.

These altruistic algorithms had to mitigate the risk of issuing loans to persons without credit scores. Safaricom devised a novel solution. M-Shwari loans are not backed by assets but by "reputational collateral" that Safaricom assembles through the analysis of a vast archive of user data that it established as having predictive potential—these data included age, repayment of emergency airtime credit, M-Pesa transactions, and daily airtime usage.[68] In explaining how this data mapped onto the relative risk for creditors, James told me that it was like "using [your] consumption of bananas to tell us something about your consumption of oranges."[69] The initial user portfolios reliably predicted repayment: Only 6.1 percent of loans were unpaid within ninety days during the initial months of the service. As more people engage with the system, Safaricom actively incorporates new data generated by M-Shwari users, iteratively improving the scoring model in the process. In so doing, it has driven the number of nonperforming loans (NPLs) down an additional 2 percent.[70]

M-Shwari has become a hugely popular service. By the end of 2016, M-Shwari boasted 16 million customers who had taken out 64 million small loans, valued at $1.4 billion.[71] Its profitability in the field of debt, too, depends in part on the company exploiting Kenya's regulatory structure, with Safaricom branding the monthly 7.5 percent premium on borrowed funds as a "facilitation fee," rather than an "interest rate," thereby getting around Kenya's interest rate caps. While the loans are small, the facilitation fee is usurious, reaching 90 percent were it to be annualized. While Safaricom frames M-Shwari as a means of deepening "financial inclusion," it has effectively translated user data—itself based on people's labors of making do—into commercially valuable insights that enable it to make claims on people's future labor.

The development of services such as M-Shwari would be impossible without the work of M-Pesa agents, who operate as critical infrastructural prosthetics, enabling users to access the platform and, in turn, facilitating Safaricom's collection of data. This data-driven business model enables Safaricom to capitalize and scale up the life courses, behaviors, and strategies of the poor. This involves work of revaluation, with customers being identified as potential designers. As Sanford Schram has argued, whereas in the past "'workers' labour was . . . a factor in production, [today] ordinary people's everyday activities are . . . commodified" in ways that add to the revenue streams of multinational corporations.[72] The poor emerge in this new landscape not merely as generators of value. Rather, their knowledge, their strategies, their patterned behaviors are themselves sources of profit. This is a more accurate accounting of what is meant by value at the "bottom of the pyramid." This profit-generating activity, which requires the daily affective and social labor of M-Pesa agents and their customers, is "treated as costless in capital's accounts, it is expropriated without compensation or replenishment and implicitly assumed to be infinite."[73]

Conclusion

Karl Polanyi argued that market societies are characterized by the disembedding of "the economy" from the total social worlds of which it is a part. This insight, he argued, was critical to understanding what is particular to the capitalist mode of production.[74] Many scholars have taken Polanyi's insights as their point of departure for understanding what is unique to capitalism under conditions of neoliberalization.[75] My discussion of Safaricom works at an angle to this narrative. While, on the one hand, Safaricom's own history is largely in line with patterns of austerity taking place on a global scale, its success has not turned on further disembedding "the economy." Indeed, today, companies like Safaricom not only work to shore up a vision of their markets as embedded, as natural extensions of existing social relationships, but depend on these relationships themselves as the foundation on which to assemble new markets. This problem is not lost on observers. Consumer protection advocates have routinely pressured regulators to intervene and break up Safaricom, arguing that its work as a financial services provider must be untethered from its work as a telecom service provider. These efforts have consistently failed. This is unsurprising given the state's own capacity for reproduction is contingent on Safaricom's success. The state, in other words, is interested in thickening the infrastructural attachments tethering it to Safaricom, for these attachments shore up the power of the state. The government, put simply, relies on this work of embedding.[76]

The work of Peninah, and those like her, sits at the center of this labor of embedding. Indeed, the success of Safaricom—its capacity to marketize existing patterns of life, of affect, of making do—turns on the relations of intimacy that are at the heart of Peninah's business. And yet, Peninah is not recognized as a knowledge-worker or a systems builder or maintainer—indeed, she is not even recognized as a generator of value—instead, she is referred to by industry insiders as a Human ATM. This representation elides the affective social work of M-Pesa agents, with material consequences. It enables companies to portray this work as merely infrastructural, all the while capitalizing on the social relations that form its basis.

By treating this as expert infrastructural work, we have a vantage from which to engage the series of erasures on which new modes of capture and subsumption turn. To do so requires recognizing the work of Peninah, and those like her, as well as the "masses" at the "bottom of the pyramid," as *both* the front and back story of contemporary modes of accumulation—their labor is exploited while their affective work is expropriated. As Elyachar has suggested, locating the relationship between "social infrastructures" and technological infrastructures offers a position from which we can see both as collectively generated resources "for which recompense should be paid or rent paid for use."[77] What is at stake in this framing is the subject of the epilogue of this book.

Epilogue

In 2016, *Nairobi Law Monthly* published a series of articles regarding the outsized role that Safaricom, the largest corporation in the region, plays in everyday life. These articles turned on the perception that the company holds a monopoly over Kenya's financial, security, and media sectors, as well as the infrastructures undergirding them. As critic ArKan Yasin wrote in an article suggestively titled "Safaricom: Empire, Kingdom or Republic?": "Safaricom is the primary arbiter of all social and economic relations for the population of the entire territory. . . . Safaricom is unique not just in the fact that it straddles commerce, banking, voice and data communications, security surveillance and entertainment, but in the depth of its dominance in terms of reach in numbers, brand and socio-political power."[1] For Yasin, Safaricom's ability to "straddle" the boundaries separating these multiple domains has entrenched the power of the corporation. It is not just in sectoral terms that Safaricom has secured its monopoly power but, as Yasin suggested, by virtue of its centralized infrastructure, which allows it to parlay "dominance" in one field into "dominance" in another.

Yasin is not alone in focusing on the work of attachment-building that he indicted in the pages of *Nairobi Law Monthly*. In explaining Safaricom's wild success, digital enthusiasts, too, celebrate the infrastructural attachments that the firm has generated, boasting that Safaricom has achieved unprecedented "stickiness." In their hands, stickiness refers to a set of features—such as non-interoperability with other financial service providers, high charges to transfer money across networks, increased per-minute calling fees between telecom companies, and the mobilization of user-data to generate new services from which profits are seized—that tether users to the corporation. Stickiness, then, refers to the infrastructural attachments that bind people, their shillings, and their data to Safaricom's expansive but ultimately centralized network.

For Yasin, though, this market dominance reflected infrastructural attachments of a more fundamental kind, pointing to the blurred lines between market and governance, the corporation and the state, the economic and the political. This has presented a fundamental conundrum, according to Yasin. As he continued: "Safaricom wields all the societal scale tools of arbitration, power and oversight without societal scale authority and culpability, and here in [sic] lies the problem." At the center of Yasin's critique was the perception that Safaricom's outsized role across the domains of its operations has rendered it state-like. As Safaricom takes on state-like functions without the attendant state-like responsibilities, it has undercut a generic notion of "the public," in the process raising questions of "the good" for which infrastructures were ostensibly constructed and services offered. These claims have empirical legs. In the 1990s, when the state began incrementally privatizing Kenya Posts and Telecommunications Corporation, leading to the birth of Safaricom, it was, among other things, a means of outsourcing the risk and costs of developing telecommunications infrastructures and services from which the state nevertheless stood to profit—first as equity holder, second as tax collector.[2] As the government of Kenya enabled Safaricom to extend its operations into realms that have putatively been the reserve of the state, it has consolidated Safaricom's status as a hybrid institutional form: the corporate-state.[3]

This distribution of infrastructural and sovereign authority has led to a viable set of concerns regarding who, ultimately, is responsible for provisioning for the Kenyan people. Despite its state-like functions, as a corporation Safaricom is not responsible to the Kenyan public writ large, but to a more discrete set of social categories: shareholders, customers, and management structures straddling the United Kingdom, Kenya, and South Africa. As Yasin wrote, "Given the sheer scale of Safaricom's dominance over the social entity, political theory dictates it should bear the burden and weight of the entity's social and economic challenges proportionately. The M-PESA transaction fee by virtue of its ubiquity and inescapability (given it is extracted from over 88% of the adult population), is a society-wide economic rent and de-facto tax." Kenyans, Yasin here argued, are functionally attached to Safaricom's infrastructures and services; coercively tethered to its centralized networks. In reframing Safaricom's fees, first as rent, then as a form of taxation, Yasin argued that the company has an obligation to provision for the public *not* as a divisible category, but as an undifferentiated one. In engaging in these regimes of seizure, Yasin argued, Safaricom is on the hook for reinvesting its profits and ameliorating the lives of Kenyans *as* Kenyans.

We would miss much if we saw Yasin's critique as one focused solely on the weakening of a state which has ceded its authority to a multinational corporation; this being one of the typical narratives of neoliberalization, which sees an inverse relationship between state authority and "neoliberalism's varied global dispensations."[4] It is true that, on the face of things, Safaricom is a publicly-traded multinational corporation. However, as I have argued (chapter 5), and as Yasin well knows, this designation obscures more than it reveals regarding the relationship between the company and the Kenyan state.

Indeed, today both the corporation and the state are poised to profit from these relations of intimacy; both are poised to benefit from these infrastructural attachments. Safaricom is not only the single largest taxpayer in the country, thereby generating revenues that line state coffers, but given that the government of Kenya is also a shareholder, when Safaricom profits, so too does the Treasury. The new "security" system, which the state single-sourced when it quietly granted Safaricom the tender in 2014, is another site where historically consolidated infrastructural attachments are poised to benefit both the corporation and the state. Not only will the corporation foot the bill in the initial instance, but as it links up new data with those gleaned through its financial and telephone services, the prospective surveillance apparatus will be more robust than anything the government could have assembled without mobilizing the infrastructural power of the corporation.[5] Finally, the massive cache of user data that Safaricom has amassed is not only entering new markets as Safaricom sells the particulars and preferences of users to other companies in a bid to generate new markets, but the Kenyan state is rumored to access this data in its rhetorically powerful, but predictably misguided, "war on terror."[6] Drawing a line between the state and the corporation, between public and private interests, is evidently no easy task in Kenya today.

The crux of Yasin's critique turned on these forms of boundary-crossing, drawing attention to how the indistinct lines between the state and the corporation, between the public and the private, are shaping the political, economic, and social landscape in Kenya. But Kenya's dispensation, which has seen the fate of the state tied to that of the corporation, did not begin with structural adjustment in the 1980s and 1990s. As it has been the task of the preceding chapters to demonstrate, contemporary corporate-state entanglements build on austere colonial inheritances. And this longer genealogy matters.

Too often, our narratives of "neoliberalization," and the forms of privatization and austerity governance with which it is associated, presume a prior set of arrangements where the domain of the economic and the domain of the political, the domain of the state and that of the corporation, were neatly bounded. This is everywhere a fiction, but it most certainly mischaracterizes settler colonial zones where state-market entanglements set the stage for colonial modes of governance whose legacies persist into the present. Indeed, the relations among state-formation and marketization, read through the infrastructures that often brought them together under the mantle of the "public good," were at the core of the British presence in the region, which saw the Crown charter the Imperial British East Africa Company (IBEA) to assert British authority in the region under the dual mandate of "commerce and civilisation" in 1888. The devolution of sovereignty to the private infrastructure firm, the Crown hoped, would enable it to execute political and commercial power in the region "on the cheap."

While the IBEA itself was a short-lived venture, the patterns of austere infrastructural statecraft that it established have been remarkably durable. Over the long twentieth century, as various iterations of the state devolved sovereign authority to private infrastructure firms, the region and its people have been subject to varying forms of company rule. The history of radio broadcasting from the 1930s to the 1950s, as I have shown, was shaped by early agreements that bound the colonial state to the networks of the private firm Cable and Wireless Ltd. (C&W). Likewise, Safaricom, while attached to the state, has come to supersede the sovereign that authorizes its operations, all the while enabling the state's reproduction. Put simply, analyses of austerity as a structuring feature of contemporary dynamics have a durable history in this postcolonial space. By focusing on the region's infrastructural history, the supposed poles of the private and the public, or the economic and the political, are revealed to be more co-constitutive than assumed dichotomies acknowledge—the political has routinely been mobilized to backstop the interests of the firm, while the corporation has repeatedly been called to the task of doing the work of governance.

However, state efforts to leverage private capital to enact infrastructural rule have not always "succeeded," even on their own terms. In the 1920s, African communities pointed to how the state was facilitating white settler accumulation, arguing that forms of extraction ostensibly seized to finance the construction of road infrastructures as "public goods" were a ruse—conflicts that led the road to emerge as a node of conflict between African subjects and the colonial state. As for radio, by the 1940s, it was widely acknowledged that C&W's monopoly in the region had unduly impeded the state's ability to develop an infrastructure of broadcasting for African listeners. The corporation, in this instance, weakened the infrastructural aspirations of the state. Over the chronology narrated here, as people debated how the state was provisioning for themselves and their communities, they engaged in comparative work, contrasting their inclusion in the infrastructural state against various others. In Kenya, in other words, the contradictory vision of the public initiated under the dual mandate of "commerce and civilisation" has been compounded by the highly stratified racial and economic order that has guided (post)colonial administrative practices and shaped everyday life. As people observed the uneven reach of infrastructures that broke down along the lines of race (chapter 2), "tribe" (chapter 4), and class refracted through the lens of ethnicity (chapter 5), they routinely called the corporate-state in its various iterations to account. The corporate-state's "failure" to smooth out these contradictions in both material and ideological terms opened up possibilities for people to claim their inclusion within the infrastructural networks of the colony and later nation. By tracking Kenya's infrastructural technopolitics as they moved out of the hands of designers and into the quotidian lives and everyday conflicts of Kenyan communities, I have shown that people mobilized these networks as platforms on which to assemble new social and political subjectivities.

If understanding how people leveraged infrastructures as objects of political and social dispute has been one core concern of this book, exploring how these networks were enacted in material terms under conditions of austerity has been the other. In each of the infrastructural tales narrated in these pages, technologists and engineers, and their plans for social transformation and accumulation, have been complicated in practice. Novel material, social, and cultural conditions have forced these men (and occasionally women) to confront the limits of their expertise. Against these material constraints, African knowledge-workers and experts have been called on to act as infrastructural prosthetics, filling the breach between the corporate-state's aspirations and material conditions. Over more than a century, these men and women have been responsible for drawing on their cultural, linguistic, and technological expertise to enact infrastructures "on the cheap." Just as men like Ambrose Wakaria were essential to making radio broadcasting "work" in the context of an infrastructurally thin state, so too have the tacit arrangements and everyday tactics of M-Pesa agents emerged in the seams of infrastructural absence. And yet infrastructural experts—like Wakaria, like Peninah—have rarely been recognized as knowledge-workers, nor as experts. Instead, systems builders refer to people like Peninah as "Human ATMs."

The effacement of expertise in Africa is durable and pernicious—and it is political. Indeed, technologists and administrators have routinely appropriated and subsumed African knowledge and expertise even as they sidelined the importance of this work. And yet, as we saw in the cases of Wakaria and

Peninah—not to mention the countless unnamed guides whose knowledge-work formed the basis of the IBEA's road network—designers have relied on this work to enact, maintain, and extend infrastructures.

And this work has been transformative. Today, the patterned exchanges of the poor and the norms that they reveal are used by Safaricom as the baseline from which to reform globally circulating technologies, generating new services and profit streams in the process. As this work is appropriated and scaled up, the infrastructures themselves are transformed and rendered "peculiarly" Kenyan. Indeed, as Safaricom's first CEO Michael Joseph was quick to learn, the peculiarly Kenyan arrangements that shape life for those at the "bottom of the pyramid" offered possibilities for "frugal innovations" that have opened up possibilities for accumulation.[7] In an earlier iteration, technologists and administrators referred to this work of embedding as "tropicalisation." Claims to the contrary notwithstanding, infrastructures are not endlessly mobile, but require localization, translation, and transformation to take hold in new material, cultural, and social zones. In each of these infrastructural tales, however, designers and technologists have reframed the work of these innovators and experts of both technics and culture as banal. Both in the past and in the present, the underwriting presupposition is that African knowledge-workers and technical experts simply *act like infrastructures*—consider the designation of M-Pesa agents as "Human ATMs."

The devaluation of the cultural and technological expertise of these workers has been uneven, and the trajectories of these knowledge-workers and experts reflect broader political-economic and social transformations of which Kenya has been a part over the long twentieth century. As explored in chapters 1 and 2, by the 1870s, Kamba of the warrior class had established themselves as central commercial players in the region, having amassed considerable wealth and prestige as they mediated commercial relations between the coastal region and inland areas. As the IBEA subsumed the topographical and social expertise of these men as the basis for the corporation's road network, administrators hoped to transform this value-generating work into rote infrastructural labor. These processes by and large succeeded, reflecting the global dynamics of exploitation and expropriation that sat at the heart of sub-Saharan Africa's colonization.

As for the radio men discussed in the middle section of the book (chapters 2 and 3), while often (though not always) waged, their contributions to the corporate-state's broadcasting network was woefully undervalued in remunerative terms, even as their cultural acumen was appropriated and put to the task of "tropicalising" Kenya's media infrastructures over the course of the 1940s and 1950s. For men like Wakaria, all of this changed following independence in 1963, a period which saw twenty-five years of economic growth.[8] By this point, as a highly educated Kikuyu man well-integrated into the state's knowledge apparatus, Wakaria was poised to join the ranks of a class of newly prosperous Kikuyu-speakers that emerged following independence under the presidency of Jomo Kenyatta.[9] His possibilities for personal enrichment and fame were augmented following liberalization in the 1990s, which saw the proliferation of vernacular-language radio stations.[10] It was from this eminence that he was finally recognized and well-compensated—though largely by private interest groups—for what he had spent his life training to become: an expert on all things Kikuyu. His ascent was not typical, and reflected the class dynamics refracted through ethnicity and gender that shaped the independent state more broadly. Indeed, following independence, many Kenyans flocked to urban centers in the hopes of securing waged employment, but, from the outset, the formal sector was unable to absorb many job seekers.[11] Women, in particular, found themselves excluded from the formal labor market.[12] These women and men turned to largely unwaged work to sustain themselves and their families.

The fates of other people featured in this book reflect further shifts in ethnicized class dynamics nationally as they articulated with global transformations. Consider Safaricom's current CEO, Peter Ndegwa (see chapter 6). He was born in 1969, six years following political independence, to Kikuyu-speakers. His father, though in the navy and traveling often, was an inspiration to Ndegwa.[13] As a teen, Ndegwa attended Starehe Boys' School—which opened in 1959 against the backdrop of the Mau Mau uprising—where he obtained his high school diploma. Following high school, he was admitted to the University of Nairobi where he was awarded a BA in Economics. As with many promising students of means, following college he traveled to the United Kingdom, there attending the London School of Business, where he received an MA in Business Administration. With his degree firmly in hand, Ndegwa entered the corporate world, working first as a consultant for PricewaterhouseCoopers, a UK-based firm, later joining East African Breweries as a director of strategy, and ultimately securing a position working for Guinness Ghana and Guinness Nigeria. His trajectory impressed many, leading to his appointment as General Manager at Diageo, a multinational spirits firm, covering Eastern and Western Europe, a position that saw him relocate from sub-Saharan Africa to Amsterdam. Ndegwa, in other words, has joined the ranks of the global corporate elite, which has ballooned since the late 1970s. This has been a period which has seen the exacerbation of inequalities that had only just begun to be ameliorated in the years following the Second World War and through the early years of independence in Kenya. But the fates of someone like Ndegwa cannot be understood outside of the trajectory of women and men like Peninah.

Born in 1987 to Kikuyu-speakers in Kiambu, Peninah completed high school and moved to Nairobi to secure a better future. There she enrolled in college, gaining a degree in hospitality. Working at a restaurant on Thika Road, she met Maina, with whom she had a child in 2012. Concerned about fulfilling her new responsibilities as a mother, Peninah quit her job and began working as an M-Pesa agent for Safaricom, the firm that Ndegwa now leads. If the trajectory of men like Wakaria and Ndegwa have been relatively exceptional, Peninah's own trajectory has been more common for the bulk of Kenya's population. As elsewhere, Kenya's unwaged economy, already large at independence, grew significantly over the course of the 1970s and 1980s, further ballooning following the enforced implementation of Structural Adjustment Policies (SAPs) in the 1990s.[14] While this afforded an opening for Wakaria, for the majority of Kenya's growing population life got harder. With a greater and greater proportion of the population operating outside of wage relations, the nature of exploitation also changed in these years.

Throughout the colonial period, Kenya's economy depended on women's socially reproductive work to offset the low wages paid to men. Today, and in the context of M-Pesa agents and digital financial services more broadly, these dynamics have been revolutionized. As explored in chapter 6, while Peninah's work within the system is exploited, the social infrastructures required to make Safaricom's network hang together is expropriated. Following Nancy Fraser, this work is "treated as costless in capital's accounts, [and] it is expropriated without compensation or replenishment and implicitly assumed to be infinite."[15] The boundaries between production and reproduction continue to blur. And it is precisely this blurring that offers a better accounting of what firms like Safaricom mean when they refer to locating value at the "bottom of the pyramid."

If we are to write histories of capitalism's present and the types of inequalities that it shores up, particularly in the face of claims to ethical capitalism—of which Safaricom, Peninah's company, is a key proponent—we must see both the present and the past differently. We must understand people like Peninah— "Human ATMs"—not simply as an effect of these infrastructural dynamics, but as, in fact, constituent to making these networks work, not simply as laboring bodies but as knowledge producers. Their work, like that of radio experts such as Katele and Wakaria, and the unnamed experts of topography whose knowledge shaped early road infrastructures, are all forms of expertise.

There are critical reasons to take this expertise and its routine devaluation seriously. James Ferguson has suggested that we need to reframe our understanding of ethics and politics if we are to take stock of an ever more ubiquitous condition. He writes:

> It is common to suppose that people make their livelihoods by being "productive"–that is, by producing, via their labor, goods and services that either meet their own needs. . . . or meet the needs of others in such a way that either their goods can be marketed . . . or their labor can be sold . . . In fact that has never been an adequate account . . . [But] recent developments in the spatial and social organization of production . . . [mean] that [the] commonsense linkage of livelihood with production [is] under considerable strain.[16]

The bulk of the population living on the African continent, he argues, is today "simply left out of the global production regime." Ferguson is not alone in making this case. Mike Davis, too, argued that much of the African continent is unlikely to be incorporated into relations of capitalist production.[17] For both authors, formal waged employment is likely to remain stagnant or decline. Writing from the perspective of South Africa, Ferguson argues that, given this dispensation, universal basic income grants offer one solution.

The conditions obtaining in South Africa are not altogether different from those in Kenya. In both places, population growth has outstripped waged employment, seeing more and more people work in the so-called informal sector. Except that in Kenya, as across much of the continent, it is unlikely that the state—a state that has rarely provisioned for its people—will intervene and "give a man a fish," in Ferguson's formulation. Moreover, we foreclose our understanding of the contemporary moment by seeing capitalist incorporation solely through the lens of waged exploitation. And these differences matter to the types of politics, the types of ethics, and the modalities of redistribution for which we advocate. Indeed, as the example of Safaricom suggests, while people are not necessarily included in today's formal wage labor economies, their activity and the data Safaricom's knowledge apparatus captures is generating wealth. As people's everyday strategies, along with the daily practices and expertise of infrastructural workers, are captured, they become the foundations on which new products are designed and new markets are assembled. This work is being mobilized by companies such as Safaricom (but in earlier iterations by the IBEA), and new modes of accumulation are sometimes the result. If we are to generate new critiques of contemporary capitalism on the continent, we need to attend to these new modes of capture *and* their histories. We need to move away from a frame that naturalizes "people as infrastructure" in order to consider the particular arrangements of capital and work, of markets and governance, that have demanded that people act infrastructurally, and we must place the transformative effects of that work front and center.

This perspective is critical for doing justice to the histories of this work and these experts. It is also essential to understanding the durable inequalities that shape our global present, one wherein private capital increasingly dictates the shape of the state, one wherein promised prosperity often looks like prolonged precarity, and one wherein the "informalization" of labor and the appropriation of everyday knowledge practices threaten to devalue work as either rote labor, or simply "making do," all the while capitalizing on the human expertise that sits at the center of the austere infrastructural state.

Notes

PREFACE

1   Hecht, "Rupture-Talk."

INTRODUCTION

1   This as measured by market capitalization. "The East Africa, Mauritius &
    Seychelles Top 30 Companies," African Financials, September 2022, accessed
    18 June 2023, https://images.africanfinancials.com/83616f9a-2022-09-east-africa
    -mauritius-top-30.pdf.
2   Fieldnotes, 7 August 2014.
3   Prahalad, *Fortune at the Bottom of the Pyramid*.
4   "Safaricom PLC Annual Report and Financial Statements," Safaricom, 2022,
    https://www.safaricom.co.ke/investor-relations-landing/reports/annual-reports.
5   Breckenridge, "Failure of the 'Single Source,'" 105.
6   This literature is vast. For some of the sources I'm thinking with, see Mamdani,
    Citizen and Subject. For Kenya on ethnicity, see Peterson, *Creative Writing*; MacArthur,
    Cartography and the Political Imagination; Berman and Lonsdale, *Unhappy Valley, Book
    One*; Berman, *Control & Crisis in Colonial Kenya*. On labor and class formation, see
    Cooper, *On the African Waterfront*. For broader accounts on ethnicity and indirect
    rule, see Ranger, "Invention of Tradition Revisited." For a response to Terence
    Ranger, see Spear, "Neo-Traditionalism and the Limits."
7   Bear, *Navigating Austerity*; Appel, *Licit Life of Capitalism*.
8   Fraser and Jaeggi, *Capitalism*; Fraser, "Behind Marx's Hidden Abode"; Wood, "Sepa-
    ration of the Economic and the Political"; Phillips-Fein, *Fear City*; Blyth, *Austerity*.
9   Ferguson, "Uses of Neoliberalism"; Capotescu, Sanchez-Sibony, and Teixeira, "Aus-
    terity without Neoliberals."
10 For exceptions, see Von Schnitzler, *Democracy's Infrastructure*; Brownell, *Gone to
    Ground*; Appel, *Licit Life of Capitalism*. For an excellent example outside of Africa,
    see Goswami, *Producing India*, especially chapter 1.

11 Special thanks to Matt Shutzer for helping me develop and sharpen this framework.

12 Bruce Berman was among the first to point to this contradiction, arguing that the colonial state in Kenya struggled to mediate between settlers, metropolitan capital, and the needs of the black African majority, as thin as officials argued these needs were. This book shares these concerns, but uses an exploration of the corporate-state as one site where the contradictions of colonial statecraft were manifest. Berman, *Control & Crisis in Colonial Kenya*.

13 Maine, *Sir Henry Maine*, 322.

14 Ciepley, "Beyond Public and Private."

15 Guyer, *Legacies, Logics, Logistics*, 157; cf. Mann, "Autonomous Power of the State."

16 Offner, *Sorting Out the Mixed Economy*.

17 For various takes on this, see Fraser and Jaeggi, *Capitalism*; Blyth, *Austerity*, 99; Wood, "Separation of the Economic and the Political"; Fraser, "Behind Marx's Hidden Abode."

18 In this, I join a growing body of literature on African technological and infrastructural expertise. See D'Avignon, *Ritual Geology*; Grace, *African Motors*; Hart, *Ghana on the Go*.

19 Grace, *African Motors*; Hart, *Ghana on the Go*; D'Avignon, *Ritual Geology*; Osseo-Asare, *Bitter Roots*; Twagira, "Introduction"; Mavhunga, *Mobile Workshop*; Mavhunga, *Science, Technology, and Innovation*; Tilley, *Africa as a Living Laboratory Empire*; Osseo-Asare, *Bitter Roots*; Jacobs, *Birders of Africa*; Mutongi, *Matatu*; Mika, *Africanizing Oncology*.

20 Goswami, *Producing India*, 33.

21 Apter, *Pan-African Nation*, 37.

22 Larkin, *Signal and Noise*; Barry, *Material Politics*.

23 For accounts that attend to the vernacularization of development, see Ferguson, "Declarations of Dependence"; Karp, "Development and Personhood"; Shipton, *Credit between Cultures*; Smith, *Bewitching Development*.

24 Larkin, "Politics and Poetics of Infrastructure"; Larkin, *Signal and Noise*; Von Schnitzler, *Democracy's Infrastructure*.

25 Mukerji, *Impossible Engineering*; Edwards, "Infrastructure and Modernity."

26 Schivelbusch, *Railway Journey*; Cronon, *Nature's Metropolis*; c.f. Collier, *Post-Soviet Social*.

27 Bagwell, *Transport Revolution*, 80–83.

28 Luxemburg, *Accumulation of Capital*; Cronon, *Nature's Metropolis*; Bagwell, *Transport Revolution*, 80–87; David Harvey, *The Limits to Capital*; Utsa Patnaik and Prabhat Patnaik, *Capital and Imperialism*.

29 The language of "state space" is drawn from Manu Goswami. However, it is worth noting that Goswami is centrally concerned with highlighting how creating "state space" in India critically turned on producing and reproducing the uneven and combined development that characterized the British Empire. Goswami, *Producing India*, 53; Hughes, *Networks of Power*; Mann, "Autonomous Power of the State."

30 Harvey, *Limits to Capital*; Klein, *Shock Doctrine*; Phillips-Fein, *Fear City*; Blyth, *Austerity*; Brown, *Undoing the Demos*; Ong, *Neoliberalism as Exception*; Bear, *Navigating Austerity*. Mattei offers an alternative to the standard chronology, arguing that aus-

terity is a tactic of statecraft that first emerged in Europe—specifically Italy and the United Kingdom—to resolve capitalism's endemic crisis tendency in the interwar period. For Mattei, austerity functions to discipline the working class and insulate capitalism from the popular demands of working people. See Mattei, *Capital Order*.

31 For exceptions, see Park et al., "Introduction"; Machava, "Reeducation Camps"; Quarshie, "Psychiatry on a Shoestring."

32 Piketty, *Capital*.

33 Piketty, *Capital*.

34 Knight and Stewart, "Ethnographies of Austerity," 2.

35 Connolly, "White Story."

36 Connolly, "White Story"; see also Harvey, *Brief History of Neoliberalism*, 11.

37 Rodney, *How Europe Underdeveloped Africa*; Amin, *Unequal Development*; Wallerstein, *Capitalist World-System*; Wallerstein, *World-Systems Analysis*; Arrighi, *Long Twentieth Century*.

38 While I share many concerns of this work, I understand austerity to be analytically and politically distinct from what Seikaly refers to as "scarcity." See especially the introduction of Seikaly, *Men of Capital*.

39 In coming to this conclusion, reading in South Asian and Middle Eastern history has been formative. See Goswami, *Producing India*; Jakes, *Egypt's Occupation*; Chaudhuri, "Lives of Value."

40 Goswami, *Producing India*.

41 Goswami, *Producing India*. The word *lumpy* better describes processes of globalization, emphasizing the necessary unevenness missed by the language of "flows," and "scapes," which suggests globalization has entailed homogenization and equal integration. See Hecht, *Being Nuclear*; cf. Appadurai, *Modernity at Large*.

42 Hecht, "Interscalar Vehicles."

43 Von Schnitzler, *Democracy's Infrastructure*, 5.

44 Berry, "Hegemony on a Shoestring." This argument shares an affinity with Berry's, but takes the argument in different directions. Whereas Berry emphasizes how policies of fiscal constraint required the use of "native authorities," I emphasize how policies of colonial austerity made the state dependent on private enterprise to enact infrastructures on the one hand, and relied on the often unpaid contributions of African knowledge-workers and experts on the other.

45 Mitchell, *Rule of Experts*.

46 "The Policy of Creating Reigning Companies," *The Economist*, September 15, 1888.

47 As d'Avignon shows based on her research in French West Africa, the colonial and postcolonial state has routinely contrasted "artisanal" mining with industrialized forms of extraction, often criminalizing the former despite the fact that industrial mining firms were reliant on Africans' knowledge of the subsoils. See d'Avignon, *Ritual Geology*.

48 The language of "milieu" is borrowed from Hunt, *Nervous State*.

49 Bear, *Navigating Austerity*, 31.

50 Haraway, *Simians, Cyborgs, and Women*, 249n7; Mackenzie, *Material Markets*. In the wake of Donna Haraway's provocative essay "The Cyborg Manifesto: Science, Technology, and Socialist Feminism in the 1980's," scholars working in a range of

disciplines have taken up the concept of "the prosthetic." Within disability studies, the proliferation of the term has come under scrutiny. At the center of these critiques is the sense that the prosthetically enhanced body has been reduced to mere metaphor, ignoring the concrete materiality of lives lived alongside and through prosthetics. For my purposes most interesting in the critiques of the figure of the cyborg–the figure of the prosthetically mediated body–is that it has rarely been used to engage in materialist critique, Haraway's initial intent notwithstanding. The approach taken here insists on the importance of the materiality of prosthetics and prosthetic work, in pursuit of a critique of capitalism's uneven, extractive, and expropriative tendencies. See Mitchell and Snyder, eds., *The Body and Physical Difference: Discourses of Disability*; Smith and Morra, eds., *The Prosthetic Impulse: From a Posthuman Present to a Biocultural Future*; Alison Kafer, *Feminist, Queer, Crip,* especially chapter 6.

51 The language of "hang together" is taken from Mol, *The Body Multiple.*

52 British National Archives [BNA] Colonial Office, "Broadcasting Cheap Medium-Wave Receivers"; Bindra, *Peculiar Kenyan*; Moorman, *Powerful Frequencies.*

53 For critiques, see Serlin, "Confronting African Histories of Technology"; Kriger, *Pride of Men*; Mavhunga, *Mobile Workshop*; Mavhunga, *Science, Technology, and Innovation.*

54 Maurer, Nelms, and Rea, "Bridges to Cash."

55 Hunt, *Colonial Lexicon.*

56 Prahalad, *Fortune at the Bottom of the Pyramid.*

57 Simone, "People as Infrastructure." I would like to thank Gabrielle Hecht for impressing on me the importance of this distinction.

58 My approach shares the concerns articulated in Fredericks, *Garbage Citizenship.*

59 Cooper, "Concept of Globalization," 190.

60 Von Schnitzler, *Democracy's Infrastructure,* 15; Collier, *Post-Soviet Social*; Elyachar, "Phatic Labor."

61 Grace, *African Motors,* 23-4. For a discussion of the shaping role of infrastructural incompleteness in Kenya, see Prince K Guma, "Incompleteness of Urban Infrastructures in Transition."

62 On the shaping capacity of what Kimari and Ernston refer to as "imperial remains," see Kimari and Ernston, "Imperial Remains and Imperial Invitations."

63 Berman, *Control & Crisis in Colonial Kenya,* 281.

64 Poggiali, "Digital Futures and Analogue Pasts?"

65 Edwards, "Infrastructure and Modernity."

66 Welker, *Enacting the Corporation,* 4.

CHAPTER 1. A DIVISIBLE SOVEREIGNTY

1 McDermott, *British East Africa,* 399.

2 Livingstone often referred to "paths," rather than "roads," when he discussed the relationship between infrastructures and "commerce and civilisation." Livingstone, *Cambridge Lectures,* 21.

3 Hunt, *A Colonial Lexicon.*

4 De Kiewiet, "History of the Imperial British East Africa Company," 17; Munro, *Maritime Enterprise and Empire*, 51.

5 Munro, *Maritime Enterprise and Empire*, 52.

6 De Kiewiet, "History of the Imperial British East Africa Company," 17.

7 Munro, *Maritime Enterprise and Empire*, 201; De Kiewiet, "History of the Imperial British East Africa Company." I follow Gabrielle Hecht's formulation of technopolitics, particularly the more expansive definition laid out in the introduction to *Entangled Geographies*, which emphasizes the unintended power effects of technopolitical assemblages.

8 Like many in the corporate world of the nineteenth century, Mackinnon's networks were not limited to commercial realms, but embraced sovereigns and those holding government posts. It was through such boundary-crossing work that he developed an intimate relationship with Belgium's King Leopold, who appealed to Mackinnon to aid in the construction of a pair of "international highway[s]," one running through the interior to Lake Nyasa and the other to Lake Victoria. In pursuit of these lofty designs, in 1876 a group of businessmen and other self-styled "philanthropists" and "humanitarians" hoping to "open up the region" met in Glasgow to discuss their dreams of infrastructural connectivity. See John Kirk to Foreign Office, 13 December 1876, British National Archives [hereafter BNA] Foreign Office [hereafter FO] 84/1454, as cited in De Kiewiet, "History of the Imperial British East Africa Company," 21. See also Munro, *Maritime Enterprise and Empire*, 194; Galbraith, *Mackinnon and East Africa*, 5.

9 First deployed by abolitionists, the concept of "legitimate commerce" was used to convince British audiences of the need to put an end to the "illegitimate" trade in enslaved people, which would be replaced by lucrative, but "legitimate," commerce. See Brown, "Origins of 'Legitimate Commerce.'"

10 Draft of the Contract to Be Entered into between Sultan of Zanzibar and Company to Be Formed for the Administration of His Domain, n.d., file 8, box 65, PP MS 1 IBEA/Mackinnon Papers, SOAS.

11 William Mackinnon to Barghash bin Said, 10 January 1878, file 60, box 77, PP MS 1 IBEA/Mackinnon Papers, SOAS.

12 McDermott, *British East Africa*, 263. The framework for the concession was initially proposed in an 1877 document; however, negotiations foundered due to pressure from the Sultan's Arab subjects and the disinterest of the Foreign Office. The document was blandly titled the "Concession Given by the Sultan of Zanzibar to the British East African Association." The 1887 Concession was followed by a secondary concession in 1888, and a third in 1889. Galbraith, *Mackinnon and East Africa*, 56, 127.

13 Trivedi, "Role of Imperial British East Africa Company," 618; "Incident in Uganda," *The Economist*, June 4, 1892.

14 McDermott, *British East Africa*, 282; Munro, *Maritime Enterprise and Empire*, 408.

15 Campbell, "Sound Finance," 9.

16 Draft Contract between Sultan and Company; "The East African Chartered Company," *The Economist*, October 3, 1891; Austen, *African Economic History*, 124.

17 Ciepley, "Beyond Public and Private."

18 Fraser, "Behind Marx's Hidden Abode," 60.

19 McDermott, *British East Africa*, 208.

20 As legal studies scholar Joshua Barkan writes, "any attempt to approach corporate power through the dichotomies of state and economy or public or private misses the ways that both corporate and state power emerge at, construct, continually transgress, and occasionally consolidate the blurred boundaries of these social spheres." See Barkan, *Corporate Sovereignty*, 113.

21 On this balancing act, see Stern, *Company-State*, 13.

22 As Barkan argues, corporations like the IBEA sought to mediate these tensions, gaining "legal recognition by arguing that their status . . . [would benefit] the common good," a commitment the IBEA tried to deliver through the financing and construction of roads. Roads, then, were not simply routes of commerce; they were both the material foundation subtending an emergent economic order and the material basis on which civilizational "uplift" depended. Barkan, *Corporate Sovereignty*, 8; Munro, *Maritime Enterprise and Empire*, 190–91; Galbraith, *Mackinnon and East Africa*, 54–55.

23 Cooper, "Alternatives to Empire," 107. On what she refers to as "degrees of imperial sovereignty," see Stoler, Duress.

24 While the literature on early colonial states in eastern Africa note in passing the central role played by corporations in laying the groundwork for formal colonial occupation, here I argue that though the IBEA's tenure as a corporate-state was short-lived, its legacy was durable. As we will see in this chapter and the one to follow, the tactics of infrastructural governance established by the IBEA formed the basic structure of colonial statecraft. This chapter, then, offers one origin story that anchors the infrastructural tales narrated in the sections to follow. Glassman, *Feasts and Riot*; Reid, "Africa's Revolutionary Nineteenth Century"; Berman, *Control & Crisis in Colonial Kenya*; Austen, *African Economic History*; Osborne, *Ethnicity and Empire in Kenya*; Ambler, *Kenyan Communities in the Age of Imperialism*. There is a re-emergent interest in the role joint-stock companies played in nineteenth-century imperial expansion. See Press, *Rogue Empires*.

25 Trivedi, "Role of Imperial British East Africa Company," 617; McDermott, *British East Africa*, 225

26 This was particularly the case in 1862 when Britain intervened in a succession dispute that saw the Sultan of Zanzibar, Sayyid Majid bin Said Al-Busaidi, assert independence from Oman. Gjerso, "Scramble for East Africa," 833.

27 Goswami, *Call of the Sea*, 133; Bishara, *Sea of Debt*; Trivedi, "Role of Imperial British East Africa Company," 616.

28 Goswami, *Call of the Sea*, 165.

29 McDermott, *British East Africa*, 3–4; Glassman, *Feasts and Riot*, 178.

30 Brennan, Burton, and Lawi, *Dar Es Salaam*, 19.

31 Trivedi, "Role of Imperial British East Africa Company," 618.

32 Musson, "Great Depression in Britain," 208. As economist A. E. Musson has argued, nineteenth-century "'imperialism' was to some extent a product of the Great Depression." Musson, "Great Depression in Britain," 228.

33 Musson, "Great Depression in Britain," 223.

34 F. D. Lugard, "The Extension of British Influence (and Trade) in Africa," *Proceedings of the Royal Colonial Institute* XXVII (1895–96), 6, as cited in Galbraith, *Mackinnon and*

East Africa, 8. Lugard was not alone in making this assessment. Critics of imperial expansion, ranging from J. A. Hobson to Vladimir Lenin to Rosa Luxemburg to W. E. B. Du Bois, understood imperial expansion in the nineteenth century as a mechanism to absorb idle capital in overseas territories, while creating new markets for manufactured goods. See Hobson, *Imperialism*; Lenin, *Imperialism*; Luxemburg, *Accumulation of Capital*; Du Bois, *Problem of the Color Line*.

35   Oliver, "British Occupation of East Africa," 56.

36   McDermott, *British East Africa*, 287–88.

37   Oliver, "British Occupation of East Africa," 56.

38   "The East African Chartered Company," *The Economist*, October 3, 1891.

39   Munro, *Maritime Enterprise and Empire*, 422.

40   Galbraith, *Mackinnon and East Africa*, 32.

41   Munro, *Maritime Enterprise and Empire*, 183–84. During his tenure, Kirk had gained considerable influence over Barghash bin Said, routinely intervening in the sovereign business of the Sultan—one of Kirk's major "successes" in this regard was convincing the Sultan to proclaim a series of anti-slavery decrees. De Kiewiet, "History of the Imperial British East Africa Company," 7.

42   Galbraith, *Mackinnon and East Africa*, 41.

43   De Kiewiet, "History of the Imperial British East Africa Company," 102.

44   Trivedi, "Role of Imperial British East Africa Company," 619.

45   Draft Contract between Sultan and Company.

46   Lugard, *Dual Mandate in British Tropical Africa*, 14.

47   "The East African Chartered Company," *The Economist*, October 3, 1891. The readership of *The Economist* were largely supporters of the Liberal party with an interest in seeing the expansion of British capital abroad. As Alexander Zevin writes: "The paper circulated . . . among the most powerful class of Victorian and Edwardian savers, the 'gentlemanly capitalists' clustered in south-east England, who made their livings in finance, banking, trade, and shipping, or as politicians, administrators, and landowners, and showed a marked preference for income derived from safe overseas assets like railway and government securities. They turned to the *Economist* not for news in the narrow sense but for political analysis to help them evaluate the risks and rewards of placing capital abroad." Zevin, *Liberalism at Large*, 120.

48   The first subscription, however, was not open to the public at large. As De Kiewiet writes, "An agreement of April 18, 1888 set forth the terms by which the IBEA Company was financed and directed. The nominal capital was set at one million pounds. The first issue was limited to 250 000 pounds in 100 pound shares. The subscribers to this issue, which was not open to the public, received special voting privileges." De Kiewiet, "History of the Imperial British East Africa Company," 99.

49   De Kiewiet, "History of the Imperial British East Africa Company," 36.

50   Trivedi, "Role of Imperial British East Africa Company," 619; The company was oversubscribed within two months of the first issue. Galbraith, *Mackinnon and East Africa*, 138.

51   Draft Contract between Sultan and Company.

52   "The Policy of Creating Reigning Companies," *The Economist*, September 15, 1888.

53   "The Danger of Trouble in East Africa," *The Economist*, April 19, 1890.

54 "The Danger of Trouble in East Africa."

55 George S. Mackenzie, "Evacuation of Uganda, Opinion of the Company," 28 September 1892, file 46, box 71, PP MS 1 IBEA/Mackinnon Papers, SOAS.

56 McDermott, *British East Africa*, 208.

57 Goswami, *Call of the Sea*, 168; Lugard, *Rise of Our East African Empire*, 325.

58 Cooper, "Alternatives to Empire," 94–123.

59 Meier, *Swahili Port Cities*, 103.

60 Sheriff, *Slaves, Spices, & Ivory*, 127.

61 McDermott, *British East Africa*, 3; Draft Contract between Sultan and Company.

62 Coupland, *Exploitation of East Africa*, 272.

63 The British secured Egypt's withdrawal. But the invasion clarified the vulnerability of the Sultanate both to the Sultan himself and to Kirk. Coupland, *Exploitation of East Africa*, 293.

64 As cited in Coupland, *Exploitation of East Africa*, 305.

65 De Kiewiet, "History of the Imperial British East Africa Company," 34. Lord Salisbury routinely opposed these labors, hoping to protect South Asian trade interests. By the time Salisbury became Foreign Minister in 1885, his position had changed, however. The treasury and taxpayers were not prepared to shoulder the costs of annexation. Given this, delegating this work to the IBEA seemed the most sensible path forward.

66 Draft Contract between Sultan and Company (emphasis mine). Some of the provisions of this document contravened the rights of other powers. Notably, the maintenance of a fixed import duty to citizens of France, Britain, Germany, and the United States. There were other stipulations. The IBEA could levy tolls on quays and jetties, but it could not prevent competitors from landing goods on the beach. Similarly, while the IBEA could levy tolls on roads it constructed, it could not stop people from using the existing routes that crisscrossed the territory. Given this, the IBEA moved quickly to build company roads and establish them as recognized tracks. Galbraith, *Mackinnon and East Africa*, 57–59.

67 John Kirk to Gerald Waller, 12 May 1878, file 60, box 77, PP MS 1 IBEA/Mackinnon Papers, SOAS.

68 John Kirk to Rev. P. Badger, 15 May 1878, file 60, box 77, PP MS 1 IBEA/Mackinnon Papers, SOAS.

69 John Kirk to Gerald Waller, 1 May 1879, file 60, box 77, PP MS 1 IBEA/Mackinnon Papers, SOAS.

70 Gerald Waller to Sayyid Barghash bin Said, Sultan of Zanzibar through Badger, Zanzibar, 11 May 1878, file 60, box 77, PP MS 1 IBEA/Mackinnon Papers, SOAS.

71 "Case to Advise on Behalf of the Imperial British East Africa Company," [1893?], file 49, box 72, PP MS 1 IBEA/Mackinnon Papers, SOAS.

72 As De Kiewiet writes: "Although the Company received powers of taxation for the coast under the Zanzibar Concession and for the interior under the Royal Charter, in neither area was it found possible to put the power into effect. In the interior where there was no system of money exchange and where the tribes were scattered over large areas, taxation was hardly feasible . . . At the coast the Zanzibar commercial treaties prevented the Company from taxing foreigners, and since

foreigners—or specifically South Asian traders—were the most eminently taxable group on the coast, the Company was effectually prevented from taxing at all." De Kiewiet, "History of the Imperial British East Africa Company," 228–29. See also McDermott, *British East Africa*, 246.

73 Gerald Waller to John Kirk, 11 May 1878, file 60, box 77, PP MS 1 IBEA/Mackinnon Papers, SOAS.

74 "Concession Given by the Sultan of Zanzibar to the British East African Association," 24 May 1887, reproduced in McDermott, *British East Africa*.

75 Galbraith, *Mackinnon and East Africa*, 150.

76 McDermott, *British East Africa*, 232; Mackenzie to the Secretary of the IBEA. 14 November 1888, file 1A, box 63, "Mombasa Letters, Sept 1888-Jan 1889," PP MS 1 IBEA/ Mackinnon Papers, SOAS.

77 McDermott, *British East Africa*, 232.

78 "Appendices to 1893 Shareholders Agreement," 29 May 1893, file 50, box 72, PP MS 1 IBEA/Mackinnon Papers, SOAS; "Case to Advise on Behalf of IBEA"; "Copy Terms of Concession of Northern Ports Submitted to Sayyid Khalifa," January 1889, file 1A, box 63, "Mombasa Letters, Sept 1888-Jan 1889," PP MS 1 IBEA/Mackinnon Papers, SOAS.

79 "Deed of Settlement of the Imperial British East Africa Company, 1889," 6 June 1889, file 51, box, 72, PP MS 1 IBEA/Mackinnon Papers, SOAS. The 1888 Concession extended the terms of the agreement "in perpetuity." Marking the incremental expansion of the sovereign authority of the Company, the IBEA gained the coastline from Wanga to Kipini. This was followed by the further expansion of company rule in 1889, which extended the territories to include Lamu and the northern islands and ports. In keeping with historical relations of indebtedness and deference, the Sultan also secured the right to take out an interest-free loan of up to £50,000 from the Company. Coupland, *Exploitation of East Africa*, 311.

80 Draft Contract between Sultan and Company.

81 Goswami, *Call of the Sea*, 179.

82 Bishara, *Sea of Debt*, 51–52.

83 Bishara, *Sea of Debt*, 53.

84 Goswami, *Call of the Sea*, 194; Sheriff, *Slaves, Spices, & Ivory*, 109.

85 Bishara, *Sea of Debt*, 52.

86 "Case to Advise on Behalf of IBEA."

87 Draft Contract between Sultan and Company.

88 "Memorandum: The Road from Dar-es-Salaam East Africa, towards the Northern End of Lake Nyanza," 20 November 1878, file 61, box 77, PP MS 1 IBEA/Mackinnon Papers, SOAS; Berman, *Control & Crisis in Colonial Kenya*, 50.

89 Draft Contract between Sultan and Company.

90 McDermott, *British East Africa*, 212. "Development" here had both economic and "civilizational" valences. Indeed, the IBEA gained its charter on the promise that the creation of new road networks would aid in the fight to end the slave trade, with newly cut roads bisecting the existing caravan routes along which slave raiders marched slaves to the coast. See Coupland, *Exploitation of East Africa*, 302.

91 "Draft of Proposed Concessions to Be Obtained from His Highness the Sultan of Zanzibar," n.d., file 2, box 64, PP MS I IBEA/Mackinnon Papers, SOAS; Galbraith, *Mackinnon and East Africa*, 48; On Barghash bin Said's impressive public works projects, see Prestholdt, *Domesticating the World*; Meier, *Swahili Port Cities*, 108.

92 "Draft of Proposed Concessions."

93 Barghash Bin Said to William MacKinnon, 3 May 1879, file 61, box 77, PP MS I IBEA/Mackinnon Papers, SOAS. See also Munro, *Maritime Enterprise and Empire*, 196.

94 Kirk to FO, 11 January 1876, BNA FO 84/1452, as cited in De Kiewiet, "History of the Imperial British East Africa Company," 11.

95 Cummings, "Aspects of Human Porterage," 128. Human porterage was an adaptation to environmental features specific to the region. While rich in flora and fauna, the region was scarce in networks of communication. By contrast to Central and West Africa, the area did not boast navigable rivers. The presence of the tsetse fly, moreover, made the use of pack animals for transport impracticable. It was thus left to human communities to facilitate trade, leading to the development of the porterage system. Rockel, *Carriers of Culture*, 4.

96 Cummings, "Aspects of Human Porterage," 104.

97 Cummings, "Aspects of Human Porterage," 193–94.

98 Cummings, "Aspects of Human Porterage," 125.

99 Cummings, "Aspects of Human Porterage," 136.

100 Cummings, "Aspects of Human Porterage," 105.

101 Maxon, *John Ainsworth*, 1, 11.

102 John Ainsworth to Chairman, Kenya Land Enquiry Commission, Nairobi, September 1931, Mss.Afr.S. 381 (2), Bodleian Library, University of Oxford. Similar "neutral zone[s]" were noted elsewhere in the territory, Reverend H.W. Innis reporting that Nandi and Luo people "were afraid to enter" one such corridor in Kavirondo. Watkins, *Report of the Kenya Land Commission*, 1132.

103 Ainsworth to Chairman, September 1931.

104 See Coronil, *Magical State*, 21–66.

105 Ainsworth to Chairman, September 1931.

106 George Wilson to Secretary, EA Scottish Mission, 23 August 1892, file 41, box 70, PP MS I IBEA/Mackinnon Papers, SOAS.

107 Wilson to EA Scottish Mission, 23 August 1892.

108 For the construction of the identity of "the Kamba" through their labor over the *longue duree*, see Osborne, *Ethnicity and Empire in Kenya*; for a discussion of Kikuyu "ethnicity" prior to 1900, see Muriuki, *History of the Kikuyu*; for a discussion of the case of Kikuyu self-fashioning as a "tribe," see Lonsdale, "Moral Economy of Mau Mau."

109 Ainsworth to Chairman, September 1931.

110 "Proclamation from Khalifa," 10 October 1888, BNA FO 84/1910, as cited in De Kiewiet, "History of the Imperial British East Africa Company," 121.

111 IBEA to Euan Smith and MB MS Agent and Consul General, 18 October 1888, file 1A, box 63, "Mombasa Letters, Sept 1888-Jan 1889," PP MS I IBEA/Mackinnon Papers, SOAS.

112 Watkins, *Report of the Kenya Land Commission*, 93.

113 Cummings, "Aspects of Human Porterage," 222.

114 Watkins, *Report of the Kenya Land Commission*, 388.

115 F. Moir to W. Mackinnon, 3 December 1877, file 9, box 65, PP MS 1 IBEA/Mackin-
     non Papers, SOAS. One 1888 report happily recounted having successfully negoti-
     ated a treaty with Sheikh Mbaruk. As he was a "man of power," this agreement
     conferred popular "confidence in our friendship and power," having a "beneficial
     effect on the Native mind." However, men of the company were well aware that
     their authority was contingent on the continued trust of the Sultan who remained
     the dominant authority in negotiations. As these matters would have to be "deli-
     cately . . . arranged through the Sultan at Zanzibar," Mbaruk was advised to await
     notice of the Sultan's approval. These cautious negotiations bore fruit for Mbaruk,
     who was subsequently employed by the company. Getting the Sultan to agree
     to these arrangements was likely a difficult task, as Mbaruk frequently revolted
     against Barghash's authority. Mackenzie to President and Court of Directors of
     IBEA, 1 December 1888, file 1A, box 63, "Mombasa Letters, Sept 1888-Jan 1889," PP
     MS 1 IBEA/Mackinnon Papers, SOAS. Galbraith, *Mackinnon and East Africa*, 135.

116 Letter dated 13 July 1877, file 3, box 64, PP MS 1 IBEA/Mackinnon Papers, SOAS.

117 "Extract of Letter Dated," Zanzibar, 24 August 1877, file 3, box 64, PP MS 1 IBEA/
     Mackinnon Papers, SOAS.

118 "Extract of Letter Dated," 24 August 1877.

119 [Illegible] to John Kirk, 9 January 1877, file 1C, box 64, PP MS 1 IBEA/Mackinnon
     Papers, SOAS. See also Glassman, *Feasts and Riot*, 182.

120 Letter dated 13 July 1877, file 3, box 64, PP MS 1 IBEA/Mackinnon Papers, SOAS.
     The threat of violence was not unusual, with company officials routinely bran-
     dishing their weapons to enforce their will. W. Moyes to William Mackinnon,
     2 October 1877, file 7, box 65, PP MS 1 IBEA/Mackinnon Papers, SOAS.

121 Beardall to Mackinnon, "Report of Progress of Work on the East African Road
     during the Past 4 Weeks," 16 August 1880, file 11, box 66, PP MS 1 IBEA/Mackinnon
     Papers, SOAS.

122 Mackenzie to Secretary of IBEA, 24 October 1888, file 1A, box, 63, "Mombasa Let-
     ters, Sept 1888-Jan 1889," PP MS 1 IBEA/Mackinnon Papers, SOAS.

123 For a detailed discussion, see Prestholdt, *Domesticating the World*.

124 Bishara, *Sea of Debt*, 52.

125 Glassman, *Feasts and Riot*, 242–44.

126 Mackenzie to Secretary of IBEA, 15 January 1889, file 1A, box 63, "Mombasa Letters,
     Sept 1888-Jan 1889," PP MS 1 IBEA/Mackinnon Papers, SOAS.

127 For competing visions of sovereignty on Zanzibar's Swahili Coast, see Brennan,
     "Lowering the Sultan's Flag."

128 W. Beardall to William MacKinnon, 14 August 1879, file 61, box 77, PP MS 1 IBEA/
     Mackinnon Papers, SOAS; W. Mayes to Kirk, September 1877, file 4, box 64, PP MS
     1 IBEA/Mackinnon Papers, SOAS.

129 N.A., 13 July 1877, file 3, box 64, PP MS 1 IBEA/Mackinnon Papers, SOAS.

130 MacKenzie to Lt. Swayne RE, 11 October 1888, file 1A, box 63, "Mombasa Letters,
     Sept 1888-Jan 1889," PP MS 1 IBEA/Mackinnon Papers, SOAS.

131 Farler, "Native Routes in East Africa," 732. See also Goswami, *Call of the Sea*, 142;
     Rockel, *Carriers of Culture*.

132 [Illegible] to Kirk, 9 January 1877.

133 Mackenzie to Sec. of IBEA, 4 January 1889, file 1A, box 63, "Mombasa Letters, Sept 1888-Jan 1889," PP MS 1 IBEA/Mackinnon Papers, SOAS; Moir to Mackinnon, 3 December 1877.

134 Lugard, *Rise of Our East African Empire*, 274.

135 J. Sangard to Sir William Mackinnon, Dagoretti, 20 October 1890, file 24, box 68, PP MS 1 IBEA/Mackinnon Papers, SOAS. Nor were blood brotherhood rites generic, with Lugard reporting that the "method of making blood-brotherhood varies slightly among the various tribes." In some instances, the blood of those party to a compact was first mixed and then placed on two small pieces of meat, which the parties then consumed. In other instances, coffee berries and salt took the place of meat. Lugard, *Rise of Our East African Empire*, 330-31.

136 Cummings, "Aspects of Human Porterage," 183.

137 For a discussion of the emergence of "local sultans" in the seams of Barghash bin Said's authority, see Goswami, *Call of the Sea*, 168.

138 Moir to Mackinnon, 3 December 1877.

139 George Mackenzie to the Hon. Sec. of IBEA, 25 September 1888, file 1A, box 63, "Mombasa Letters, Sept 1888-Jan 1889," PP MS 1 IBEA/Mackinnon Papers, SOAS.

140 Mackenzie to IBEA, 25 September 1888.

141 Letter to His Highness, Sayd Khalifa Said Seyyd, Sultan of Zanzibar, 20 January 1889, file 1A, box 63, "Mombasa Letters, Sept 1888-Jan 1889," PP MS 1 IBEA/Mackinnon Papers, SOAS; MacKenzie to Lt. Swayne, 11 October 1888. See also Stern, *Company-State*, 14.

142 De Kiewiet, "History of the Imperial British East Africa Company," 110. For a discussion of the centrality of treaties in nineteenth-century European expansionism, see Press, *Rogue Empires*.

143 Galbraith, *Mackinnon and East Africa*, 135.

144 De Kiewiet, "History of the Imperial British East Africa Company," 110.

145 Press, *Rogue Empires*, 10.

146 Mackenzie to Secretary of IBEA, 15 January 1889.

147 Watkins, *Report of the Kenya Land Commission*, 90-91.

148 [Illegible] to Kirk, 9 January 1877.

149 [Illegible] to Kirk, 9 January 1877.

150 James Stuart, Edinburgh, May 1874, file 1C, box 64, PP MS 1 IBEA/Mackinnon Papers, SOAS.

151 McDermott, *British East Africa*, 114.

152 McDermott, *British East Africa*, 111-13.

153 McDermott, *British East Africa*, 109-10; Galbraith, *Mackinnon and East Africa*, 112.

154 The Kabaka's authority was vulnerable. Zanzibari "Arab" traders had made inroads into the interior over the course of the century, procuring ivory and slaves, which they exchanged for guns, cotton, and other manufactured goods. Goods were not the only thing on offer, the traders bringing with them new cultural forms and a new religion, Islam, to which people close to the court began converting. Following closely on the heels of traders came British explorers, including Henry Morton Stanley. Like the Sultan, the Buganda Kabaka, Mutesa I (1856-1884), hoped to draw

on the power associated with Europeans to consolidate his territorial control. It was on this premise that the Kabaka encouraged Stanley to prompt Christian missionaries to settle in his kingdom. Anglican missionaries began arriving in 1877, followed by the Roman Catholic White Fathers in 1879. This competitive world was complicated by the position of the Kabaka, who refused to align himself with a single faith. This did not stop his courtiers, however, some of whom began converting to the two Christian creeds in the early 1880s. See Low, *Fabrication of Empire*, 2–4.

155 McDermott, *British East Africa*, 115.

156 Low, *Buganda in Modern History*, 86; Frederick D. Lugard to John Kirk, 4 February 1891, file 25, box 68, PP MS 1 IBEA/Mackinnon Papers, SOAS.

157 Galbraith, *Mackinnon and East Africa*, 231.

158 De Kiewiet, "History of the Imperial British East Africa Company," 249.

159 W. Mackinnon to the Marquis of Salisbury, Foreign Office, 17 December 1890, file 25, box 68, PP MS 1 IBEA/Mackinnon Papers, SOAS.

160 Mackinnon to Marquis of Salisbury, 17 December 1890; Mackinnon was not being naïve in making these requests, having received government subsidies for his ventures in the years before the establishment of the IBEA. Munro, *Maritime Enterprise and Empire*.

161 "The East African Chartered Company," *The Economist*, October 3, 1891. This letter was first published in the *Times* and was apparently penned by someone close to the Company.

162 De Kiewiet, "History of the Imperial British East Africa Company," 271.

163 In exchange, Germany gained the Heligoland archipelago located in the North Sea.

164 Munro, *Maritime Enterprise and Empire*.

165 Munro, *Maritime Enterprise and Empire*, 465.

166 Gerald Portal to the Marquis of Salisbury, 29 October 1891, file 1B, box 63, PP MS 1 IBEA/Mackinnon Papers, SOAS.

167 Portal to Marquis of Salisbury, 29 October 1891.

168 Gerald Portal to Earl Rosebery, 3 November 1892, FO 881/6362, as cited in Munro, *Maritime Enterprise and Empire*, 469.

169 "Uganda," *The Economist*, October 8, 1892.

170 "Uganda," *The Economist*, October 8, 1892.

171 "Uganda," *The Economist*, October 8, 1892.

172 "Case to Advise on Behalf of IBEA."

173 "Case to Advise on Behalf of IBEA."

174 "Case to Advise on Behalf of IBEA."

175 "Case to Advise on Behalf of IBEA."

176 McDermott, *British East Africa*, 241.

177 See Stern, *Company-State*, 25.

178 "Case to Advise on Behalf of IBEA."

179 Lewis Pelly, "Comments on Draft Letter concerning the Operations of the IBEA Being Prepared for Lord Salisbury," 16 January 1892, file 1B, box 63, PP MS 1 IBEA/Mackinnon Papers, SOAS.

180 Henry James and Montague Muir Mackenzie, "Opinion," 1 February 1893, file 49, box 72, PP MS 1 IBEA/Mackinnon Papers, SOAS.

181 James and Mackenzie, "Opinion," 1 February 1893.
182 As Marie J. De Kiewiet writes: "The Foreign Office, on the other hand, while admitting that the Company should be recompensed for the surrender of its concession and its Charter, was unwilling to pay more than dignity and the pressure of public opinion demanded. The Undersecretaries did not admit the justice of the Uganda argument, nor did they concede to Buxton's 'curious idea' that the Company should receive compensation for 'pure philanthropy.'" De Kiewiet, "History of the Imperial British East Africa Company," 303. And these negotiations were protracted with Makenzie spending months "in tedious and increasingly ill-tempered negotiations with Foreign Office officials and in frequent meetings and consultations with the shareholders. The Company's right to compensation was admitted from the start, but the amount and source of payment presented serious difficulties. The Directors' original claim—which was, after consideration and reconsideration, finally accepted two years later—was for compensation on the basis of ten shillings and sixpence in the pound." De Kiewiet, "History of the Imperial British East Africa Company," 301.
183 Galbraith, *Mackinnon and East Africa*, 230–34.
184 *Correspondence Respecting the Retirement of the Imperial British East Africa Company*, 16–25; Galbraith, *Mackinnon and East Africa*, 230–34. The penultimate deal stipulated that the Zanzibar treasury pay £150,000. However, in subsequent negotiations, it was decided that Zanzibar would pay £200,000.
185 Great Britain Parliament House of Commons, *Parliamentary Papers*, 51–52.
186 As one observer remarked: "The path here was quite lost and overgrown in the rich green grass, as the people remarked: 'It had killed the road.'" "Mr. Fitzgerald's Reports," September 1891, file 49, box 72, PP MS 1 IBEA/Mackinnon Papers.
187 Yonge, "Uganda Road," 23.

CHAPTER 2. THE POLITICS OF VALUATION

1 Ernest J. L. Berkeley to Sec. London, 28 January 1892, file 49, box 72, PP MS 1 IBEA/Mackinnon Papers, SOAS.
2 This concept emerged out of many long conversations with Kevin P. Donovan. My first published piece working with the concept was Emma Park, "The Right to Sovereign Seizure?" For Donovan's elaboration of the concept, see Donovan, *The Moneychanger State: Economic Sovereignty and Citizenship in Decolonizing East Africa*.
3 Hecht makes a similar argument; see Hecht, *Being Nuclear*, 16.
4 "Track to Tambach," 23 January 1933, Kenya National Archives [hereafter KNA] PC/RVP.GA/8/1.
5 A similar argument can be found in Cummings, "Aspects of Human Porterage."
6 "Track to Tambach," 23 January 1933. For the most thorough account of imperial finance and taxation in Africa to date, see Gardner, *Taxing Colonial Africa*. See also Woker, "A Taxing Empire: Power and Taxation in the French Colonial Empire, 1850s–1950s."
7 Livingstone, *Dr. Livingstone's Cambridge Lectures*, 14.
8 McDermott, *British East Africa*, 208.

9 Lt. Swayne R. S. to the Administrator-in-Chief, Imperial British East Africa Company, 23 November 1888, file 1A, box 63, "Mombasa Letters, Sept 1888–Jan 1889," PP MS 1 IBEA/Mackinnon Papers, SOAS.

10 George S. Mackenzie to the Secretary, Imperial British East Africa Company, 26 November 1888, file 1A, box 63, "Mombasa Letters, Sept 1888–Jan 1889," PP MS 1 IBEA/Mackinnon Papers, SOAS; Ernest L. Bentley to the Administrator, 7 October 1892, file 49, box 72, PP MS 1 IBEA/Mackinnon Papers, SOAS.

11 For more on colonial representations of "native paths" and a discussion of vernacular networks of mobility, see Grace, *African Motors*, chapter 1.

12 Lugard, *Rise of Our East African Empire*, 247.

13 Lugard, *Rise of Our East African Empire*, 248; Joseph Thomson, *Through Masai Land*, 310, as cited in Watkins, *Report of the Kenya Land Commission*, 547.

14 Macdonald, *Soldiering and Surveying*, 10; "Memorandum: The Road from Dar-es-Salaam East Africa, towards the Northern End of Lake Nyassa," 20 November 1878, file 61, box 77, PP MS 1 IBEA/Mackinnon Papers, SOAS; W. Beardall to William Mackinnon, 16 June 1879, file 59, box 76, PP MS 1 IBEA/Mackinnon Papers, SOAS; James Weaver to Administrator, 7 January 1893, file 49, box 72, PP MS 1 IBEA/Mackinnon Papers, SOAS. See also Lugard, *Rise of Our East African Empire*, 277.

15 Emphasis on the "straight" lines of newly cut roads is ubiquitous in the company archive. See "Jelori Road," n.d., file 49, box 72, PP MS 1 IBEA/Mackinnon Papers, SOAS.

16 "Memorandum: Road from Dar-es-Salaam," 20 November 1878.

17 Swayne to IBEA, 23 November 1888; Macdonald, *Soldiering and Surveying*, 17; Beardall to Mackinnon, 27 February 1879, file 61, box 77, PP MS 1 IBEA/Mackinnon Papers, SOAS.

18 Lugard, *Rise of Our East African Empire*, 244.

19 F. Moir to W. Mackinnon, 3 December 1877, file 9, box 65, PP MS 1 IBEA/Mackinnon Papers, SOAS.

20 "Extract of Letter," 24 August 1877, file 3, box 64, PP MS 1 IBEA/Mackinnon Papers, SOAS.

21 Lugard, *Rise of Our East African Empire*, 266.

22 Moir to Mackinnon, 3 December 1877.

23 Lugard, *Rise of Our East African Empire*, 322.

24 Lugard, *Rise of Our East African Empire*, 300.

25 Lugard, *Rise of Our East African Empire*, 325.

26 Lugard, *Rise of Our East African Empire*, 300, 324.

27 As Stanley wrote of *mpagazi*, or porters, this was a "useful person" who was "the camel, the horse, the mule, the ass, the train, the wagon and the cart of East and Central Africa" without whom "Salem would not obtain her ivory, Boston and New York their African ebony, their frankincense, myrrh and gum copal." Rockel, *Carriers of Culture*, 4.

28 Lugard, *Rise of Our East African Empire*, 342.

29 W. Moyes to William Mackinnon, 2 October 1877, file 7, box 65, PP MS 1 IBEA/Mackinnon Papers, SOAS.

30 Lugard, *Rise of Our East African Empire*, 338.

31 Lugard, *Rise of Our East African Empire*, 353.

32 Lugard, *Rise of Our East African Empire*, 244.

33 "Memorandum: Road from Dar-es-Salaam," 20 November 1878; Beardall to Mackinnon, 16 June 1879; C. Craufurd, "Report on Gulu-Gulu & Mbungo," 16 November 1888, file 1A, box 63, "Mombasa Letters, September 1888-January 1889," PP MS 1 IBEA/Mackinnon Papers, SOAS; Mackenzie to Lt. Swayne, Zanzibar, 11 October 1888, file 1A, box 63, "Mombasa Letters, September 1888-January 1889," PP MS 1 IBEA/Mackinnon Papers, SOAS.

34 John Ainsworth, "East Africa 'Kenya' Reminiscences," n.d., MSS.Afr.s.380, Bodleian Library, University of Oxford.

35 Moir to Mackinnon, 29 January 1877, file 1C, box 64, PP MS 1 IBEA/Mackinnon Papers, SOAS.

36 Beardall to Mackinnon, 16 June 1879.

37 Smith, "Road-Making and Surveying," 274.

38 Smith, "Road-Making and Surveying," 274.

39 Smith, "Road-Making and Surveying," 272.

40 Beardall to Waller, 12 January 1880, file 61, box 77, PP MS 1 IBEA/Mackinnon Papers, SOAS. Note that the Kingani [Ruvu] River is in present day Tanzania, and was adjacent to Bagamoyo, a trade entrepot in the region.

41 Beardall to Waller, 12 January 1880; Beardall to W. Mackinnon, 14 August 1879, file 61, box 77, PP MS 1 IBEA/Mackinnon Papers, SOAS; Beardall to Mackinnon, 10 October 1879, file 61, box 77, PP MS 1 IBEA/Mackinnon Papers, SOAS.

42 Beardall to Mackinnon, 14 August 1879.

43 Beardall to Mackinnon, 14 August 1879.

44 Beardall to MacKinnon, 14 August 1879. Though often company roads remained "winding," as this was the only way to "obtain suitable grades for waggon [*sic*] traffic." Austin, *Macdonald in Uganda*, 252. For discussions of the importance of the politics of angular aesthetics, see Adas, *Machines as the Measure of Men*; Mrázek, *Engineers of Happy Land*; Bissell, *Urban Design, Chaos, and Colonial Power*.

45 Macdonald, *Soldiering and Surveying*, 31.

46 Macdonald, *Soldiering and Surveying*, 21.

47 George Wilson to Secretary, EA Scottish Mission, 23 August 1892, file 41, box 70, PP MS 1 IBEA/Mackinnon Papers, SOAS.

48 Macdonald, *Soldiering and Surveying*, 26, 35. See also Lugard, *Rise of Our East African Empire*, 272.

49 Wilson to EA Scottish Mission, 23 August 1892.

50 C. Craufurd, "Report on Gulu-Gulu & Mbungo," 16 November 1888, file 1A, box 63, "Mombasa Letters, September 1888-January 1889," PP MS 1 IBEA/Mackinnon Papers, SOAS.

51 Macdonald, *Soldiering and Surveying*, 66.

52 Smith, "Road-Making and Surveying," 269; H. H. Austin, "Diaries: E.A. Railway Survey - Mombasa-Lake Victoria Nyanza, 16th October 1891–21st May 1892," Book I (Nairobi: Macmillan Library), 61, cited in Cummings, "Aspects of Human Porterage," 145.

53 Ainsworth, "East Africa 'Kenya' Reminiscences."

54 Wilson to EA Scottish Mission, 23 August 1892.

55 Wilson to EA Scottish Mission, 23 August 1892.

56 Wilson to EA Scottish Mission, 23 August 1892.

57 Guyer and Belinga, "Wealth in People."

58 There is very good scholarship on the construction of ethnicity in Kenya. See Osborne, *Ethnicity and Empire in Kenya*, 19–20 who offers an account of the construction of the Kamba through their labor over the *longue duree*. For a discussion of Kikuyu "ethnicity" prior to 1900, see Muriuki, *History of the Kikuyu*. For a discussion of the case of Kikuyu self-articulation as a "tribe," see Lonsdale, "Moral Economy of Mau Mau."

59 Osborne, *Ethnicity and Empire in Kenya*, 22–23.

60 Cummings, "Aspects of Human Porterage," 155–56.

61 Cummings, "Aspects of Human Porterage," 185.

62 Wilson to EA Scottish Mission, 23 August 1892.

63 Wilson to EA Scottish Mission, 23 August 1892.

64 Beardall to Waller, 11 December 1879, file 61, box 77, PP MS 1 IBEA/Mackinnon Papers, SOAS.

65 Wilson to EA Scottish Mission, 23 August 1892.

66 Cummings, "Aspects of Human Porterage," 191–92.

67 Beardall to William Mackinnon, "Report of Progress of Work on the East African Road during the Past 4 weeks," 16 August 1880, file 11, box 66, PP MS 1 IBEA/Mackinnon Papers, SOAS.

68 Cummings, "Aspects of Human Porterage," 273.

69 Ambler, *Kenyan Communities*; Muriuki, *History of the Kikuyu*.

70 See Thompson, "Time, Work-Discipline, and Industrial Capitalism," 85. For examples of how Thompson's ideas surrounding time have been taken up by scholars of Africa, see Cooper, *Plantation Slavery*, 181; Martin, *Leisure and Society*. Keletso Atkins works with a concept akin to E. P. Thompson's notion of "moral economy" in her discussion of incommensurate visions of work and time. On this, see Atkins, *The Moon Is Dead! Give Us Our Money!* For a discussion of the limits of the concept, see Glassman, *Feasts and Riot*.

71 Beardall to Mackinnon, 27 February 1879.

72 For a discussion of this phenomenon in West Africa, see Guyer, *Marginal Gains*. For a discussion of this phenomenon in East Africa, see Pallaver, "What East Africans Got."

73 Ainsworth, "East Africa 'Kenya' Reminiscences."

74 Mackenzie to Swayne, 11 October 1888; "Memorandum: Road from Dar-es-Salaam," 20 November 1878. For a wonderful discussion of the importance of consumer "taste" in eastern Africa and how it shaped global markets in commodities for consumption, see Prestholdt, *Domesticating the World*.

75 Mackenzie to F. J. Jackson, Esq., 10 November 1888, file 1A, box 63, "Mombasa Letters, Sept 1888-Jan 1889," PP MS 1 IBEA/Mackinnon Papers, SOAS.

76 Mackenzie to Swayne, 11 October 1888.

77 H. Leakey, "Kenya Land Commission," *Evidence*, 1865, as cited in Ambler, *Kenyan Communities*, 126. On beads as stores of value, see Prestholdt, *Domesticating the World*, 74. G. S. Mackenzie wrote to the Secretary of the IBEA, reporting on the

delivery of the gift of maxim guns to the Sultan. G. S. Mackenzie to Hon. Secretary IBEA, 23 December 1888, file 1A, box 63, "Mombasa Letters, Sept 1888–Jan 1889," PP MS 1 IBEA/Mackinnon Papers, SOAS.

78 Lugard, *Rise of Our East African Empire*, 274.

79 A. G. Swayne, "Report on Gulu-Gulu (Ukamba Country)," 17 November 1888, file 1A, box 63, "Mombasa Letters, Sept 1888-Jan 1889," PP MS 1 IBEA/Mackinnon Papers, SOAS.

80 Smith, "Road-Making and Surveying," 272.

81 Wilson to EA Scottish Mission, 23 August 1892.

82 Eagleton, "Currency as Commodity," 115.

83 Eagleton, "Currency as Commodity," 115; Kirk to Granville, 7 April 1884, FO 1677/55, British National Archives.

84 Galbraith, *Mackinnon and East Africa*, 20; Mwangi, "Of Coins and Conquest," 770. While this is not Mwangi's framing, to look to these histories of currencies offers a different trajectory for our understanding of the history of capitalism in the British Empire, one which moves not from metropole to colony, but one that traversed colonial spaces, connecting India to eastern Africa. As one 1890 report noted: "Rupees and Pice have been pretty well circulated and are being disposed of quite freely." Mackenzie to W. P. Alexander, Esq., 9 January 1890, file 31, box 69, PP MS 1 IBEA/Mackinnon Papers, SOAS.

85 Mr. Dick to Mackinnon, 2 December 1890, file 24, box 68, PP MS 1 IBEA/Mackinnon Papers, SOAS.

86 Mwangi, "Of Coins and Conquest," 778.

87 Dick to Mackinnon, 2 December 1890.

88 George S. Mackenzie to Sec. IBEA, 23 October 1888, file 1A, box 63, "Mombasa Letters, September 1888–January 1889," PP MS 1 IBEA/Mackinnon Papers, SOAS.

89 Mackenzie to IBEA, 23 October 1888.

90 Mackenzie to the Secretary, 13 May 1889, file 16, box 67, PP MS 1 IBEA/Mackinnon Papers, SOAS.

91 For a discussion of coins and flags as "ideological texts," see Stern, *Company-State*, 14.

92 As Catherine Eagleton writes, by "the late 1880s, the issue of coins was . . . seen as important in order to facilitate wage payments—so that roads could be cleared and building works undertaken." Eagleton, "Currency as Commodity," 127.

93 Cummings, "Aspects of Human Porterage," 229–35.

94 W. Moyes to Mackinnon, 26 September 1877, file 7, box 65, PP MS 1 IBEA/Mackinnon Papers, SOAS.

95 Letter, Second camp near village Keswire, 14 July 1877, file 3, box 64, PP MS 1 IBEA/Mackinnon Papers, SOAS.

96 Letter, n.d., file 3, box 64, PP MS 1 IBEA/Mackinnon Papers, SOAS.

97 Letter, 10 July 1877, file 3, box 64, PP MS 1 IBEA/Mackinnon Papers, SOAS.

98 Moyes to Mackinnon, 26 September 1877.

99 Smith, "Road-Making and Surveying," 278.

100 Ernest L. Bentley to the Administrator, Mombasa, 10 November 1891, file 49, box 72, PP MS 1 IBEA/Mackinnon Papers, SOAS.

101 Ernest L. Bentley to the Administrator, 17 November 1891, file 49, box 72, PP MS 1 IBEA/Mackinnon Papers, SOAS.

102 Roitman, *Fiscal Disobedience*.

103 "Track to Tambach," 23 January 1933.

104 "The Policy of Creating Reigning Companies," *The Economist*, September 15, 1888.

105 Goswami, *Producing India*; Jakes, *Egypt's Occupation*.

106 Ernest J. L. Berkeley to Sec. London, 28 January 1892, file 49, box 72, PP MS 1 IBEA/Mackinnon Papers, SOAS; "Draft of the Contract to Be Entered into between Sultan of Zanzibar and Company to Be Formed for the Administration of His Domain," n.d., file 8, box 65, PP MS 1 IBEA/Mackinnon Papers, SOAS; John Kirk to Gerald Waller, 12 May 1878, file 60, box 77, PP MS 1 IBEA/Mackinnon Papers, SOAS.

107 Bentley to Administrator, 17 November 1891.

108 Mwangi, "Of Coins and Conquest," 771.

109 Mwangi, "Of Coins and Conquest," 781.

110 Berman, *Control & Crisis*, 51.

111 Peterson, *Creative Writing*, 14; Muriuki, *History of the Kikuyu*, 136.

112 Watkins, *Report of the Kenya Land Commission*, 289.

113 Ambler, *Kenyan Communities*, 123–24, 126.

114 Ambler, *Kenyan Communities*, 141.

115 Hopkins, *Greatest Killer*. See also Davis, *Late Victorian Holocausts*, especially chapter 1.

116 Ambler, *Kenyan Communities*, 141.

117 Boyes, *King of the Wa-Kikuyu*, 191. See also, Watt and Watt, *Heart of Savagedom*, 360.

118 De Kiewiet, "History of the Imperial British East Africa Company," 249; Leggett, "Economics and Administration," 398. The construction of the railroad began in 1896.

119 Ainsworth, "East Africa 'Kenya' Reminiscences"; Muriuki, *A History of the Kikuyu*, 137.

120 Ambler, *Kenyan Communities*, 137. For a discussion of this period of devastation in relation to Kamba communities, see Osborne, *Ethnicity and Empire in Kenya*, 43–51.

121 J. W. Stauffacher, "History of the AIM, 1895–1913," box 17, *Mission History*, Africa Inland Mission (AIM), as cited in Peterson, *Creative Writing*, 15.

122 Boyes, *King of the Wa-Kikuyu*, 191.

123 Grimes, *Life out of Death*, 22.

124 Watt and Watt, *Heart of Savagedom*, 358.

125 R. Blayden-Taylor, "Kenya Land Commission," *Evidence*, 759, as cited in Ambler, *Kenyan Communities*, 128.

126 Ambler, *Kenyan Communities*, 139.

127 Barlow, *English-Kikuyu Dictionary*, 334. I'm very grateful to Derek Peterson for sharing his notes on Kikuyu words for the routes of mobility, and for his insights regarding the connections Kikuyu-speakers drew between the paths they traversed and the social work their mobility was undertaken to secure. Julie Livingston makes a similar point regarding the social functions of precolonial road networks and how they were effaced by the "tarred road." See Julie Livingston, *Self-Devouring*

Growth, 86. For a fulsome analysis of vernacular networks of mobility, see Grace, *African Motors*.

128 Ambler, *Kenyan Communities*, 124.

129 Ambler, *Kenyan Communities*, 142.

130 Boyes, *King of the Wa-Kikuyu*, 142, 189.

131 Peterson, *Creative Writing*, 15.

132 Ambler, *Kenyan Communities*, 138–39.

133 Watt and Watt, *Heart of Savagedom*, 342.

134 Atieno-Odhiambo, "Colonial Government," 97; Kitching, *Class and Economic Change*, 213.

135 Atieno-Odhiambo, "Colonial Government," 97.

136 Cummings, "Aspects of Human Porterage," 248.

137 R. D. Wolff, *Britain and Kenya, 1870-1930: The Economics of Colonialism* (Nairobi, 1974), 55, as cited in Tarus, "History of Direct Taxation," 44.

138 For an account of the racialized forms of predatory inclusion in the context of the US housing market, see Taylor, *Race for Profit*.

139 Hindlip, *British East Africa*, 66.

140 Mwangi, "Of Coins and Conquest," 779–80.

141 While not spelled out in the colonial record, it is reasonable to assume that these fiscal techniques were simultaneously geared toward social transformation, the goal being to encourage people to pursue monogamous unions.

142 Hindlip, *British East Africa*, 68.

143 Tarus, "History of Direct Taxation," 62.

144 Berman, *Control & Crisis*, 63.

145 Lowe, *Intimacies of Four Continents*.; Aaron Jakes, "Infrastructures of Occupation" in *Egypt's Occupation*, 32–57.

146 Berman, *Control & Crisis*, 79.

147 Berman, *Control & Crisis*, 51.

148 Berman, *Control & Crisis*, 54.

149 "Total Revenue Collected in Kikuyu Province from 1902-1907," KNA PC/CP/1/1/2, as cited in Mwangi, "Order of Money," 56.

150 Watkins, *Report of the Kenya Land Commission*, 495.

151 Berman and Lonsdale, *Unhappy Valley*.

152 "Fort Hall District Record Book, 1919," KNA PC/CP/1/7/1.

153 "Track to Tambach," 23 January 1933.

154 Gardner, *Taxing Colonial Africa*; Jakes, "Boom, Bugs, Bust."

155 Cummings, "Aspects of Human Porterage," 248.

156 Berry, "Hegemony on a Shoestring"; Forstater, "Taxation and Primitive Accumulation"; Dodd, *Social Life of Money*, 63.

157 Hindlip, *British East Africa*, 68. By 1912, while it was agreed that an increase in taxation could not be pursued with the sole goal of increasing the labor supply, it was "recognised that, in the first instance, taxation" brought "natives into the labour market." East Africa Protectorate, *Native Labour Commission, 1912-13—Evidence and Report*, (Nairobi: Printed by the Government Printer, n.d.), 329, Kenya National Archives.

158 "Track to Tambach," 23 January 1933.

159   Reply of Under-Secretary of State for the Colonies, Winston Churchill to Edward Sassoon, 29 July 1907, "Official: Colonial Office: Kikuyu Hut Tax," CHAR/10/35, Churchill Archive; Fredrick Lugard, *Political Memoranda: Revision of, Instructions to Political Officers on Subjects of Chiefly Political and Administrative* (London, 1968), as cited in Tarus, "History of Direct Taxation," 46. For a discussion of the role of taxation as both a fiscal and political technology, see Tilly, *Coercion, Capital, and European States*, 49. Brenda Chalfin takes this up in her study of customs duties in Ghana. See Chalfin, *Neoliberal Frontiers*, 25.

160   Hindlip, *British East Africa*, 66.

161   Isaac Tarus, "The Early Colonial History of the Keiyo of Kenya, 1900–1939" (MA thesis, University of Nairobi, 1944), as cited in Tarus, "History of Direct Taxation," 62–63.

162   C. W. Hobley, *Kenya from Chartered Company to Crown Colony* (London, 1929), 123–24, as cited in Tarus, "History of Direct Taxation," 63.

163   W. McGregor Ross, *Kenya from Within: A Short Political History* (London, 1927), 199, as cited in Tarus, "History of Direct Taxation," 47.

164   Ainsworth, "East Africa 'Kenya' Reminiscences."

165   Provincial Commissioner, Central Province to Chief Secretary, 17 August 1939, KNA DC/NY/2/15/2.

166   Reply of Churchill to Sassoon, 29 July 1907.

167   Meinhertzhagen, *Kenya Diary*, 64.

168   van Zwanenberg, *Colonial Capitalism and Labour*, 77.

169   van Zwanenberg, *Colonial Capitalism and Labour*, 80.

170   Tarus, "History of Direct Taxation," 22.

171   Assistant District Commissioner, Machakos, to the Chief Native Commissioner, 1920s, KNA PC/CENT/2/4/1, Roads in Native Reserves, 1913–27.

172   Machakos to the Chief Native Commissioner, 1920s.

173   The politics of mobility did not affect all equally. Indeed, as a robust literature suggests, young men and women often used mobility to assert their autonomy against a gerontocratic, patriarchal and rural order. Berman and Lonsdale, *Unhappy Valley*; White, *Comforts of Home*; Thomas, *Politics of the Womb*; Bujra, "Women 'Entrepreneurs.'"

174   van Zwanenberg, *Colonial Capitalism and Labour*, 111.

175   "Political Records Book—A Protest," KNA PC/CP/1/4/1, Kikuyu District Political Record.

176   Berman, *Control & Crisis*, 147.

177   "Race Relations Papers," KNA MSS/3/215, 1937–38.

178   "A Note on the Principle of Paid Labour as against Free Labour on Roads," 1924, KNA PC/CENT/2/4/1, Roads in Native Reserves, 1913–27.

179   Berman and Lonsdale, *Unhappy Valley*, 13.

180   van Zwanenberg, *Colonial Capitalism and Labour*, 112.

181   John Ainsworth, "The Question of Segregation," n.d., MSS.Afr.s.380, Bodleian Library, University of Oxford.

182   "Fort Hall Record Book, 1926," KNA DC/FHI/6.

183   "Fort Hall Record Book, 1926."

184   "Fort Hall Record Book, 1927," KNA DC/FHI/7.

CHAPTER 3. "TROPICALISING" TECHNOLOGIES

1   Samwel Okwako Libuko, "Essays by Schoolchildren," 1944, KNA HAKI/13/298, Information and Propaganda.
2   Vetus Oduor, "Soil Erosion," 1944, KNA HAKI/13/298, Information and Propaganda.
3   "Cinematograph; Mobile Cinema Van; Reports," 1940, KNA HAKI/13/328, Information Committee.
4   "Broadcasting Cheap Medium-Wave Receivers for Colonial Peoples," BNA CO/875/59/8.
5   Arthur Champion, "Report of the Government Cinema Unit for the Week Ending Sunday, September 22nd," KNA HAKI/13/328.
6   Arthur Champion, "Report of the Activities of Government Cinema Unit up to Week Ending August 25th 1940," KNA HAKI/13/328.
7   Champion, "Report of Activities up to Week Ending August 25th 1940."
8   British Broadcasting Corporation (BBC) to Reiss, 4 July 1952, KNA AHC/18/7, Broadcasting in the Colonies, Broadcasting Policy.
9   M. F. Hill, Information Officer, "Draft Memorandum on the Broadcasting Service Arranged by the Kenya Information Office with the Cooperation of Messrs. Cable & Wireless, Ltd., with Certain Proposals for the Improvement of the Nairobi Broadcasting Station," 23 January 1941, KNA HAKI/13/276 or CS/2/8/121, Information and Propaganda.
10 The language of broadcasting development is ubiquitous in the colonial archive. For example, see Blackburne to Clark, 1949, CO 875/68/6, Broadcasting, East Africa.
11 Committee on Broadcasting in the Colonies, as quoted in Hill, "Draft Memorandum," 23 January 1941.
12 E. R. Davies, Acting Information Officer to the Chief Secretary, Secretariat, Nairobi, "The Kenya Information Office after the War," 8 June 1943, KNA HAKI/13/136, Information and Propaganda: Organisation and Establishment, Postwar Organization, 1943-47.
13 Boyce, "Origins of Cable and Wireless Limited," 87.
14 M. F. Hill, Information Officer, "Memorandum of the Broadcasting Service Arranged by the Kenya Information Office with the Cooperation of Messrs. Cable & Wireless, Ltd., with Certain Proposals for the improvement of the Nairobi Broadcasting Station," 11 February 1941, British National Archives [hereafter BNA] Colonial Office [hereafter CO] 875/2/14, Broadcasting, East Africa.
15 H. L. G. Gurney, Chief Secretary to the Governors' Conference to the Hon. Chief Secretary, 14 January 1943, KNA HAKI/13/276; BBC to Reiss, 4 July 1952.
16 Gurney to Hon. Chief Secretary, 14 January 1943; BBC to Reiss, 4 July 1952.
17 Governor Henry Moore to the Information Office, n.d., KNA HAKI/13/328.
18 Hunt, *Colonial Lexicon.*
19 Lonsdale, "Moral Economy of Mau Mau"; Peterson, *Creative Writing;* Glassman, *Word of Words, War of Stones.*
20 Englund, "Anti Anti-Colonialism"; For a discussion of how radio appeals to publics across multiple scales—"local," national, transnational—see Gunner, Ligaga, and Moyo, *Radio in Africa;* Moorman, *Powerful Frequencies.*

21   Heath, "Broadcasting in Kenya," 61.

22   Heath, "Broadcasting in Kenya," 61.

23   Heath, "Broadcasting in Kenya," 62.

24   "The Cable and Wireless Merger," *The Economist*, 4 August 1928.

25   Boyce, "Origins of Cable and Wireless Limited," 91–96.

26   Roland Belfort, "The Empire: Cable v. Radio, the Coming Struggle and How to Avert It," London, 1898, MSS. Marconi 240, Marconi Archives, Bodleian Library, University of Oxford; Hill, "Draft Memorandum," 23 January 1941; Boyce, "Origins of Cable and Wireless Limited," 95; Heath, "Broadcasting in Kenya," 62.

27   Heath, "Broadcasting in Kenya," 60.

28   Baglehole, *A Century of Service*, 15.

29   Berman, *Control & Crisis in Colonial Kenya*, 130–31.

30   Heath, "Broadcasting in Kenya," 60. For a discussion of the centrality of telegraphic coordination in producing what she refers to as "imperial state space," see Goswami, *Producing India*, 53.

31   Norman E. Walter and Allan S. Ker, "Some Reflections of the Early Days of Broadcasting in Kenya," Cable and Wireless Archive [hereafter CW] DOC/12/135, Telegraph Museum, Porthcurno.

32   *Annual Report for Kenya, 1927*, 44, as cited in Heath, "Broadcasting in Kenya," 57.

33   Walter and Ker, "Early Days of Broadcasting in Kenya."

34   Great Britain, PRO, CO 533/389 048664, Kenya Original Correspondence, as cited in Heath, "Broadcasting in Kenya," 57.

35   Walter and Ker, "Early Days of Broadcasting in Kenya"; Great Britain, *Handbook on Broadcasting Services*.

36   Moorman, *Powerful Frequencies*, 35.

37   Great Britain, PRO, CO 533/389/8 048664, "Broadcasting Licence and Agreement between the Post Master General of the Colony and Protectorate of Kenya and the British East African Broadcasting Company," 1927, as cited in Heath, "Broadcasting in Kenya," 65.

38   Heath, "Broadcasting in Kenya," 56.

39   Walter and Ker, "Early Days of Broadcasting in Kenya."

40   "Broadcasting Licence and Agreement," as cited in Heath, "Broadcasting in Kenya," 57.

41   Walter and Ker, "Early Days of Broadcasting in Kenya."

42   Walter and Ker, "Early Days of Broadcasting in Kenya."

43   Walter and Ker, "Early Days of Broadcasting in Kenya."

44   For a discussion of what she refers to as "hegemony on a shoestring," see Berry, "Hegemony on a Shoestring."

45   Heath, "Broadcasting in Kenya," 62.

46   Heath, "Broadcasting in Kenya," 65.

47   Heath, "Broadcasting in Kenya," 66.

48   Walter and Ker, "Early Days of Broadcasting in Kenya."

49   Walter and Ker, "Early Days of Broadcasting in Kenya."

50   Walter and Ker, "Early Days of Broadcasting in Kenya."

51 Walter and Ker, "Early Days of Broadcasting in Kenya."

52 Walter and Ker, "Early Days of Broadcasting in Kenya."

53 D. H. Williams, District Commissioner, Kiambu to the Hon. Provincial Commissioner, Central Province, Nyeri, "Experimental Broadcasts to Africans," 20 September 1939, KNA AE/34/62, Posts and Telegraphs, Wireless, 1938-39.

54 Williams, "Experimental Broadcasts to Africans"; In this instance, the scheme was funded by a combination of the colonial administration and the Kiambu Local Native Council, each contributing £25. "Draft: Memorandum on Broadcasting to Natives," n.d., KNA AE/34/62, Posts and Telegraphs, Wireless, 1938-39.

55 "Broadcasting Listener Research," 9 May 1950, Broadcasts for Africans, Vol. 1, Uganda National Archives [hereafter UNA].

56 Berman, *Control & Crisis in Colonial Kenya*, 163.

57 D.H. Williams, District Commissioner, Kiambu to the Hon. Provincial Commissioner, Central Province, Nyeri. "A Report on a Series of Ten Experimental Broadcasts to Africans Held from August 18th to 20th October, 1939," 4 November 1939, KNA AE/34/62, Posts and Telegraphs, Wireless, 1938-39. Harry Thuku was an interesting selection here. His pan-Africanist activism in the 1920s advocated for a politics of unity that transcended not simply the lines of race and ethnicity, but colonial boundaries themselves. This work led to his arrest and subsequent exile in 1922. His time in exile seems to have moderated the radical, and upon his release in the 1930s, his political work was centered on the concerns of Kikuyu moderates. It was perhaps this change in focus that made Thuku appear to the administration to be a reliable interlocutor in the field of broadcasting. Gabay, "Decolonizing Interwar Anticolonial Solidarities." Interestingly, Thuku does not reference his involvement with radio in his highly detailed autobiography. See Thuku, *Harry Thuku: An Autobiography*.

58 "Draft: Memorandum on Broadcasting to Natives."

59 Williams, "Experimental Broadcasts to Africans."

60 M. F. Hill, "Memorandum on Broadcasting to Africans in Kenya," n.d., KNA AHC/10/63.

61 Ambrose Wakaria, in discussions with the author, 2014.

62 Peterson, *Creative Writing*.

63 Wakaria, in discussions with the author, 2014.

64 Wakaria, in discussions with the author, 2014.

65 A. M. Champion, "Native Welfare in Kenya," September 1944, Anglican Church of Kenya Archives [hereafter ACKA], Nairobi.

66 Champion, "Native Welfare in Kenya."

67 W. F. P. Kelly, District Officer, 1936-63, "Chapter V: The Mobile Cinema," in *Kenya Memoir*, ed. W. F. P. Kelly, MSS.Afr.s.2229:1, Bodleian Library, University of Oxford.

68 Arthur Champion, "Report of Tour," 18 February-2 March 1941, KNA HAKI/13/128.

69 Arthur M. Champion, "Report of the Government Cinema Unit for the Week Ending September 29, 1940," KNA HAKI/13/32. At other times, films were accompanied by running commentary in Kiswahili alone. Kelly, "The Mobile Cinema."

70 Ambrose C. Coghill, "Mambo Leo: A Memorandum on the Education of East African Natives By Means of the Mobile Cinema Projection Units," n.d., DC/MUR/3/10/2, Cinema and Stage.

71 Coghill, "Mambo Leo."
72 Arthur M. Champion, "Report of Activities of the Government Cinema Unit Week-Ending August 18, 1940," KNA HAKI/13/328.
73 Kelly, "The Mobile Cinema."
74 Kelly, "The Mobile Cinema."
75 Bartholomew Orwa, "Composition," 17 October 1940, KNA HAKI/13/328.
76 J. E. Obala, "Composition," 17 October 1940, KNA HAKI/13/328.
77 Report from Committee Appointed by the Secretary of State to Report on Broadcasting in the Colonies, excerpted in Hill, "Draft Memorandum," 23 January 1941.
78 Kelly, "The Mobile Cinema."
79 Lewis, *Empire State-Building*, 258-60.
80 "Government Mobile Cinema Annual Report, 1944," KNA HAKI/13/309.
81 Willis, "Men on the Spot."
82 "Government Mobile Cinema Annual Report, 1944."
83 Brian Larkin has referred to this aspiration as "the colonial sublime." Larkin, *Signal and Noise*, 7.
84 "Government Mobile Cinema Annual Report, 1944."
85 cf. Larkin, *Signal and Noise*.
86 "Government Mobile Cinema Annual Report, 1944."
87 "Government Mobile Cinema Annual Report, 1944."
88 "Government Mobile Cinema Annual Report, 1944."
89 "Government Mobile Cinema Annual Report, 1944."
90 Hunt, *Colonial Lexicon*.
91 W. Sellers, "Mobile Cinema Shows in Africa," *Colonial Cinema*, December 1951 (emphasis added).
92 "Government Mobile Cinema Annual Report, 1944."
93 Mukerji, *Impossible Engineering*. For an interesting take on the centrality of breakdown and how it shapes African technological action, see Grace, *African Motors*. For a more general take on the analytic and political possibilities of thinking with breakdown and repair, see Jackson, "Rethinking Repair."
94 Arthur Champion, "The Government Mobile Cinema Unit Report for Week Ending October 20 1940," KNA HAKI/13/328.
95 F. H. Knight, "Projector Maintenance," *Colonial Cinema*, December 1949.
96 "Replacement of Prints," *Colonial Cinema*, March 1945.
97 "Replacement of Prints."
98 "Replacement of Prints."
99 Champion, "Report for Week Ending September 22."
100 For a discussion of the importance of studying maintenance and repair, see Jackson, "Rethinking Repair."
101 "Replacement of Prints."
102 "Replacement of Prints."
103 E. F. Twining, W. K. Brasher, and C. A. L. Richards, *Broadcasting Investigations* (Entebbe: Government Printer, Uganda, 1939), KNA AHC/34/62.
104 Dickson, Captain, General Staff Intelligence, "Mobile Propaganda Safari in Uganda," 25 March 1943, BNA CO/875/8/8.

105 Knight, "Projector Maintenance."
106 A similar point is made by Joshua Grace in the context of vehicles. As Grace notes, African mechanics adapted these technologies, often under conditions of shortage, reworking them into vehicles well-suited to conditions in Tanzania. On this, see chapter 2 of Grace, *African Motors*.
107 Sellers, "Mobile Cinema Shows in Africa."
108 R. W. Harris, A.R.P.S. of the Colonial Film Unit, "16 mm. Bell and Howell-Gaumont 601," *Colonial Cinema*, June 1950.
109 Knight, "Projector Maintenance."
110 Knight, "Projector Maintenance."
111 Arthur Champion, "Report of Mobile Cinema Unit up to Week Ending October 6, 1940," KNA HAKI/13/328.
112 Dickson, "Mobile Propaganda Safari."
113 Peterson, *Creative Writing*.
114 "Religious Broadcasting: A Communication from Miss Wrong on his Subject," n.d., ACKA CMO/CMS/1. See also Leonard Beecher to Hill, December 1940, KNA AE/34/64.
115 "Religious Broadcasting: A Communication from Miss Wrong on his Subject"; "Memorandum of the Translation Bureau Attached to the Information Office," KNA AHC/10/68.
116 "Memorandum of the Translation Bureau."
117 "Memorandum of the Translation Bureau."
118 "Memorandum of the Translation Bureau."
119 "Memorandum of the Translation Bureau."
120 Latour, "Why Has Critique Run Out of Steam?"
121 As one report read: "A man's native speech is almost like his shadow, inseparable from his personality. In our way of speech we must each, as the old saying runs, drink water out of our own cistern. For each one of us is a member of a community. . . . It is the mother tongue which gives to the adult mind the relief and illumination of utterance, as it clutches after the aid of words when new ideas or judgements spring from the wordless recesses of thought or feeling under the stimulus of physical experience or of emotion." Advisory Committee on Education in the Colonies, "Memorandum on Language in African School Education," November 1943, ACKA ET/AEC/1, African Education Council, 1920–1956.
122 Excerpt from "Memorandum on Language in African School Education."
123 "Memorandum on Language in African School Education."
124 "Memorandum on Language in African School Education."
125 "Memorandum on Language in African School Education."
126 "Memorandum on Language in African School Education."
127 "Memorandum on Language in African School Education." Whether the use of Kiswahili in broadcasting was fundamentally reshaping the linguistic landscape of the Kenya colony during the war years is hard to ascertain. It is clear, however, that people's experiences fighting in the war had convinced them of the role that language could play in binding together community. As members of the Local Na-

tive Council (LNC), Fort Hall wrote: "Swahili [simply] does not exist outside East Africa." "Memorandum on Language in African School Education."

128 As a member of the Translation Bureau noted: "The preparation of African broadcasts does not only involve their molding into a form of English capable of proper translation into the vernaculars . . . It also requires that the broadcasts should be attractive and informative." "Memorandum of the Translation Bureau."

129 E. R. Davies copied to J. L. Beecher, Nairobi, 24 April 1941, KNA AHC/10/68.

130 Patrick Jubb, Director of Broadcasting, "Mount Kenya Station, Nyeri," 15 July 1960, KNA AHC/18/34, Regional Mount Kenya Station; Dickson, "Mobile Propaganda Safari."

131 John Gitonga, for example, and like Wakaria, was the son of a committed Christian. In Gitonga's case, his father was a clergyman of the Church of England. Gitonga had been educated in mission schools, spending three years at a government school. He later attended Alliance High School, from whence he proceeded to Makerere College in Uganda, thereafter attending King's College. His elder brother was a District Officer, with another brother having secured the position of Chief Inspector of Police in Kakamega, a third brother acting as a medical assistant, and a fourth as a "farm manager." As Gitonga pointed out, in defending himself against accusations of insubordination, "most of us are all in Govt. Service." These men, in terms of education and ascension within the colonial hierarchy, were exceptional. John Gitonga to the Accountant and Broadcasting Officer, Nairobi, 1 May 1956, KNA HAKI/13/84, "Information and Propaganda."

132 In 1940, the Advisory Committee on Information noted: "The Information Officer . . . was dissatisfied with the Vernacular Broadcasts. He said that on the previous day he had caused the Kikuyu script to be translated back into English and this had revealed serious discrepancies with the English script, which itself was not good." "Extract from Minutes of Meeting of Advisory Committee on Information Held on the 14th December, 1940," KNA HAKI/13/276 or CS/2/8/121, Information and Propaganda.

133 Hill, "Memorandum," 11 February 1941.

134 J. Wilson, "Memorandum: The Use of 'Pidgin' and Vernacular for Propaganda Purposes in the Colonies," n.d., BNA CO 875/5/13. This comment was made in reference to the use of a "pidgin" in Nigeria. One of the interesting features of the world of broadcasting is that expertise in this period often did not rely on connections between colonies and the metropole, but networks of knowledge work that were being generated by colonial officials working at sites across the British Empire.

135 Miss Wrong, Secretary of International Committee on Colonial Literature for Africa, 1941, ACKA CMO/CMS/2. As another report recounted, while some knowledge-workers "caught on to the idea at once," one administrator reported others had "no nose for news." Transcript of broadcast by E. R. Davies, "Some Problems of Broadcasting to Africans," Nairobi, 8th August 1941, KNA AHC/10/68.

136 Champion, "Report for Week Ending October 20 1940."

137 Champion, "Report for Week Ending October 20 1940."

138 G. G. S. Hutchinson to the Provincial Commissioner, Nyanza Province, "African Broadcasts," 1 December 1941, KNA AHC/10/68.

139   During the war years, "well-known chiefs" were invited to the microphone to counter the feeling that broadcasts to African troops had become "too impersonal, coming as they do from an unknown announcer in Nairobi." Geoffrey Northcote, Principal Information Officer, Ministry of Information, East Africa Command to Information Officers, 28 August 1942, BNA CO 875/8/21, Propaganda, Conference of Information Officers in Nairobi, 1942–43. Other observers noted that "the chief attraction . . . [for audiences was a speaker with a] well known voice, and personality, irrespective of the matter on which he speaks." Hutchinson to DC, Central Kavirondo, n.d. KNA DC/KSM1/28/52, Publications and Records, Propaganda and Wireless Sets. Similarly, as early as 1940, people on the coast reported that Africans showed a great interest in talks "given by Africans whom they know." The goal here was to "introduce a personal element into the vernacular broadcasts by means of short talks by men who are well known to the tribe to which they are talking . . . This gave a domestic touch to the programmes." Broadcasting autobiographies was a move in the same direction. Indeed, Davies readily acknowledged that a "good deal more credence is given to the broadcasts if the voices of well known natives are occasionally heard by the listeners." E. R. Davies, as cited in Information Officer, Kampala to Honorable Chief Secretary, Entebbe, "Broadcasting," 11 June 1941, Broadcasts for Africans, Vol. I, UNA; Transcript of broadcast by Davies, "Some Problems of Broadcasting to Africans."

140   Transcript of broadcast by Davies, "Some Problems of Broadcasting to Africans" (emphasis added). Others were of the same opinion, arguing that the "psychology and political views" of Africans would "have to be taken seriously" when programming was designed. DC, CP, Kenya, n.d., likely 1936–38, KNA DC/MUR/3/10/2, Stage Plays and Cinematography, 1934–52.

141   E. R. Davies, Officer in Charge of Africa Section, "African Propaganda, 1942," KNA HAKI/13/229.

142   Transcript of broadcast by Davies, "Some Problems of Broadcasting to Africans."

143   Dickson, "Mobile Propaganda Safari."

144   This is well-illustrated by Louis Leakey's treatment of knowledge-workers. In 1940, for example, James Beauttah was cut from the African Information Services (AIS) roster. He had been scheduled to broadcast on soil conservation efforts that would become part and parcel of the spate of expertly administered reforms that typified the "second colonial occupation" of the postwar period. Beauttah was already in the employ of the state, working as a propaganda officer of the Local Native Council (LNC). He was also experienced in the world of information, having worked as a commentator with Arthur Champion in the early 1940s. Leakey emphatically rejected Beauttah's appointment, arguing he "felt very strongly on the matter," as Beauttah had "taken an oath that he was the enemy of the British and would probably be under detention within a week." The chairman of the meeting countered this position, stating that he had known "James Beauttah for many years and that the latter had been a member of the first Native Council to be elected in the native locations of Nairobi in 1926 when he himself had been District Commissioner" there. E. R. Davies, too, challenged Leakey, arguing that Beauttah had been "sent into broadcast by the District Commissioner at Fort Hall." Even the Chief

Native Commissioner (CNC), the Chairman noted, he had "approved of his giving a talk on the wireless." "Broadcast Programmes Sub-Committee, Eighth Meeting," 12 July 1940, KNA AE/34/63. Leakey's rejection of Beauttah likely had less to do with Beauttah's disdain for the British than with Beauttah's membership in the Kikuyu Central Association (KCA), an organization that Leakey felt unduly challenged the authority of conservative Kikuyu elders, such as Chief Waruhiu, with whom he was close. In this instance, Leakey used his position as a "cloak for *fetina*," working to settle old scores with foes. Leakey "considered that announcers should be selected with great care and some time in advance so that something would be found out regarding them." Berman and Lonsdale, "Louis Leakey's Mau Mau," 178.

145 Peterson, *Creative Writing*. Myles Osborne notes that observers were impressed by the reach of what they conceived of as "Mau Mau propaganda." See Osborne, "The Rooting Out of Mau Mau," 93.

146 Perhaps unsurprisingly, then, it would be missionaries—people who saw themselves as having unique insights into the local—who argued for ever more stringent protocol when it came to recruiting knowledge-workers. Louis Leakey, for his part, argued that, despite evidence that local voices would augment the success of broadcasts, "any such talks by chiefs and other prominent natives" were not "advisable in war time." "Broadcast Programmes Sub-Committee, Eighth Meeting," 12 July 1940.

147 As the report continued, this "impromptu" character left administrators with no "idea what they [African knowledge-workers]" were "going to say next." "Broadcasting," n.d., BNA CO 875/3/6. This threat was not limited to newsmen, however. "We could never guarantee," wrote another administrator, "that a singer would not convey remarks about the weather to the enemy." Miss Wrong, ACKA CMO/CMS/2. As another report recounted, there was uncertainty regarding "whether *any* native is a fit and proper person to broadcast or as to whether a chief or other native cuts any ice or no." "Broadcast Programmes Sub-Committee, Eight Meeting," 12 July 1940.

148 Transcript of broadcast by Davies, "Some Problems of Broadcasting to Africans."

149 D. C. S. to G., 31 December 1946, KNA HAKI/13/229.

150 Officer In Charge, Kenya Information Office, "African Translators and Announcers," 30 August 1940, KNA AE/34/64, Posts and Telegraph, 1940-41.

151 "African Translators and Announcers," 30 August 1940.

152 "Careers in Government Service, Recruitment Pamphlet No. 2," The Department of Information, Kenya (Nairobi: Government Printer), KNA HAKI/13/85.

153 "Staff Functions," n.d., KNA HAKI/13/262.

154 "African Translators and Announcers," 30 August 1940.

155 Director of Information to All Provincial Information Officers, "Points from Mr. Senior's Address," 14 June 1956, KNA HAKI/13/262.

156 "Points from Mr. Senior's Address," 14 June 1956.

157 "Points from Mr. Senior's Address," 14 June 1956.

158 "Points from Mr. Senior's Address," 14 June 1956.

159 Berman, *Control & Crisis in Colonial Kenya*, 381-83.

160 Guyer and Belinga, "Wealth in People."

161 Berman and Lonsdale, *Unhappy Valley: Violence & Ethnicity*.

162 Berman, *Control & Crisis in Colonial Kenya*, 349.

163 Berman and Lonsdale, *Unhappy Valley: Violence & Ethnicity*.

164 Lonsdale, "Authority, Gender & Violence"; Peterson, "Wordy Women."

165 Peterson, *Ethnic Patriotism*.

166 Berman and Lonsdale, *Unhappy Valley: Violence & Ethnicity*.

167 Osborne, "The Rooting Out of Mau Mau," 93.

168 For a more general account of this strategy, see Cooper, *On the African Waterfront*.

169 The sources I work with in this section are partial at best. They are based on the few remaining audio files from the period, which today are housed at the Kenya Broadcasting Corporation (KBC). Holding true to colonial representational practices, rarely are the voices of knowledge-workers attached to personalities, nor are they named. Like other forms of African expertise, these unnamed men are, instead, represented as easily replaced non-experts. However, as the preceding sections have demonstrated, these men were anything but replaceable: they were actively courted experts of both technics and culture. Put differently, this anonymizing impulse notwithstanding, a close reading of these broadcasts demonstrates how the cultural expertise of African knowledge-workers was actively brought to bear in efforts to convince Kenya's diverse communities to side with the government in its fight against Mau Mau. My concern here is less with elucidating whether these strategies "succeeded" or "failed." Instead, my efforts are directed to understanding how these notions of development were actively comingled in this period, and to elucidating the place of vernacular knowledge-workers, as well as the infrastructure of radio broadcasting, in advancing this twinned vision.

170 Berman and Lonsdale, *Unhappy Valley: Violence & Ethnicity*, 327, 333.

171 Berman and Lonsdale, *Unhappy Valley: Violence & Ethnicity*, 333.

172 Berman and Lonsdale, *Unhappy Valley: Violence & Ethnicity*, 341.

173 Berman and Lonsdale, *Unhappy Valley: Violence & Ethnicity*, 356.

174 Peterson, *Ethnic Patriotism*, 178–94.

175 Leakey, *The Southern Kikuyu Before 1903*.

176 "Broadcast #7," n.d., Tape A4, Kenya Broadcasting Corporation [hereafter KBC] Audio Archive. Translations were done by Vivian Wanjiku, to whom I am tremendously indebted for her work and her insights.

177 "Broadcast #2," n.d., Tape A5, KBC Audio Archive.

178 "Broadcast #2," Tape A5.

179 "Broadcast #7," Tape A4; Clark, "Louis Leakey as Ethnographer," 385; Berman and Lonsdale, "Louis Leakey's Mau Mau," 178; Osborne, "The Rooting Out of Mau Mau," 86.

180 "Broadcast #7," Tape A4.

181 Berman and Lonsdale, *Unhappy Valley: Violence & Ethnicity*, 344. It is here that John Lonsdale discusses the labor that went into dividing the space of the wild from the space of domestic life.

182 "Broadcast #2," n.d., Tape A2, KBC Audio Archive.

183 "Broadcast #9," n.d., Tape A7, KBC Audio Archive.

184 Lonsdale, "Moral Economy of Mau Mau," 344.

185 "Broadcast #2," Tape A5.

186 The language of the "bush telegraph" is ubiquitous in this archive. See "Information and Propaganda," KNA HAKI/13/81, Information and Propaganda, 1950–55.

187 "Information and Propaganda," KNA HAKI/13/81.

188 "Broadcast #4," n.d., Tape A4, KBC Audio Archive.

189 John Gitonga to the Director of Establishment and the Director of Information, Nairobi, 12 June 1956, KNA HAKI/13/84, Information and Propaganda.

190 Wakaria, in discussion with the author, 22 June 2015.

191 One story was recalled by many interlocutors: the story of Mburu Matemo. Matemo was Wakaria's senior in the field of information work. The two were close, with Wakaria continuing to visit Matemo's widow following his death. Like other information workers, Matemo had to negotiate this fine line between intimacy and estrangement as he reported on Mau Mau. Matemo's story, like all good stories, is subject to variation, but in broad strokes it goes as follows. African knowledge-workers reporting on Mau Mau were prohibited from publicizing the deaths of Europeans; instead, Mau Mau losses were to be foregrounded. One day, Matemo recounted a battle that had taken place at Brackenhurst. Deviating from policy, he reported on the deaths of Europeans. He "used a proverb," Wakaria explained, describing Europeans using the term *thukumu*. "That's what he did," Wakaria continued; it "was as simple as that." *Thukumu* was a Kikuyu concept used to describe foreign trading partners in the period prior to colonial occupation. Matemo, perhaps, thought that by tapping into this longer linguistic genealogy of otherness, he would escape the watchful eye of Leakey. He was wrong. Leakey, himself, was concerned with locating culture in language. Indeed, it was based on interviews with elite Kikuyu elders that he had set himself the task of writing *The History of the Southern Kikuyu*, which ostensibly documented precolonial Kikuyu culture and social life. In this text, Leakey put to page his knowledge of the term *thukumu*. In his rendering, it was a Kikuyu concept used at the turn of the century to describe foreign trading partners—first, Swahili, and later, European. Leakey caught Matemo in his vernacular insubordination and "chased him," summarily firing the broadcaster. They were all vulnerable in the face of Leakey and this vulnerability stemmed from years of intimacy that had transformed Leakey into "a Kikuyu," as Wakaria explained. Ambrose Wakaria, in conversation with the author, 22 June 2015; Leakey, *Southern Kikuyu Before 1903*, 498.

192 White, *Comforts of Home*, 188.

CHAPTER 4. BROADCASTING THE FUTURE

1 J. Grenwell Williams, R. W. P. Cockburn, and W. A. Roberts of the BBC, "Report of Kenya Broadcasting Commission, June 1954," KNA JA/12/115.

2 See Cooper, *Decolonization and African Society*.

3 Young, *Closer Union of the Dependencies*, 130–31.

4 See Moorman, *Powerful Frequencies*, especially chapter 2.

5 Colonial Office Advisory Committee on Education in the Colonies, *Mass Education in African Society*, 8.

6 Cooper, *Decolonization and African Society*.

7 Colonial Office Advisory Committee on Education in the Colonies, *Mass Education in African Society*, 7. These transformations had been thoroughgoing, the report argued, writing that "another phenomenon perhaps not wholly unrelated to the changes of which we have just spoken is the changing attitude in the face of evil conditions. In the past that attitude might be described as one of social fatalism which led the people to tolerate with patience conditions of misfortune and even injustice. They were accepted with the same fatalistic resignation as fire and flood and earthquake." With this coming to an end, administrators complained, people were blaming the government for "misfortune," which, of course, administrators thought was completely misguided.

8 [Illegible] to Hon. C. S. and Hon. C. N. C., 11 March 1940, KNA AE/34/63, Posts and Telegraphs, Wireless, Broadcasting, To Natives.

9 Colonial Office Advisory Committee on Education in the Colonies, *Mass Education in African Society*.

10 Ambrose C. Coghill, "Mambo Leo: A Memorandum on the Education of East African Natives by Means of Mobile Cinema Projection Units," n.d., KNA DC/MUR/3/10/2.

11 Colonial Office Advisory Committee on Education in the Colonies, *Mass Education in African Society*, 8.

12 The report continued: "We must not omit consideration of the political aspirations which have emerged in some parts of the Colonial Empire in vigorous form and are spreading over far wider areas. The force of those aspirations has been accentuated by the magnitude of the struggle in which we are now involved and also by the certainty that the issue of that struggle will decide the common destiny. There is a universal sense of a common interest in the decision, which makes all men responsive to the same forces even though the form of response may differ widely. Thus the stage is set for the conception of a common citizenship which begins to acquire reality and expression. This common citizenship is not to be envisaged as a benevolent concession from above: it is the practical outcome of common trial and common effort. This ideal and all that it entails may well be the keynote of all progress within the colonies as well as in their political relations with the people of Britain." And so people had to be guided. "The principles of local government, understood and practised by many African communities in the past, will have to be 'modernised' to meet new conditions, and the wisdom of the old men will have to be interwoven with the new learning to make a real and not an artificial basis for local democracy. In many rural areas the traditional forms of recreation have lost their attraction and new ones must be devised to lighten the dullness of village life." Colonial Office Advisory Committee on Education in the Colonies, *Mass Education in African Society*, 8.

13 M. F. Hill, Information Officer, Kenya, excerpted in "Proceedings of the Conference of Information Officers," 21–24 May 1943, KNA HAKI/13/170 or CS/2/8/3.

14 Colonial Office Advisory Committee on Education in the Colonies, *Mass Education in African Society*, 8.

15 Colonial Office Advisory Committee on Education in the Colonies, *Mass Education in African Society*, 8.

16   G. G. S. Hutchinson, "Broadcasting in the Colonies," 31 August 1948, KNA HAKI/13/276. He continued: "And it should not . . . be debated whether or not a demand now exists where there are no facilities but whether or not there is a need to increase the medium through which African population[s] can be educated."

17   "Provision of Studies and Communal Listening Facilities, Grant of 27,130," 7 December 1951, BNA CO 875/68/6, Broadcasting, East Africa, 1949.

18   Arthur Creech Jones, Confidential Circular, "Broadcasting in the Colonies," BNA CO 875/31/1, Receivers for Colonial Peoples: Policies, 1948. See also Armour, "BBC and the Development of Broadcasting," 359, for a discussion of how Creech Jones emphasized the importance of expanding broadcasting across Anglophone Africa.

19   "Provision of Studies and Communal Listening Facilities." As one administrator argued, "the purposes of broadcasting to Africans can be described as educational, administrative and strategic."

20   "Provision of Studies and Communal Listening Facilities"; see also BNA CO/875/64/3, Broadcasting, Kenya.

21   There were a number of such commissions in British-held African colonies in these years. See Armour, "BBC and the Development of Broadcasting." Indeed, in the 1940s, a survey was conducted to assess the viability of establishing broadcasting on an East African basis. However, these plans were abandoned due to resistance from Tanganyika and Uganda. See W. E. C. Varley, Engineering Division, British Broadcasting Corporation, "Broadcasting Survey of the British East African Territories," April 1946, KNA HAKI/13/276 or CS/2/8/121, Information and Propaganda.

22   Atmospheric noise refers to natural variations caused by shifting weather patterns, specifically lightning discharges. "Cheap Wireless Receivers for Colonial Peoples," n.d., British National Archives [hereafter BNA] Colonial Office [hereafter CO] 875/31/1, Receivers for Colonial Peoples: Policies, 1948.

23   W. E. C. Varley, Engineering Division, British Broadcasting Corporation, "The Broadcasting Survey of the British East African Territories, 1946," British Broadcasting Corporation Written Archive [hereafter BBCWA] E1/21, Countries African East African Broadcasting, 1943–1947.

24   "Report of Kenya Broadcasting Commission, June 1954," Cable and Wireless Archive [hereafter CW] DOC/5 1774/1, Telegraph Museum, Porthcurno.

25   "Report of Kenya Broadcasting Commission, June 1954."

26   "Report of Kenya Broadcasting Commission, June 1954."

27   "Broadcasting, East Africa," 1949, BNA CO 875/68/6.

28   "Cheap Wireless Receivers for Colonial Peoples."

29   For a discussion of the distinction between "matters of concern" and "matters of fact," see Latour, "Why Has Critique Run Out of Steam?," 225–48.

30   "Report of Kenya Broadcasting Commission, June 1954."

31   G. G. S. Hutchison to J. H. Reiss, 27 October 1949, KNA AHC/10/63, Information and Propaganda, Wireless Broadcasting, 1949.

32   Minute of J. B. Millar, 20 November 1951, BNA CO 875/64/3, Broadcasting, Kenya.

33   It is not clear how subscribers broke down along the lines of race. "Kenya Broadcasting: History of Cable and Wireless Limited Activities," CW DOC/6/326/1, Telegraph Museum, Porthcurno.

34 Heath, "Broadcasting in Kenya," 65.
35 J. B. Miller, 21 September 1950, BNA CO 875/68/6.
36 Philip Mitchell to Geoffrey Northcote, 13 January 1945, KNA HAKI/13/245, Information and Propaganda, 1944–46.
37 Millar, 21 September 1950.
38 M. F. Hill, Information Officer, "Draft Memorandum on the Broadcasting Service Arranged by the Kenya Information Office with the Cooperation of Messrs. Cable & Wireless, Ltd., with Certain Proposals for the Improvement of the Nairobi Broadcasting Station," 23 January 1941, KNA HAKI/13/276 or CS/2/8/121, Information and Propaganda.
39 Chief Broadcasting Engineer to the Clerk of the Aberdare County Council, March 1960, KNA AHC/18/18.
40 Heath, "Broadcasting in Kenya," 73.
41 Hutchinson, "Broadcasting in the Colonies."
42 It is worth noting that the contours of this agreement were unique within the empire *and* were anomalous for C&W. Among Britain's colonial holdings, Kenya was the first to have a regular wireless broadcasting service directed specifically to Europeans and South Asians. As for C&W, while it was the single largest telecommunications entity in the world, it was in Kenya alone that the entity offered "a completely programmed broadcasting service." "Report of Kenya Broadcasting Commission, June 1954."
43 If the Government of Kenya was going to challenge C&W's monopoly, it would have to demonstrate that C&W was unable or unwilling to provide services at a reasonable cost. Such as they were made, efforts to open up the informational landscape to other commercial providers failed. Governor's Deputy, Kenya to Secretary of State for the Colonies, 6 July 1943, BNA CO 875/4/30.
44 The BBC took this responsibility very seriously. It often involved "the broadcasting of programmes for which there is no active demand, or which . . . [ran] counter to the wishes of the majority. A readiness to undertake programmes for which there is no active demand," the BBC argued, was "essential for any broadcasting organization with ideals and a determination not to stagnate . . . such programmes must be heard, perhaps many times, before opinions are formed . . . this inevitably means that supply must constantly precede demand." Broadcasting as a "public good" mitigated against the application of market principles, such as consumer desire. As the report continued, if people "disliked a series of broadcasts . . . that would not in itself be considered a valid reason why the policy should be reversed or the programmes withdrawn." In this regard, the path taken in the early years by the colonial administration was in line with those of the BBC. See "BBC, A Review of Listener Research Methods," n.d., BNA CO 875/70/6.
45 O. J. Whitley, Assistant Head of Colonial Services to James Millar, Colonial Office, 29 August 1950, BBC, BNA CO 875/64/3, Broadcasting, Kenya.
46 Whitley to Millar, 29 August 1950.
47 "Cheap Wireless Receivers for Colonial Peoples." Department of Information, "Annual Report, 1955," KNA HAKI/13/251.

48 The first indication that there was an Imperial need for "cheap wireless receivers" was made known to manufacturers at an interdepartmental meeting convened under the War Office in 1946. "Cheap Wireless Receivers for Colonial Peoples."

49 Kenya & Uganda Customs Collector, Kampala to C. M. A. Gayer, Public Relations and Social Welfare Dept., Kampala, 19 May 1949, BNA CO 875/68/3, Broadcasting, Propaganda, 1949–51. This, it was acknowledged, would always be "a danger when the ordinary short or medium-wave frequencies are used."

50 Broadcast Relay Service Limited, "Rediffusion," n.d., BNA CO 875/40/3.

51 Broadcast Relay Service Limited, "Rediffusion."

52 Broadcast Relay Service Limited, "Rediffusion."

53 "Cheap Wireless Receivers for Colonial Peoples."

54 "Cheap Wireless Receivers for Colonial Peoples."

55 Varley, "Broadcasting Survey, 1946."

56 Varley, "Broadcasting Survey, 1946." One way of managing situations where "electricity . . . [was] not available," would be to establish "central charging point[s]" where "accumulator batteries" could be repowered. In more remote locales, dry batteries, which lasted six months, seemed a good solution. A. H., "Receivers for Use in Villages," 13 January 1946, Broadcasts for Africans, Vol. 1, Uganda National Archives [hereafter UNA].

57 "Cheap Wireless Receivers for Colonial Peoples."

58 "Cheap Wireless Receivers for Colonial Peoples."

59 "Cheap Wireless Receivers for Colonial Peoples."

60 "Broadcasting: Cheap Medium-Wave Receivers for Colonial Peoples," BNA CO/875/59/8; "Cheap Wireless Receivers for Colonial Peoples."

61 "Cheap Wireless Receivers for Colonial Peoples."

62 Harry Franklin, "Draft Report on the Saucepan Special: The Poor Man's Radio for Rural Populations," BNA CO 875/60/1, Broadcasting: Cheap Short-Wave Receivers for Colonial Peoples. Franklin claimed to have embedded these considerations in his design, describing himself as "having [had] the vision and enterprise to see the vast potential market for the poor man's radio and the part which radio could play in the development of the colonies." Franklin, *Report on 'the Saucepan Special'*.

63 Franklin, "Draft Report on the Saucepan Special."

64 The Postmaster General to the Honorable Chief Native Commissioner, Kenya, 21 January 1950, KNA HAKI/13/293, Wired Broadcasts, 1950.

65 "Broadcasting Services: Note by Kenya," May 1956, KNA AHC/10/65; G. G. S. Hutchison to Sir Geoffrey Northcote, 20 August 1945, KNA AHC/10/68. Others were more concerned by the possibility that "irresponsible and shady outfits" would try to enter the market posing as respectable private firms. "Ref. (30) and (31)," 26 October 1946, KNA HAKI/13/296, Information and Propaganda, Use of Cinema, 1941–52.

66 "Report of Kenya Broadcasting Commission, June 1954."

67 "Report of Kenya Broadcasting Commission, June 1954."

68 "Extract from Report on Information Services Kenya, 1949," BNA CO 875/64/3.
69 Berec Battery Export Company to Mr. Miller, Esq., 1 June 1951, BNA CO 875/60/2.
70 Peter Lister, "Report on the Saucepan Special," 7 July 1951, BNA CO 875/60/2; Peter
    Lister, "'The Saucepan Special': A Light in Darkest Africa," newspaper article, n.d.,
    BNA CO 875/60/2; See also Harry Franklin, "A New World Market for Britain" in
    "Draft Report on the Saucepan Special: The Poor Man's Radio for Rural Populations,"
    BNA CO 875/60/1, Broadcasting: Cheap Short-Wave Receivers for Colonial Peoples.
71 Letter from J. E. King, n.d., BNA CO 875/60/2; Department of Information, "An-
    nual Report, 1955"; Osborne, *Ethnicity and Empire in Kenya*, 90.
72 "Broadcasting Services: Note by Kenya," May 1956.
73 Cooper, *On the African Waterfront*.
74 "Mombasa District Annual Report, 1951–59," KNA CQI/19/29 or DC/KMSA/I/6.
75 "Mombasa District Annual Report, 1951–59."
76 "Mombasa District Annual Report, 1953," PC/COAST/2/I; Brennan, "Lowering the
    Sultan's Flag," 840.
77 "Report on the Use of Vernaculars in Broadcasting with Special Reference to
    Kikamba," 16 September 1953, KNA AHC/10/68; Hutchison to Northcote, 20
    August 1945; The Governor's Deputy to the Secretary of State for the Colonies,
    "CONFIDENTIAL," 25 August 1950, BNA CO 875/64/3; "Provision of Studies and
    Communal Listening Facilities."
78 K. J. A. Hunt, District Commissioner, Central Nyanza to the District Commission-
    ers of North Nyanza, Kakamega, South Nyanza, Kisii, and Kericho, 18 Decem-
    ber 1950, KNA DC/KSM/I/28/54, Wireless Sets.
79 A. D. Shirreff, DC, Central Nyanza to the PC, Nyanza Province, "Provision of
    Wireless Sets for Africans," 1952, KNA DC/KSM/I/28/54; C. H. Williams, PC, Nyanza
    Province, to the CNC, "Wireless Sets—Nyanza Province," 28 October 1952, KNA.
    Luo communities were not the only groups affected by this change in policy. By
    1949, broadcasts in Guisii and Nandi were discontinued: "The reason for the ces-
    sation of these programmes being that listener attendance figures tended to show
    that their expense was not justified." "Extract from Report on Information Service
    Kenya, 1949," BNA CO 875/64/3.
80 R. M. Woodford to the District Commissioners, African Schools, Municipal
    African Affairs Officers, All Information Rooms, "Outside Speakers for Vernacular
    Broadcasts During 1949," 13 January 1949, KNA DC/KSM/I/28/54.
81 Hunt to District Commissioners, 18 December 1950; "Report on the Use of
    Vernaculars."
82 The administration amped up its informational work on the coast in general fol-
    lowing the 1947 strike. "Extract from Report on Information Service Kenya, 1949";
    "Record of a Meeting Held in the Secretariat on 3rd December, 1954 at 2.30 p.m,"
    KNA AHC/10/66; Heath, "Broadcasting in Kenya," 103.
83 Monthly Reports, November 1948, KNA UY/13/I.
84 Nasser, *Philosophy of the Revolution*, 60.
85 Brennan, "Radio Cairo." As James Brennan notes, however, the aspirations of
    Third World solidarity were complicated in practice by local dynamics of race,
    religion, and nationalism, which impeded Afro-Asian solidarity.

86 Nasser, *Philosophy of the Revolution*, 70.

87 Colin Legum, "Cairo Bid to Lead African Nationalists," *The Observer*, 27 November, 1955, 4.

88 Osgood Carruthers, "'Voice of Arabs' Stirs Mideast: Broadcasts Are Now Most Potent Propaganda Aimed at West and the U.S. Cairo Relationship against METO Request from Jordan Links Undefined Arab Radio," *New York Times*, 15 January 1956, E5.

89 Brennan, "Radio Cairo."

90 "The American Influence in East and Southern Africa," 5 April 1958, KNA BB/1/204; "Colonialism and Imperialism in Somaliland i.e. British Somaliland, French Somaliland, Ethiopian Somaliland and N. F. P.," 6 November 1958, KNA BB/1/204; "Somaliland Demand for Freedom & Unity," 13 September 1958, KNA BB/1/204; "A Brief History of Slave Trade in Africa," 11 September 1958, KNA BB/1/204.

91 "Africa Awake," 24 August 1958, KNA BB/1/204.

92 "Africa Awake," 24 August 1958. This sentiment appears to have been widespread. As one broadcast to Somali listeners pleaded: "Stop quarrelling and cursing each other. Instead be one hand and one voice and fight against imperialism, occupation and supporters of imperialism." "Daily Report on the Programmes of Cairo Radio Monitored by Special Branch, Coast Province, Kenya," 24 August 1958, KNA BB/1/204.

93 The Governor to the Chief Secretary, n.d., KNA AHC/10/66.

94 The Provincial Commissioner to the Administrative Secretary, Coast Province, 28 March 1955, KNA AHC/10/66.

95 Governor to Chief Secretary, n.d., KNA AHC/10/66.

96 Watkins Pritchford, Director of Information to the Chief Secretary, "Strengthening of Information Services in Coast and Nyanza Provinces," 22 August 1955, KNA AHC/18/70, Nyanza and Coast Province Broadcasting.

97 Hutchinson, "Broadcasting in the Colonies."

98 "Notes on Mr. Hutchinson's Visit to the Coast," Mombasa, June 1942, KNA HAKI/13/128.

99 "Kenya Intelligence Committee: Appreciation of the Arab Situation at the Coast," October 1956, BNA CO/822/804/2, as cited in Brennan, "Radio Cairo," 176.

100 "Mombasa District Annual Report, 1954," KNA PC/COAST/2/1/90.

101 Kindy, *Life and Politics in Mombasa*, 164.

102 Kindy, *Life and Politics in Mombasa*, 164–65.

103 Kindy, *Life and Politics in Mombasa*, 164–65.

104 Kindy, *Life and Politics in Mombasa*, 164.

105 "Mombasa District Annual Report, 1953," KNA PC/COAST/2/1/84.

106 This tension comes up many times in Kindy's autobiography. See Kindy, *Life and Politics in Mombasa*.

107 "Mombasa District Annual Report, 1957," KNA PC/COAST/2/1/102.

108 Kindy, *Life and Politics in Mombasa*, 173.

109 Governor to Chief Secretary, 22 November 1954, KNA AHC/10/66.

110 J. A. H. Wolff, DC, "Annual Report, Mombasa, 1957," KNA PC/COAST/2/1/102.

111 Osgood Caruthers, "Cairo Broadcasts Irk London, Paris: Inflammatory Talks to Africa Incite Natives to Revolt against 'Imperialists,'" *New York Times*, March 1, 1956, 2.

112 Brennan, "Radio Cairo," 177.

113 Pritchford, "Strengthening of Information Services"; G. W. Jamieson for the Assistant Chief Secretary, "Broadcasting from Mombasa," 16 November 1956, KNA AHC/10/66; See also Brennan, "Radio Cairo," 176.

114 Jamieson, "Broadcasting from Mombasa."

115 Jamieson, "Broadcasting from Mombasa."

116 James Brennan notes that colonial officials were beset by fears that, as some broadcasts claimed, this programme was being broadcast from Kenya's White Highlands. Brennan, "Radio Cairo," 178.

117 Osgood Caruthers, "Cairo Radio Stirs Restive Peoples: Air Barrage Covers a Broad Area," Special to the NYT, *New York Times*, 6 July 1958, E5.

118 Seaghan Maynes, "British Jam Cairo Radio to Halt Rumors," *Daily Boston Globe*, November 17, 1956, 2.

119 Rumors did not just travel from the outside in. There were also rumors moving in the other direction, from Kenya to places as far away as California, USA, where in 1940, there was a rumor reported that colonial subjects in East Africa were punished by death for owning radios. H. Gurney, Information Officer to the Honorable Chief Secretary, 23 September 1940, KNA AE/34/64, Posts and Telegraphs, 1940–41. The language of "colonial lexicon" comes from Hunt, *Colonial Lexicon*.

120 Notes of Ag. A. S., Chief Secretary, n.d., KNA AHC/10/66.

121 See "Political Struggle and the Crisis of the Colonial State," in Berman, *Control & Crisis*, 300–346; Cooper, *Decolonization and African Society*.

122 Ray Vicker, "Cairo Radio: It Promotes Nasserism with Lies, Hate, and Obscenity," *Wall Street Journal*, 14 August 1958, 6.

123 Vicker, "Cairo Radio."

124 Hill, "Draft Memorandum."

125 Young, *Closer Union of the Dependencies*, 130–31.

126 Governor to the Chief Secretary, n.d., KNA AHC/10/66.

127 "The Activities of Cairo Radio and Their Impact on Territories to Which They Are Directed, Appendix," n.d., BNA CO 1035/22. The government claimed Egypt had fourteen transmitters: three high-power medium wave, six low-power medium wave, and five high-power short wave. See also Caruthers, "Cairo Radio Stirs Restive Peoples"; Department of Information to Secretariat, 15 July 1954, KNA AHC/10/68, African Broadcasts.

128 Caruthers, "Cairo Radio Stirs Restive Peoples."

129 Caruthers, "Cairo Radio Stirs Restive Peoples."

130 Remarks by Governor of Kenya on His tour of the Swahili Coast, [1956?], KNA AHC/10/66.

131 Provincial Commissioner to Administrative Secretary, 28 March 1955.

132 "Extension Made in Radio Time," *East African Standard*, 15 November 1954, KNA AHC/10/68.

133 Michael Blundell, "African Broadcasting," 15 July 1954, KNA AHC/10/68; "Memorandum—Request for Funds to Purchase Two Transmitters," n.d., KNA AHC/10/68.

134 From Department of Information to Secretariat, 15 July 1954, KNA AHC/10/68, African Broadcasts.

135 KNA AHC/10/68 or 66.

136 J. H. Reiss, Director of Information to Secretariat, 15 July 1954, KNA AHC/10/68, African Broadcasts.

137 Pritchford, "Strengthening of Information Services."

138 Ag. A. S. to Chief Secretary, 1 April 1955, KNA AHC/10/66. The installation of a medium wave transmitter in Mombasa had a number of politically expedient affordances. Not only would a medium wave transmitter strengthen the reach of signals over the area, but it would circumvent the "problems" thrown up by the increased ownership of shortwave receivers, which made it "all too easy for [listeners] to pick up the more powerful transmissions from Cairo, All India Radio and Omdurman." It was in this context that Emergency monies, presumably allocated to defeat Mau Mau, were used to erect new transmitting sites both in Kisumu and Mombasa. J. H. Reiss, Director of Information, "A Review of the Functions, Scope and Organisation of the Kenya Department of Information," KNA CA/16/117.

139 Ministry of African Affairs to Secretariat, "Egyptian Propaganda," 25 July 1954, KNA AHC/10/68, Information and Propaganda, Wireless Broadcasts to Africans, Nairobi.

140 Lonsdale, "Moral Economy of Mau Mau."

141 In the preceding chapter, I addressed how the government mobilized African knowledge-workers—and vernacular notions of individual progress or *wiathi* (self-mastery)—as it sought to use radio as a tool in the fight against Mau Mau. In this section, I pursue a different tack. By considering the 1950s through the lens of the contested politics of broadcasting, I argue, we can situate Mau Mau as but one project that sought to redraw the boundaries of belonging with which the colonial state had to contend in these years. To this end, in the following, I bracket Mau Mau (and Central Kenya more generally) to focus on other spaces where projects of competitive scale-making were in the offing: specifically, Kenya's Swahili Coast.

142 See Peterson, "Patriotism and Dissent in Western Kenya."

143 Anderson, "Yours in Struggle for Majimbo," 549.

144 "Photographic, Broadcasting, Publications, Provincial Offices," KNA HAKI/13/251 or CS/2/8/247, Information and Propaganda, 1955–56.

145 While there is good reason to be skeptical of this publication and the editorial decisions made regarding what was allowed to grace the pages of the magazine, there are equally good reasons to take this seriously as a source. As Harri Englund writes, while government publications "may not furnish uplifting examples" of resistance narratives "we must examine them as cultural and political artifacts in their own right." Englund, "Anti Anti-Colonialism," 222.

146 "Yaliyosemwa (The Things They Say)," *Habari za Radio*, June-July 1954.

147 *Habari za Radio*, October 1956.

148 "Minutes of the Meeting of the Provincial Information Officers," [1957?], KNA AHC/8/73.

149   "Saucepan Radios," n.d., KNA DC/TAMB/3/18/4, Saucepan Radios, 1949–57.

150   "Photographic, Broadcasting, Publications, Provincial Offices," n.d., KNA HAKI/13/251 or CS/2/8/247, Information and Propaganda, 1955–56.

151   "Photographic, Broadcasting, Publications, Provincial Offices."

152   Suleiman Abdullah, "Msaada Wa Kununua Radio (Help to Purchase Radio)," *Habari za Radio*, February 1955.

153   "Yaliyosemwa," *Habari za Radio*.

154   Angelo Daudi, "Maoni ya Wasikilizaji (Listeners Views)," *Habari za Radio*, October 1954.

155   "Letter," *Habari za Radio*, 1959.

156   Ole Aomo, "Programs in Maasai," *Habari za Radio*, February 1955.

157   The language of the "bush telegraph" is ubiquitous in this archive. See KNA HAKI/13/81, Information and Propaganda, 1950–55.

158   J. H. Reiss, 1954, KNA AHC/10/66.

159   This was paid for by Colonial Welfare and Development Funds. [Illegible] Rose to Commissioners of Nyanza, Kakamega, Nyanza, Kisumu, and Nanzya, Kisii, "Reference my Letter 16/A of 11th and 29th April," 21 May 1953, KNA DC/KSM/1/28/54.

160   [Illegible] Rose, Executive Officer, African Information Services to the District Commissioners of North Nyanza, Central Nyanza, South Nyanza, Kericho, "Saucepan Radios," 29 April 1953, KNA DC/KSM/1/28/54.

161   Debate on Information Office Vote, Legislative Council Debates, Vol. 72, 12 June 1957, cols. 1276–87, as cited in Heath, "Broadcasting in Kenya," 122. As noted by Heath: "The AIS was responsible for explaining . . . [constitutional] reform and procedures for . . . [the] election to the public. However, African candidates were not permitted to make political speeches on the radio nor were the subsequently elected members permitted to make their opinions regarding government policies heard on the air even though Europeans had that privilege."

162   Heath, "Broadcasting in Kenya," 122.

163   Kenya, Legislative Council Debates, col. 1539, as cited in Heath, "Broadcasting in Kenya," 157.

164   Momanyi, *Ronald Ngala*.

165   Wolff, "Annual Report, Mombasa, 1957."

166   Berman and Lonsdale, "Labors of 'Muigwithania'"; Berman and Lonsdale, *Unhappy Valley: Violence & Ethnicity*.

167   Wolff, "Annual Report, Mombasa, 1957."

168   Wolff, "Annual Report, Mombasa, 1957."

169   M. N. Edwards to Chief Secretary, "Legislative Council Question No. 39," 2 November 1957, KNA AHC/10/67, Information and Propaganda.

170   E. C. Eggins to J. H. Reiss, Director of Information, 1957, KNA AHC/18/71, Mombasa Broadcasting Scheme, Policy.

171   E. C. Eggins to John H. Reiss, Esq., Director of Information, 22 June 1957, KNA AHC/18/7.

172   Mboya, "Kenya's Constitutional Crisis."

173   "Minute from Informations Services for Somalis, Convened December 1957," January 1958, KNA BB/1/204.

174 Dean, Broadcasting Officer to the Acting Director of Information, "Introduction," 30 August 1955, KNA AHC/18/70.
175 Pritchford, "Strengthening of Information Services."
176 Jamieson, "Broadcasting from Mombasa."
177 A. M. Dean, Broadcasting Officer, Department of Information, Nairobi, "A Scheme for the Future Development of Broadcasting in Kenya," August 1955, KNA AHC/18/70.
178 Letter from Hunt, Provincial Information Officer, Coast Province, "Secret," n.d., KNA AHC/18/70.

CHAPTER 5. THE POLITICS OF DIVISIBILITY

1 This rebranding was contested in Parliament, with opponents arguing that it was an affront to the image of Daniel arap Moi and an effacement of "national heritage." See Jeremiah Kiplang'at, "MPs Oppose Bid to Rename Kasarani Stadium," *Nation*, 4 December 2013.
2 For a discussion of the importance of the visual presence of the figure of the "dictator"/authoritative elder in asserting political authority in Kenya, see Blunt, *For Money and Elders*. For broader perspectives, see Mbembe, *On the Postcolony*; Bayart, *The State in Africa*.
3 Tej Kohli, "Why Kenya, Home to Africa's 'Silicon Valley', Is Set to Be the Continent's Ultimate Tech Hub," *Mail & Guardian Africa*, 21 February 2015; Jake Bright and Aubrey Hruby, "The Rise of Silicon Savannah and Africa's Tech Movement," *Tech Crunch*, 23 July 2015.
4 Poggiali, "Digital Futures and Analogue Pasts? Citizenship and Ethnicity in Techno-Utopian Kenya"; Nyabola, *Digital Democracy, Analogue Politics*.
5 Ferguson, "Declarations of Dependence."
6 Welker, *Enacting the Corporation*.
7 Haugerud, *Culture of Politics in Modern Kenya*.
8 Breckenridge, "Failure of the 'Single Source.'"
9 Mitchell, "Society, Economy, and the State Effect."
10 Jenkins, *Bonds of Inequality*.
11 Bear, "Alternatives to Austerity," 5.
12 For a discussion of the heterogeneous effects of policies of austerity, see Ferguson, "Uses of Neoliberalism"; Von Schnitzler, *Democracy's Infrastructure*; Welker, *Enacting the Corporation*. See also Hibou, "Privatising the Economy to Privatising the State"; Collier, *Post-Soviet Social*, 27.
13 Guyer, *Legacies, Logics, Logistics*, 157. See also Callon and Muniesa, "Economic Markets."
14 Macharia, "Clientelism, Competition and Corruption," 47.
15 Macharia, "Clientelism, Competition and Corruption," 49.
16 Macharia, "Clientelism, Competition and Corruption," 50.
17 Macharia, "Clientelism, Competition and Corruption," 62.
18 Kenyatta, *Harambee!*, 7.
19 Berman, *Control & Crisis in Colonial Kenya*, 425–28.

20 Tyler, Hughes, and Renfrew, "Kenya: Facing the Challenges."
21 Macharia, "Clientelism, Competition and Corruption," 54–5.
22 Macharia, "Clientelism, Competition and Corruption," 73.
23 Macharia, "Clientelism, Competition and Corruption," 72–3; Republic of Kenya, *Committee of Review of Statutory Boards Report*, 22, as cited in Himbara, "Myths and Realities of Kenyan Capitalism," 98.
24 Kenya, though, was well ahead of other countries in sub-Saharan Africa, leading in telecommunications penetration until 1992. See Macharia, "Clientelism, Competition and Corruption," 64.
25 Macharia, "Clientelism, Competition and Corruption," 61.
26 Macharia, "Clientelism, Competition and Corruption," 67.
27 Republic of Kenya, *Report and Recommendations of the Working Party on Government Expenditures*, as cited in Himbara, "Myths and Realities of Kenyan Capitalism," 98.
28 Macharia, "Clientelism, Competition and Corruption," 85.
29 "Kenya Telecom Reform: A Good Story Gone Bad?" Wikileaks, 10 April 2005, https://wikileaks.org/plusd/cables/05NAIROBI1488_a.html; *Kenya National Assembly Official Record* (Hansard), 27 September 2007.
30 *Kenya National Assembly Official Record* (Hansard), 20 June 2000. The language of "unbundling" is an actor category used by international financial institutions to describe the process of privatization.
31 For a discussion of the distinction between "matters of concern" and "matters of fact," see Latour, "Why Has Critique Run Out of Steam?"
32 Leys, Borges, and Gold, "State Capital in Kenya," 316.
33 For a discussion, see Pitcher, "Entrepreneurial Governance."
34 *Kenya National Assembly Official Record* (Hansard), 10 June 2004.
35 *Kenya National Assembly Official Record* (Hansard), 10 June 2004.
36 *Kenya National Assembly Official Record* (Hansard), 18 June 2003.
37 "Kenya Telecom Reform: A Good Story Gone Bad?"
38 "Kenya Telecom Reform: A Good Story Gone Bad?"
39 "Kenya Telecom Reform: A Good Story Gone Bad?"
40 *Kenya National Assembly Official Record* (Hansard), 31 October 2001; "Kenya Telecom Reform: A Good Story Gone Bad?"
41 *Kenya National Assembly Official Record* (Hansard), 23 October 2001.
42 *Kenya National Assembly Official Record* (Hansard), 23 October 2001.
43 Guyer and Belinga, "Wealth in People."
44 Kenya National Assembly Public Investments Committee, *Fifteenth Report*.
45 Ewan Sutherland, "A Short Note on Corruption in Telecommunications in Kenya," SSRN, 31 January 2012, http://dx.doi.org/10.2139/ssrn.1996429.
46 "Kenya Telecom Reform: A Good Story Gone Bad?"
47 Chalfin, *Neoliberal Frontiers*; Bear, "Alternatives to Austerity."
48 On the dubious concept of technological "leapfrogging," see Bill Maurer, who argues: "Industry and academic researchers usually attribute the rapid dissemination of the mobile phone [and mobile money] to infrastructural leapfrogging: there is no need to lay cables, no large scale infrastructure projects required." Maurer, "Mobile Money," 593.

49   Goggin, *Cell Phone Culture*, 1.

50   "ICT penetration rates per 100 inhabitants, 2007," International Telecommunication Union (website), 2007, https://web.archive.org/web/20221217195738/http://www.itu.int/ITU-D/ict/statistics/ict/graphs/ICT_penetration_2007.jpg. Last accessed 2017.

51   FSD Kenya, "Financial Access in Kenya: Results of the 2006 National Survey," FinDev Gateway (website), October 2007, https://www.fsdkenya.org/finaccess/financial-access-in-kenya-results-of-the-2006-national-survey/.

52   "Kenya Telecom Reform: A Good Story Gone Bad?"

53   *Kenya National Assembly Official Record* (Hansard), 15 December 1999; *Kenya National Assembly Official Record* (Hansard), 27 September 2007.

54   While these debates largely broke down along partisan lines, with opponents of privatization associated with the opposition party of Raila Odinga and supporters associated with the administration of Mwai Kibaki, here I direct attention to the discursive frameworks through which these debates took place.

55   *Kenya National Assembly Official Record* (Hansard), 28 April 2004.

56   *Kenya National Assembly Official Record* (Hansard), 5 July 2005.

57   *Kenya National Assembly Official Record* (Hansard), 27 April, 2004.

58   *Kenya National Assembly Official Record* (Hansard), 5 July, 2005.

59   *Kenya National Assembly Official Record* (Hansard), 2 April, 2002.

60   *Kenya National Assembly Official Record* (Hansard), 28 April, 2004.

61   *Kenya National Assembly Official Record* (Hansard), 5 July, 2005.

62   Geschiere, *Perils of Belonging*; D'Avignon, *Ritual Geology*.

63   Emphasis added in the first and second instance. Phil, "The Coming Two Tier Global Economy: Which Is Thy Tier?" *Kumekucha* (blog), 24 May 2011, https://web.archive.org/web/20111011115214/http://kumekucha.blogspot.com/2011/05/coming-two-tier-global-economy-which-is.html. Phil's profile lists social worker as his profession, and Kibera, a large "informal" settlement on Nairobi's periphery, as his location.

64   *Kenya National Assembly Official Record* (Hansard), 27 April 2004.

65   *Kenya National Assembly Official Record* (Hansard), 27 April 2004.

66   *Kenya National Assembly Official Record* (Hansard), 27 April 2004.

67   *Kenya National Assembly Official Record* (Hansard), 27 April 2004.

68   This analysis benefited from two very different discussions of the relationship between qualities and quantities; see Guyer, *Marginal Gains*; Nelson, *Who Counts?*

69   This subtitle was taken from language used in Kenyan Parliamentary debates. *Kenya National Assembly Official Record* (Hansard), 27 April 2004; *Kenya National Assembly Official Record* (Hansard), 27 September 2007. *Nyama choma* is roasted meat, a favorite meal in Kenya. Note that the name of this organization has been changed to protect the identities of our interlocutors.

70   Prahalad, *Fortune at the Bottom of the Pyramid*, 1. For an analysis of this rationale at work, see Elyachar, *Markets of Dispossession*; Elyachar, "Next Practices"; Elyachar, "Phatic Labor."

71   "Kenya's Big Push for ICT Reform and Infrastructure," Wikileaks, 25 October 2007, https://web.archive.org/web/20230619200437/https://wikileaks.org/plusd/cables/07NAIROBI4202_a.html. Last accessed 2017.

72 "Kenya's Big Push for ICT Reform and Infrastructure."

73 Kenya National Assembly Public Investments Committee, *Fifteenth Report*.

74 "Kenya's Big Push for ICT Reform and Infrastructure," https://web.archive.org /web/20230619200437/https://wikileaks.org/plusd/cables/07NAIROBI4202_a .html. Last accessed 2017.

75 Since 2019, and for the first time, the CEO of Safaricom is a Kenyan national, Peter Ndegwa. Following the death of Bob Collymore in 2019, there were lively debates over who his successor would be. Interestingly, Kenyans were not uniformly supportive of the new CEO being a Kenyan, arguing that this would lead to "corruption" in the firm's top management. Others argued that this was long overdue, celebrating the further nationalization of the firm's top leadership.

76 Kenya National Assembly Public Investments Committee, *Fifteenth Report*.

77 Kenya National Assembly Public Investments Committee, *Fifteenth Report*. These questions were raised in spite of the fact that TKL's contributions had been in kind, not capital.

78 Kenya National Assembly Public Investments Committee, *Fifteenth Report*.

79 Kenya National Assembly Public Investments Committee, *Fifteenth Report*.

80 Guyer, *Legacies, Logics, Logistics*, 154.

81 Guyer, *Legacies, Logics, Logistics*, 154.

82 Kenya National Assembly Public Investments Committee, *Fifteenth Report*.

83 Kenya National Assembly Public Investments Committee, *Fifteenth Report*.

84 Kenya National Assembly Public Investments Committee, *Fifteenth Report*.

85 *Kenya National Assembly Official Record* (Hansard), 27 September 2007.

86 Emphasis added. *Kenya National Assembly Official Record* (Hansard), 27 September 2007.

87 *Kenya National Assembly Official Record* (Hansard), 27 September 2007.

88 Emphasis added. *Kenya National Assembly Official Record* (Hansard), 27 September 2007.

89 Holmes and Marcus, "Fast Capitalism," 33–57.

90 *Kenya National Assembly Official Record* (Hansard), 2 October 2007.

91 *Kenya National Assembly Official Record* (Hansard), 2 October 2007.

92 *Kenya National Assembly Official Record* (Hansard), 2 October 2007.

93 Chris, "Why You Support Corruption If You Buy Safaricom Shares," *Kumekucha* (blog), 26 March 2008, http://kumekucha.blogspot.com/2008/03/why-you-support -corruption-if-you-buy.html.

94 *Kenya National Assembly Official Record* (Hansard), 13 October 2010.

95 *Kenya National Assembly Official Record* (Hansard), 2 October 2007 (emphasis added).

96 *Kenya National Assembly Official Record* (Hansard), 14 June 2007.

97 Guyer, *Legacies, Logics, Logistics*, 60.

98 Branch, *Kenya: Between Hope and Despair*.

99 Kroll Associates UK Limited, "KTM Report," Wikileaks, 27 April 2004, https://web .archive.org/web/20220630124551/https://wikileaks.org/wiki/KTM_report. Last accessed 2017.

100 "Kenya: Safaricom IPO Gets Green Light," Wikileaks, 18 March 2008, https:// wikileaks.org/plusd/cables/08NAIROBI762_a.html. Last accessed 2017.

101 Branch, *Kenya: Between Hope and Despair.*

102 Phil, "Safaricom Now Joins Partisan Politics," *Kumekucha* (blog), January 1, 2008, https://web.archive.org/web/20080105042320/http://kumekucha.blogspot.com/2008/01/safaricom-now-joins-partisan-politics.html.

103 "Kenya: Safaricom IPO Gets Green Light."

104 "Cabinet Anxiety at a High!" *Cribarworld* (blog), 7 April 2008, https://web.archive.org/web/20230410181459/https://cribarworld.wordpress.com/2008/04/07/cabinet-anxiety-at-a-high/(emphasis added).

105 Chris, "The 'Sina Makosa' Grand Coalition and the Mouth-Watering Safaricom IPO," *Kumekucha* (blog), 23 March 2008, https://web.archive.org/web/20111010164443/http://kumekucha.blogspot.com/2008/03/sina-makosa-grand-coalition-and-mouth.html.

106 "The 'Sina Makosa' Grand Coalition."

107 "The 'Sina Makosa' Grand Coalition."

108 Comments Section, responding to "The 'Sina Makosa' Grand Coalition."

109 Comments Section, responding to "The 'Sina Makosa' Grand Coalition."

110 Blunt, *For Money and Elders.*

111 Comments Section, responding to "The 'Sina Makosa' Grand Coalition." *Matatus* are privately owned public service vehicles used by the majority of Kenyans. In everyday discourse, they are used to mark the difference between the people (*wananchi*) and elites, who famously travel in Mercedes Benzes, gaining for themselves the title *wabenzi.*

112 *Kenya National Assembly Official Record* (Hansard), 8 September 2009.

113 This term comes from Lonsdale, "Moral Economy of Mau Mau."

114 Rajak, *In Good Company*, 92.

115 Shika-Msa, "Safaricom IPO Goes Tribal," *Kumekucha* (blog), 28 March 2008, https://web.archive.org/web/20111010162043/http://kumekucha.blogspot.com/2008/03/safaricom-ipo-goes-tribal.html.

116 Rebecca Wanjiku, "Kenya's Safaricom IPO Disappoints Small Investors," NetworkWorld (website), 4 June 2008, https://web.archive.org/web/20230104193411/https://www.networkworld.com/article/2280772/kenya-s-safaricom-ipo-disappoints-small-investors.html.

117 "A Happy Roar from Kenya," *The Economist*, June 9, 2008.

118 Duncan Miriri, "UPDATE 2—Kenya's Safaricom IPO Oversubscribed by 532 pct." *Reuters*, 30 May 2008; International Finance Corporation, "Public-Private Partnership Stories; Kenya: Telkom Kenya," January 2013, https://www.reuters.com/article/business/media-telecom/kenyas-safaricom-ipo-oversubscribed-by-532-pct-idUSL30071239/. Oversubscription occurs when initial demand exceeds the amount of shares offered in an IPO, requiring an increase in share price or an issue of additional shares. In Safaricom's case, the excess demand for shares was so great that it was forced to refund a significant portion of the purchases.

119 Interview with James Oweno, 21 July 2015.

120 *Kenya National Assembly Official Record* (Hansard), 4 November 2008.

121 *Kenya National Assembly Official Record* (Hansard), 16 July 2008.

122 Safaricom, Annual Report, 2014, Safaricom (website), https://www.safaricom.co.ke
/investor-relations-landing/reports/annual-reports.

123 "Safaricom Foundation," Safaricom (website), archived 3 December 2005,
https://web.archive.org/web/20151203215916/http://www.safaricom.co.ke/about-us
/transforming-lives/safaricom-foundation.

124 Rajak, *In Good Company*, 99–100; Welker, *Enacting the Corporation*, 17.

125 "The Bitter Option Indeed!" *Kumekucha* (blog), 15 October 2011, https://web.archive
.org/web/20120119090244/http://kumekucha.blogspot.com/2011/10/bitter-option
-indeed.html. Author wrote of "Safaricom strategy at the expense of Kenyans:
Sustain big fat profits at all costs!!"

126 Bindra, *Peculiar Kenyan*, 90.

127 Bear and Mathur, "Remaking the Public Good," 18.

CHAPTER 6. SAFARICOM'S AUSTERE LABOR REGIME

1 Central Organization of Trade Unions (COTU-K), "Press Statement: New Safaricom CEO Dangerous for Workers' Rights," 26 April 2021, https://web.archive.org
/web/20210427173640/https://cotu-kenya.org/new-safaricom-ceo-dangerous-for
-workers-rights/.

2 Donald Kipkorir, "Why M-Pesa Agents Are Safaricom Employees," *Business Daily*, 16
March 2021.

3 For a discussion on digital rentiership, see Kean Birch and D. T. Cochrane, "Big
Tech: Four Emerging Forms of Digital Rentiership."

4 The language of the "bottom of the pyramid" was popularized by management
professor and consultant, C. K. Prahalad. See Prahalad, *The Fortune at the Bottom of
the Pyramid*.

5 Fraser, "Behind Marx's Hidden Abode," 63. Their labour is akin to "phatic
labour"—those forms of uncompensated work that generate, as Julia Elyachar has
noted, a "social infrastructure on which other projects oriented around the pursuit
of profit . . . [can] be constructed." Elyachar, "Phatic Labor," 453.

6 Maurer, "Data-Mining for Development," 129; Maurer, "Mobile Money," 592.

7 Meagher, "Cannibalizing the Informal Economy," 18.

8 Guyer, *Marginal Gains*.

9 "Safaricom PLC Annual Report and Financial Statements," 2022, Safaricom
(website), https://www.safaricom.co.ke/investor-relations-landing/reports/annual
-reports.

10 Fieldnotes, 7 August 2014.

11 Fieldnotes, 7 August 2014.

12 Hochschild, *Managed Heart*, 7.

13 Hochschild, *Managed Heart*, 7.

14 Fieldnotes, 7 August 2014.

15 Fraser, "Behind Marx's Hidden Abode," 66.

16 For a discussion of the relationship between the rise of the ATM and labor relations,
which documents how technological developments reduced what was initially a

large outlay of capital required to install and network these devices, see Batiz-Lazo, "ATMs," 198.

17 "Kenya Telecom Reform: A Good Story Gone Bad?" Wikileaks, 10 April 2005, https://wikileaks.org/plusd/cables/05NAIROBI1488_a.html.

18 Fraser, "Behind Marx's Hidden Abode," 64.

19 For a sensitive account for people's responses to the collapse of these promises in Zambia, see Ferguson, *Expectations of Modernity*.

20 Roy, *Poverty Capital*.

21 The "knowledge problem" and the lack of faith in planning that characterizes neoliberalism is broadly attributable to Friedrich Hayek. See Slobodian, *Globalists*; Elyachar, "Next Practices"; Von Schnitzler, *Democracy's Infrastructure*, 11.

22 Elyachar, "Next Practices," 117.

23 Akin Oyebode, "M-PESA and Beyond—Why Mobile Money Worked in Kenya and Struggles in Other Markets," *TechCabal*, accessed 1 September 2016.

24 Jack and Suri, "Mobile Money."

25 Breckenridge, "The Failure of the 'Single Source,'" 105.

26 Breckenridge, "The Failure of the 'Single Source,'" 96.

27 Klein and Mayer, "Mobile Banking and Financial Inclusion," 6.

28 Klein and Mayer, "Mobile Banking and Financial Inclusion," 7–8.

29 For a discussion of the discursive construction of agents as "intermediaries" rather than "mediators," see Maurer, Nelms, and Rea, "Bridges to Cash." It is worth noting that there is, in fact, an ongoing debate regarding the relative "agency" of "agents" within the mobile money industry. For the purposes of this chapter, however, my concern is mainly with the question of how the social work of M-Pesa agents is appropriated and revalued, and where the resultant profits end up.

30 Jack and Suri, "Mobile Money."

31 For a discussion of the ideological effects of the language of self-employment in contexts where large firms retain managerial oversight, see Crouch, "Long-Term Perspective on the Gig Economy."

32 Fieldnotes, 8 February 2015.

33 Klein and Mayer, "Mobile Banking and Financial Inclusion," 5.

34 See "Experience M-PESA - Safaricom," Safaricom (website), 5 October 2019, https://web.archive.org/web/20191005043736/https://www.safaricom.co.ke/personal/M-Pesa/getting-started/experience-M-Pesa.

35 Mol, *Body Multiple*.

36 Fieldnotes, 5 February 2015.

37 Fieldnotes, 13 June 2015.

38 Fieldnotes, 28 June 2015.

39 Fieldnotes, 13 June 2015.

40 Fieldnotes, 15 December 2014.

41 Fieldnotes, 7 August 2014.

42 Fieldnotes, 7 August 2014.

43 Elyachar, "Phatic Labor," 457.

44 Fraser, "Behind Marx's Hidden Abode," 59.

45 Miller and Rose, *Governing the Present*, 98.
46 This is the language Safaricom generated to market M-Pesa. For a treatment of how M-Pesa fits into Kenyans' fiduciary lives and alternative strategies of provisioning through mobilizing social networks, see Kusimba, *Reimagining Money*.
47 Murphy, *Sick Building Syndrome*.
48 Fieldnotes, 28 June 2015.
49 Fieldnotes, 8 February 2015.
50 Ferguson, "Uses of Neoliberalism," 172.
51 Jane Guyer uses the term "turbulence" (as opposed to "uncertainty," "risk," etc., which is the language of financial markets) as a way to mark how the lived experiences of financialization are distinct from the "uncertainty" that characterizes financial markets themselves. See Guyer, *Legacies, Logics, Logistics*.
52 Fieldnotes, 8 February 2015.
53 Keane, "Market, Materiality and Moral Metalanguage."
54 See "We Innovate," Safaricom (website), archived 16 March 2017, https://web.archive.org/web/20170316232908/http://www.safaricom.co.ke/about-us/innovation.
55 "Michael Joseph Was Spot On over Our 'Peculiar Habits,'" *Standard*, December 6, 2012, https://web.archive.org/web/20221004010309/http://www.standardmedia.co.ke/article/2000072201/michael-joseph-was-spot-on-over-our-peculiar-habits.
56 Bindra, *Peculiar Kenyan*.
57 Rajak, *In Good Company*.
58 Keane, "Market, Materiality and Moral Metalanguage."
59 Latour, *Science in Action*.
60 Tsing, "Sorting Out Commodities," 36.
61 Interview with James Oweno, 21 July 2015.
62 Hughes and Lonie, "M-PESA."
63 Berry, "Hegemony on a Shoestring."
64 Berry, "Hegemony on a Shoestring."
65 Interview with James Oweno, 21 July 2015.
66 "Safaricom Foundation Strategy 2014–2017," Safaricom Foundation (website), archived 30 November 2021, https://web.archive.org/web/20211130125931/https://www.safaricomfoundation.org/wp-content/uploads/2021/11/2014-2017-SUMMARY-VERSION-OF-STRATEGY-1.pdf.
67 For a more extended discussion, see Donovan and Park, "Knowledge/Seizure"; Donovan and Park, "Algorithmic Intimacy."
68 Breckenridge, "Failure of the 'Single Source,'" 95.
69 Fieldnotes, 23 June 2015.
70 FSD Kenya and FSD Africa, "The Growth of M-Shwari in Kenya—A Market Development Story," FSD Africa (website), November 2016, 4, https://web.archive.org/web/20210825044019/https://www.fsdafrica.org/wp-content/uploads/2019/08/M-Shwari_Briefing-final_digital.pdf-9.pdf. Last accessed 2017.
71 Breckenridge, "Failure of the 'Single Source,'" 105.
72 Schram, *Return of Ordinary Capitalism*, 26.
73 Fraser, "Behind Marx's Hidden Abode," 63.
74 Polanyi, *Great Transformation*.

75 Slobodian, *Globalists*. See especially the introduction.
76 Omar Mohammed, "Safaricom Will Not Be Forced to Loosen Its Dominant Hold on Kenya's Mobile-Money Market," *Quartz Africa*, 2 August 2015, https://web.archive .org/web/20200610134744/https://qz.com/africa/472886/safaricom-will-not-have-to -give-up-its-nearly-80-hold-of-kenyas-mobile-money-market/; Joseph Purnell, "Safaricom Pushes Back against Monopoly Complaints," *Telcotitans*, 9 December 2021, https://www.telcotitans.com/vodafonewatch/safaricom-pushes-back-against -monopoly-complaints/4189.article.
77 Elyachar, "Next Practices," 121

EPILOGUE

1   ArKan Yasin, "Safaricom: Empire, Kingdom or Republic?" *Nairobi Law Monthly*, 7 June 2017, https://nairobilawmonthly.com/safaricom-empire-kingdom-republic.
2   Breckenridge, "Failure of the 'Single Source.'"
3   Ciepley, "Beyond Public and Private."
4   Chalfin, *Neoliberal Frontiers*, 37.
5   Beatrice Obwocha, "Government, Safaricom Sign Deal for Sh15 Billion Security Surveillance System," Daily Nation, 25 November 2014, https://nation.africa/kenya /news/government-safaricom-sign-deal-for-sh15-billion-security-surveillance -system-1046276.
6   "Kenya: Safaricom Partners with Msurvey to Measure Kenya's Digital Economy," All Africa, 29 June 2017, https://allafrica.com/stories/201706290869.html; Donovan and Park, "Knowledge/Seizure."
7   Meagher, "Cannibalizing the Informal Economy."
8   Sandgren, *Mau Mau's Children*.
9   Sandgren, *Mau Mau's Children*. Wakaria's trajectory was not dissimilar from the men interviewed by Sandgren. Though these men were somewhat younger, they too were Kikuyu men who had an edge following independence, both by virtue of the density of schools in Kenya's Central Province and because of the ethnic partisanship of the government of Jomo Kenyatta, which favored Kikuyu jobseekers in staffing administrative and public sector posts.
10 Ogola, "Political Economy of the Media."
11 Federation of Kenyan Employers, "The Informal Economy in Kenya," March 2021, 12, https://www.ilo.org/wcmsp5/groups/public/---ed_emp/---emp_ent/documents /publication/wcms_820312.pdf.
12 Mwatha, "Study of Women Small Scale Enterprise."
13 "Corporate Hack with Leader's Soul," *Business Daily*, 20 August 2015, https://web .archive.org/web/20211124095128/https://www.businessdailyafrica.com/Corporate -hack-with-leader-s-soul/539444-2840024-ysgymiz/index.html.
14 Federation of Kenyan Employers, "The Informal Economy in Kenya"; King, *Jua Kali Kenya*.
15 Fraser, "Behind Marx's Hidden Abode," 63.
16 Ferguson, *Give a Man a Fish*.
17 Davis, *Planet of Slums*.

Bibliography

Adas, Michael. *Machines as the Measure of Men: Science, Technology, and Ideologies of Western Dominance*. Ithaca, NY: Cornell University Press, 2015. https://doi.org/10.7591/9780801455261.

Ambler, Charles H. *Kenyan Communities in the Age of Imperialism: The Central Region in the Late Nineteenth Century*. New Haven, CT: Yale University Press, 1988.

Amin, Samir. *Unequal Development: An Essay on the Social Formations of Peripheral Capitalism*. Translated by Brian Pierce. New York: Monthly Review Press, 1976.

Anderson, David M. "'Yours in Struggle for Majimbo.' Nationalism and the Party Politics of Decolonization in Kenya, 1955–64." *Journal of Contemporary History* 40, no. 3 (July 2005): 547–64.

Appadurai, Arjun. *Modernity at Large: Cultural Dimensions of Globalization*. Minneapolis: University of Minnesota Press, 1996.

Appel, Hannah. *The Licit Life of Capitalism: U.S. Oil in Equatorial Guinea*. Durham, NC: Duke University Press, 2019.

Apter, Andrew H. *The Pan-African Nation: Oil and the Spectacle of Culture in Nigeria*. Chicago: University of Chicago Press, 2005.

Armour, Charles. "The BBC and the Development of Broadcasting in British Colonial Africa 1946–1956." *African Affairs* 83, no. 332 (July 1984): 359–402.

Arrighi, Giovanni. *The Long Twentieth Century: Money, Power and the Origins of Our Times*. New York: Verso, 2010.

Atieno-Odhiambo, E. S. "The Colonial Government, the Settlers and the 'Trust' Principle in Kenya 1939." *Transafrican Journal of History* 2, no. 2 (1972): 94–113.

Atkins, Keletso E. *The Moon Is Dead! Give Us Our Money! The Cultural Origins of an African Work Ethic, Natal, South Africa, 1843–1900 (Social History of Africa)*. Portsmouth, NH: Heineman J. Currey, 1993.

Austen, Ralph. *African Economic History: Internal Development and External Dependency*. London: James Currey, 1987.

Austin, Herbert Henry. *With Macdonald in Uganda: A Narrative Account of the Uganda Mutiny and Macdonald Expedition in the Uganda Protectorate and the Territories to the North*. London: Edward Arnold, 1903. http://archive.org/details/withmacdonaldinuooaust.

Baglehole, Kenneth Charles. *A Century of Service: A Brief History of Cable and Wireless Ltd. 1868–1968.* London: Bournehall Press, 1970.

Bagwell, Philip Sidney. *The Transport Revolution.* London: Routledge, 1988. http://site .ebrary.com/id/10060774.

Barkan, Joshua. *Corporate Sovereignty: Law and Government under Capitalism.* Minneapolis: University of Minnesota Press, 2013.

Barlow, Arthur. *English-Kikuyu Dictionary.* Edited by Thomas Godfrey Benson. London: Clarendon Press, 1975.

Barry, Andrew. *Material Politics: Disputes along the Pipeline.* Chichester, West Sussex: Wiley-Blackwell, 2013.

Batiz-Lazo, Bernardo. "ATMs." In *Paid: Tales of Dongles, Checks, and Other Money Stuff,* edited by Bill Maurer and Lana Swartz, 197–210. Cambridge, MA: MIT Press, 2017.

Bayart, Jean-Francois. *The State in Africa: The Politics of the Belly.* Boston: Polity, 2009.

Bear, Laura. "Alternatives to Austerity: A Critique of Financialized Infrastructure in India and Beyond." *Anthropology Today* 33, no. 5 (October 2017): 3–7.

Bear, Laura. *Navigating Austerity: Currents of Debt along a South Asian River.* Anthropology of Policy. Stanford, CA: Stanford University Press, 2015.

Bear, Laura, and Nayanika Mathur. "Introduction: Remaking the Public Good: A New Anthropology of Bureaucracy." *Cambridge Journal of Anthropology* 33, no. 1 (2015): 18–34.

Berman, Bruce. *Control & Crisis in Colonial Kenya: The Dialectic of Domination.* London: James Currey, 1996.

Berman, Bruce, and John Lonsdale. "The Labors of 'Muigwithania': Jomo Kenyatta as Author, 1928–45." *Research in African Literatures* 29, no. 1 (1998): 16–42.

Berman, Bruce, and John Lonsdale. "Louis Leakey's Mau Mau: A Study in the Politics of Knowledge." *History and Anthropology* 5, no. 2 (1991): 143–204.

Berman, Bruce, and John Lonsdale. *Unhappy Valley: Conflict in Kenya & Africa, Book One: State and Class.* Eastern African Studies. London: James Currey, 1992.

Berman, Bruce, and John Lonsdale. *Unhappy Valley: Conflict in Kenya & Africa, Book Two: Violence & Ethnicity.* Eastern African Studies. Oxford: James Currey, 1992.

Berry, Sara. "Hegemony on a Shoestring: Indirect Rule and Access to Agricultural Land." *Africa: Journal of the International African Institute* 62, no. 3 (1992): 326–55.

Bindra, Sunny. *The Peculiar Kenyan: Tongue-in-Cheek Writings from the Sunday Nation 2003–2009.* Nairobi: Storymoha, 2013.

Birch, Kean and D. T. Cochrane, "Big Tech: Four Emerging Forms of Digital Rentier-ship," *Science as Culture* 31, no. 1 (January 2, 2022): 44–58.

Bishara, Fahad Ahmad. *A Sea of Debt: Law and Economic Life in the Western Indian Ocean, 1780–1950.* Asian Connections. Cambridge: Cambridge University Press, 2017.

Bissell, William Cunningham. *Urban Design, Chaos, and Colonial Power in Zanzibar.* Bloomington: Indiana University Press, 2011.

Blunt, Robert W. *For Money and Elders: Ritual, Sovereignty, and the Sacred in Kenya.* Chicago: University of Chicago Press, 2019.

Blyth, Mark. *Austerity: The History of a Dangerous Idea.* Oxford: Oxford University Press, 2015.

Boyce, Robert. "The Origins of Cable and Wireless Limited, 1918–1939: Capitalism, Imperialism, and Technical Change." In *Communications under the Seas: The Evolving*

Cable Network and Its Implications, edited by Bernard Finn and Daqing Yang, 81–114. Cambridge, MA: MIT Press, 2009.

Boyes, John. *John Boyes, King of the Wa-Kikuyu: A True Story of Travel and Adventure in Africa.* London: Methuen, 1912.

Branch, Daniel. *Kenya: Between Hope and Despair, 1963–2011.* New Haven, CT: Yale University Press, 2011.

Brantley, Cynthia. *Giriama and Colonial Resistance in Kenya, 1800–1920.* Berkeley: University of California Press, 2018.

Breckenridge, Keith. "The Failure of the 'Single Source of Truth about Kenyans': The NDRS, Collateral Mysteries and the Safaricom Monopoly." *African Studies* 78, no. 1 (January 2019): 105.

Brennan, James R. "Lowering the Sultan's Flag: Sovereignty and Decolonization in Coastal Kenya." *Comparative Studies in Society and History* 50, no. 4 (2008): 831–61. https://doi.org/10.1017/S0010417508000364.

Brennan, James R. "Radio Cairo and the Decolonization of East Africa, 1953–1964." In *Making a World after Empire: The Bandung Moment and Its Political Afterlives*, edited by Christopher J. Lee, 173–95. Athens: Ohio University Press, 2010.

Brennan, James R., Andrew Burton, and Yusufu Qwaray Lawi, eds. *Dar es Salaam: Histories from an Emerging African Metropolis.* Dar es Salaam: Mkuki na Nyota, 2007.

Brown, Christopher Leslie. "The Origins of 'Legitimate Commerce.'" In *Commercial Agriculture: The Slave Trade and Slavery in Atlantic Africa*, edited by Robin Law, Susan Schwarz, and Silke Strickrodt, 138–57. Suffolk, UK: Boydell & Brewer, 2013.

Brown, Wendy. *Undoing the Demos: Neoliberalism's Stealth Revolution.* New York: Zone Books, 2017.

Brownell, Emily. *Gone to Ground: A History of Environment and Infrastructure in Dar es Salaam.* Pittsburgh: University of Pittsburgh Press, 2020.

Bujra, Janet. "Women 'Entrepreneurs' of Early Nairobi." *Canadian Journal of African Studies* 9, no. 2 (1975): 213–34.

Callon, Michel, and Fabiane Muniesa. "Economic Markets as Calculative Collective Devices." *Organization Studies* 26, no. 8 (2005): 1229–50.

Campbell, Todd C. "Sound Finance: Gladstone and British Government Finance, 1880–1895." PhD diss., London School of Economics and Political Science, 2004.

Capotescu, Cristian, Oscar Sanchez-Sibony, and Melissa Teixeira. "Austerity without Neoliberals: Reappraising the Sinuous History of a Powerful State Technology." *Capitalism: A Journal of History and Economics* 3, no. 2 (2022): 379–420.

Chalfin, Brenda. *Neoliberal Frontiers: An Ethnography of Sovereignty in West Africa.* Chicago: University of Chicago Press, 2010.

Chaudhuri, Meghna. "Lives of Value: Land, Labor, and Agrarian Finance in South Asia, 1830–1950." PhD diss., New York University, 2020.

Ciepley, David. "Beyond Public and Private: Toward a Political Theory of the Corporation." *American Political Science Review* 107, no. 1 (2013): 139–58.

Clark, Carolyn. "Louis Leakey as Ethnographer: On the Southern Kikuyu before 1903." *Canadian Journal of African Studies* 23, no. 3 (1989): 380–98.

Collier, Stephen J. *Post-Soviet Social: Neoliberalism, Social Modernity, Biopolitics.* Princeton, NJ: Princeton University Press, 2011.

Colonial Office Advisory Committee on Education in the Colonies. *Mass Education in African Society*. London: His Majesty's Stationery Office, 1944.

Connolly, N. D. B. "A White Story." *Dissent Magazine*, 22 January 2018.

Cooper, Frederick. "Alternatives to Empire: France and Africa after World War II." In *The State of Sovereignty: Territories, Laws, Populations*, 21st Century Studies, v. 3, edited by Douglas Howland and Luise White, 94–123. Bloomington: Indiana University Press, 2009.

Cooper, Frederick. "Conflict and Connection: Rethinking Colonial African History." *The American Historical Review* 99, no. 5 (1994): 1516–45.

Cooper, Frederick. *Decolonization and African Society: The Labor Question in French and British Africa*. African Studies Series 89. Cambridge: Cambridge University Press, 1996.

Cooper, Frederick. *On the African Waterfront: Urban Disorder and the Transformation of Work in Colonial Mombasa*. New Haven, CT: Yale University Press, 1987.

Cooper, Frederick. *Plantation Slavery on the East Coast of Africa*. New Haven, CT: Yale University Press, 1977.

Cooper, Frederick. "What Is the Concept of Globalization Good For? An African Historian's Perspective." *African Affairs* 100, no. 399 (2001): 189–213.

Cooper, Frederick, and Randall M. Packard, eds. *International Development and the Social Sciences: Essays on the History and Politics of Knowledge*. Berkeley: University of California Press, 1997.

Coronil, Fernando. *The Magical State: Nature, Money, and Modernity in Venezuela*. Chicago: University of Chicago Press, 1997.

Correspondence Respecting the Retirement of the Imperial British East Africa Company. London: Her Majesty's Stationery Office, 1895.

Coupland, Reginald. *The Exploitation of East Africa 1856–1890: The Slave Trade and the Scramble*. Evanston, IL: Northwestern University Press, 1967.

Cronon, William. *Nature's Metropolis: Chicago and the Great West*. New York: W. W. Norton, 1992.

Crouch, Colin. "A Long-Term Perspective on the Gig Economy." *American Affairs* 2, no. 2 (2018): 51–64. https://americanaffairsjournal.org/2018/05/a-long-term-perspective-on-the-gig-economy.

Cummings, Robert J. "Aspects of Human Porterage with Special Reference to the Akamba of Kenya: Towards an Economic History, 1820–1920." PhD diss., University of California, Los Angeles, 1975. ProQuest Dissertations & Theses Global (302758396). https://login.libproxy.newschool.edu/login?url=https://www.proquest.com/dissertations-theses/aspects-human-porterage-with-special-reference/docview/302758396/se-2?accountid=12261.

D'Avignon, Robyn. *A Ritual Geology: Gold and Subterranean Knowledge in Savanna West Africa*. Durham, NC: Duke University Press, 2022.

Davis, Mike. *Late Victorian Holocausts: El Niño Famines and the Making of the Third World*. London: Verso, 2001.

Davis, Mike. *Planet of Slums*. London: Verso, 2006.

De Kiewiet, Marie J. "History of the Imperial British East Africa Company 1876–1895." PhD diss., King's College London, 1955.

Dodd, Nigel. *The Social Life of Money*. Princeton, NJ: Princeton University Press, 2016. https://doi.org/10.1515/9781400880867.

Donovan, Kevin P. *The Moneychanger State: Economic Sovereignty and Citizenship in Decolonizing East Africa.* Cambridge: Cambridge University Press, 2024.

Donovan, Kevin P., and Emma Park. "Algorithmic Intimacy: The Data Economy of Predatory Inclusion in Kenya." *Social Anthropology* 30, no. 2 (2022): 120–39.

Donovan, Kevin P., and Emma Park. "Knowledge/Seizure: Debt and Data in Kenya's Zero Balance Economy." *Antipode: A Radical Journal of Geography* 54, no. 4 (2022): 1063–85.

Du Bois, W. E. B. *The Problem of the Color Line at the Turn of the Twentieth Century: The Essential Early Essays.* Edited by Nahum Dimitri Chandler. New York: Fordham University Press, 2015.

Eagleton, Catherine. "Currency as Commodity, as Symbol of Sovereignty and as Subject of Legal Dispute: Henri Greffulhe and the Coinage of Zanzibar in the Late Nineteenth Century." In *Currencies of the Indian Ocean World*, edited by Steven Serels and Gwyn Campbell, 113–140. Cham, CH: Palgrave Macmillan, 2019. https://doi.org/10.1007/978-3-030-20973-5.

Edwards, Paul N. "Infrastructure and Modernity: Force, Time, and Social Organization in the History of Sociotechnical Systems." In *Modernity and Technology*, edited by Thomas J. Misa, Philip Brey, and Andrew Feenberg, 185–225. Cambridge, MA: MIT Press, 2002.

Elyachar, Julia. *Markets of Dispossession: NGOs, Economic Development, and the State in Cairo.* Durham, NC: Duke University Press, 2005.

Elyachar, Julia. "Next Practices: Knowledge, Infrastructure, and Public Goods at the Bottom of the Pyramid." *Public Culture* 24, no. 1 (2012): 109–29.

Elyachar, Julia. "Phatic Labor, Infrastructure, and the Question of Empowerment in Cairo." *American Ethnologist* 37, no. 3 (2010): 452–64.

Englund, Harri. "Anti Anti-Colonialism: Vernacular Press and Emergent Possibilities in Colonial Zambia." *Comparative Studies in Society and History* 57, no. 1 (2015): 221–47.

Farler, J. P. "Native Routes in East Africa from Pangani to the Masai Country and the Victoria Nyanza." *Proceedings of the Royal Geographical Society and Monthly Record of Geography* 4, no. 12 (1882): 730–42.

Ferguson, James. "Declarations of Dependence: Labour, Personhood, and Welfare in Southern Africa." *Journal of the Royal Anthropological Institute* 19, no. 2 (June 2013): 223–42.

Ferguson, James. *Expectations of Modernity: Myths and Meanings of Urban Life on the Zambian Copperbelt.* Oakland: University of California Press, 1999.

Ferguson, James. *Give a Man a Fish: Reflections on the New Politics of Distribution.* The Lewis Henry Morgan Lectures. Durham, NC: Duke University Press, 2015.

Ferguson, James. "The Uses of Neoliberalism." *Antipode* 41, no. S1 (2009): 166–84.

Forstater, Mathew. "Taxation and Primitive Accumulation: The Case of Colonial Africa." *Research in Political Economy* 22 (January 1, 2005): 51–64. https://doi.org/10.1016/S0161-7230(04)22002-8.

Franklin, Harry. *Report on 'the Saucepan Special,' The Poor Man's Radio for Rural Populations.* Lusaka: Northern Rhodesia Information Department, 1950.

Fraser, Nancy. "Behind Marx's Hidden Abode: For an Expanded Conception of Capitalism." *New Left Review* 86 (2014): 55–72.

Fraser, Nancy, and Rahel Jaeggi. *Capitalism: A Conversation in Critical Theory*. Medford, MA: Polity, 2018.

Fredericks, Rosalind. *Garbage Citizenship: Vital Infrastructures of Labor in Dakar, Senegal*. Durham, NC: Duke University Press, 2018.

Freed, Libbie. "Networks of (Colonial) Power: Roads in French Central Africa after World War I." *History and Technology: An International Journal* 26, no. 3 (2010): 203–23.

Gabay, Clive. "Decolonizing Interwar Anticolonial Solidarities: The Case of Harry Thuku." *International Journal of Postcolonial Studies* 20, no. 4 (2018): 549–66.

Galbraith, John S. *Mackinnon and East Africa 1878–1895: A Study in the "New Imperialism."* Cambridge Commonwealth Series. Cambridge: Cambridge University Press, 1972.

Gardner, Leigh. *Taxing Colonial Africa: The Political Economy of British Imperialism*. Oxford Historical Monographs. Oxford: Oxford University Press, 2012.

Geschiere, Peter. *The Perils of Belonging: Autochthony, Citizenship, and Exclusion in Africa and Europe*. Chicago: University of Chicago Press, 2009.

Gjerso, Jonas Fossli. "The Scramble for East Africa: British Motives Reconsidered, 1884–95." *Journal of Imperial and Commonwealth History* 43, no. 5 (2015): 831–60.

Glassman, Jonathon. *Feasts and Riot: Revelry, Rebellion, and Popular Consciousness on the Swahili Coast, 1856–1888*. Social History of Africa. Portsmouth, NH: Heinemann, 1995.

Glassman, Jonathon. *War of Words, War of Stones: Racial Thought and Violence in Colonial Zanzibar*. Bloomington: Indiana University Press, 2011.

Goggin, Gerard. *Cell Phone Culture: Mobile Technology in Everyday Life*. New York: Routledge, 2006.

Goswami, Chhaya. *The Call of the Sea: Kachchhi Traders in Muscat and Zanzibar, c. 1800–1880*. New Delhi: Orient BlackSwan, 2011.

Goswami, Manu. *Producing India: From Colonial Economy to National Space*. Chicago Studies in Practices of Meaning. Chicago: University of Chicago Press, 2004.

Grace, Joshua. *African Motors: Technology, Gender, and the History of Development*. Durham, NC: Duke University Press, 2021.

Great Britain. *Handbook on Broadcasting Services in the Colonies, etc.* London: Information Department, Colonial Office, 1953.

Great Britain Parliament House of Commons. *Parliamentary Papers: 1850–1908*. London: His Majesty's Stationery Office, 1907.

Grimes, Mable S. *Life out of Death or the Story of the Africa Inland Mission*. London: Africa Inland Mission, 1917.

Guma, Prince K. "Incompleteness of Urban Infrastructures in Transition: Scenarios from the Mobile Age in Nairobi." *Social Studies of Science* 50, no. 5 (October 1, 2020): 728–50.

Gunner, Liz, Dina Ligaga, and Dumisami Moyo, eds. *Radio in Africa: Publics, Cultures, Communities*. Johannesburg: Wits University Press, 2011.

Guyer, Jane I. *Legacies, Logics, Logistics: Essays in the Anthropology of the Platform Economy*. Chicago: University of Chicago Press, 2016.

Guyer, Jane I. *Marginal Gains: Monetary Transactions in Atlantic Africa*. The Lewis Henry Morgan Lectures 1997. Chicago: University of Chicago Press, 2004.

Guyer, Jane I. "Representation without Taxation: An Essay on Democracy in Rural Nigeria, 1952–1990." *African Studies Review* 35, no. 1 (1992): 41–79.

Guyer, Jane I., and Samuel M. Eno Belinga. "Wealth in People as Wealth in Knowledge: Accumulation and Composition in Equatorial Africa." *Journal of African History* 36, no. 1 (March 1995): 91–120. https://doi.org/10.1017/S0021853700026992.

Haraway, Donna. *Simians, Cyborgs, and Women: The Reinvention of Nature*. New York: Routledge, 1991.

Harrison, Graham. "Debt, Development and Intervention in Africa: The Contours of a Sovereign Frontier." *Journal of Intervention and Statebuilding* 1, no. 2 (June 1, 2007): 189–209.

Hart, Jennifer. *Ghana on the Go: African Mobility in the Age of Motor Transportation*. Bloomington: Indiana University Press, 2016.

Harvey, David. *A Brief History of Neoliberalism*. Reprint, Oxford: Oxford University Press, 2011.

Harvey, David. *The Limits to Capital*. London: Verso, 2018.

Haugerud, Angelique. *The Culture of Politics in Modern Kenya*. Cambridge: Cambridge University Press, 1997.

Heath, Carla Wilson. "Broadcasting in Kenya: Policy and Politics, 1928–1984." PhD diss., University of Illinois Urbana-Champaign, 1986.

Hecht, Gabrielle. "Africa and the Nuclear World: Labor, Occupational Health, and the Transnational Production of Uranium." *Comparative Studies in Society and History* 51, no. 4 (2009): 896–926.

Hecht, Gabrielle. *Being Nuclear: Africans and the Global Uranium Trade*. Cambridge, MA: MIT Press, 2014.

Hecht, Gabrielle, ed. *Entangled Geographies: Empire and Technopolitics in the Global Cold War*. Inside Technology. Cambridge, MA: MIT Press, 2011.

Hecht, Gabrielle. "Interscalar Vehicles for an African Anthropocene: On Waste, Temporality, and Violence." *Cultural Anthropology* 33, no. 1 (February 22, 2018): 109–41.

Hecht, Gabrielle. "Rupture-Talk in the Nuclear Age: Conjugating Colonial Power in Africa." *Social Studies of Science* 32, no. 5–6 (December 1, 2002): 691–727.

Hibou, Béatrice. "From Privatising the Economy to Privatising the State: An Analysis of the Continual Formation of the State." In *Privatizing the State*, edited by Béatrice Hibou, 1–46. New York: Columbia University Press, 2004.

Himbara, David. "Myths and Realities of Kenyan Capitalism." *Journal of Modern African Studies* 31, no. 1 (1993): 93–107.

Hindlip, Charles Allsopp. *British East Africa: Past, Present, and Future*. London: T. Fisher Unwin, 1905.

Hobson, J. A. *Imperialism: A Study*. London: James Nisbet, 1902.

Hochschild, Arlie Russel. *The Managed Heart: Commercialization of Human Feeling*. Berkeley: University of California Press, 2012.

Holmes, Douglas R., and George E. Marcus. "Fast Capitalism: Para-Ethnography and the Rise of the Symbolic Analyst." In *Frontiers of Capital: Ethnographic Reflections on the New Economy*, edited by Melissa S. Fisher and Greg Downey, 33–57. Durham, NC: Duke University Press, 2006.

Hopkins, Donald R. *The Greatest Killer: Smallpox in History*. Chicago: University of Chicago Press, 2002.

Howell, John. "What's Wrong with Managers?" *Development Policy Review* 11, no. 1 (1978): 53–69.

Hughes, Nick, and Susie Lonie. "M-PESA: Mobile Money for the 'Unbanked' Turning Cellphones into 24-Hour Tellers in Kenya." *Innovations* 2, no. 1–2 (2007): 63–81.

Hughes, Thomas Parke. *Networks of Power: Electrification in Western Society, 1880–1930.* Softshell Books History of Technology. Baltimore, MD: John Hopkins University Press, 1993.

Hunt, Nancy Rose. *A Colonial Lexicon of Birth Ritual, Medicalization, and Mobility in the Congo.* Durham, NC: Duke University Press, 1999.

Hunt, Nancy Rose. *A Nervous State: Violence, Remedies, and Reverie in Colonial Congo.* Durham, NC: Duke University Press, 2016.

Jack, William, and Tavneet Suri. "Mobile Money: The Economics of M-PESA." NBER Working Paper Series W16721. Cambridge, MA: National Bureau of Economic Research, 2011.

Jackson, Steven. "Rethinking Repair." In *Media Technologies: Essays on Communication, Materiality, and Society*, edited by Tarleton Gillespie, Pablo Boczkowski, and Kirsten Foot, 221–239. Cambridge, MA: MIT Press, 2014.

Jacobs, Nancy. *Birders of Africa: History of a Network.* New Haven, CT: Yale University Press, 2016.

Jakes, Aaron. "Boom, Bugs, Bust: Egypt's Ecology of Interest, 1882–1914." *Antipode* 49, no. 4 (February 2016). https://doi.org/10.1111/anti.12216.

Jakes, Aaron. *Egypt's Occupation: Colonial Economism and the Crises of Capitalism.* Stanford, CA: Stanford University Press, 2020.

Jenkins, Destin. *The Bonds of Inequality: Debt and the Making of the American City.* Chicago: University of Chicago Press, 2021.

Kafer, Alison. *Feminist, Queer, Crip.* Bloomington: Indiana University Press, 2013.

Karp, Ivan. "Development and Personhood." In *Critically Modern: Alternatives, Alterities, Anthropologies*, edited by Bruce M. Knauft, 82–104. Bloomington: Indiana University Press, 2002.

Karuka, Manu. *Empire's Tracks: Indigenous Nations, Chinese Workers, and the Transcontinental Railroad.* Oakland: University of California Press, 2019.

Keane, Webb. "Market, Materiality and Moral Metalanguage." *Anthropological Theory* 8, no. 27 (2008): 27–42.

Kennedy, P. M. "Imperial Cable Communications and Strategy 1870–1914." *English Historical Review* 86, no. 341 (1971): 728–52.

Kenya National Assembly Public Investments Committee. *Fifteenth Report of the Public Investments Committee on the Accounts of State Corporations.* Nairobi: National Assembly, 2007.

Kenyatta, Jomo. *Harambee! The Prime Minister of Kenya's Speeches 1963–1964.* Nairobi: Oxford University Press, 1964.

Kimari, Wangui, and Henrik Ernston. "Imperial Remains and Imperial Invitations: Centering Race within the Contemporary Large-Scale Infrastructures of East Africa." *Antipode* 52, no. 3 (May 2020): 825–46.

Kindy, Hydar. *Life and Politics in Mombasa.* Nairobi: East Africa Broadcasting House, 1972.

King, Kenneth. *Jua Kali Kenya: Change and Development in an Informal Economy 1970–95.* London: James Currey, 1996.

Kitching, G. N. *Class and Economic Change in Kenya: The Making of an African Petite Bourgeoisie 1905–1970.* New Haven, CT: Yale University Press, 1980.

Klein, Michael, and Colin Mayer. "Mobile Banking and Financial Inclusion: The Regulatory Lessons." Policy Research Working Papers 5664. World Bank, June 22, 2013.

Klein, Naomi. *The Shock Doctrine: The Rise of Disaster Capitalism*. New York: Picador, 2008.

Knight, Daniel, and Charles Stewart. "Ethnographies of Austerity: Temporality, Crisis and Affect in Southern Europe." *History and Anthropology* 27, no. 1 (January 2016): 1–18.

Kriger, Colleen. *Pride of Men: Ironworking in 19th Century West Central Africa*. Portsmouth, NH: Heinemann, 1999.

Kusimba, Sibel. *Reimagining Money: Kenya in the Digital Finance Revolution*. Stanford, CA: Stanford University Press, 2021.

Larkin, Brian. "The Politics and Poetics of Infrastructure." *Annual Review of Anthropology* 42 (2013): 327–43.

Larkin, Brian. *Signal and Noise: Media, Infrastructure and Urban Culture in Nigeria*. Durham, NC: Duke University Press, 2008.

Latour, Bruno. *Science in Action: How to Follow Scientists and Engineers through Society*. New York: Harvard University Press, 1987.

Latour, Bruno. "Why Has Critique Run Out of Steam? From Matters of Fact to Matters of Concern." *Critical Inquiry* 30, no. 2 (2004): 225–48.

Leakey, Louis. *The Southern Kikuyu Before 1903*. London: Academic Press, 1977.

Leggett, E. Humphrey. "Economics and Administration in British East Africa." *Geography* 14, no. 5 (1928): 395–405.

Lenin, V. I. *Imperialism: The Highest Stage of Capitalism*. London: Penguin Classics, 2010.

Lewis, Joanna. *Empire State-Building: War & Welfare in Kenya, 1925–52*. Eastern African Studies. Athens: Ohio University Press, 2000.

Leys, Colin. *Underdevelopment in Kenya: The Political Economy of Neo-Colonialism 1964–1971*. Berkeley: University of California Press, 1975.

Leys, Colin, Jane Borges, and Hyam Gold. "State Capital in Kenya: A Research Note." *Canadian Journal of African Studies* 14, no. 2 (1980): 307–17.

Livingstone, David. *Dr Livingstone's Cambridge Lectures: Together with a Prefatory Letter by the Rev. Professor Sedgwick*. Cambridge, UK: Deighton, Bell, 1858.

Livingston, Julie. *Self-Devouring Growth: A Planetary Parable as Told from Southern Africa*. Durham, NC: Duke University Press, 2019.

Lonsdale, John. "Authority, Gender & Violence: The War within Mau Mau's Fight for Land and Freedom." In *Mau Mau and Nationhood*, edited by E. S. Atieno-Odhiambo and John Lonsdale, 46–75. Oxford: James Currey, 2003.

Lonsdale, John. "The Moral Economy of Mau Mau: Wealth, Poverty and Civic Virtue in Kikuyu Political Thought." In *Unhappy Valley: Conflict in Kenya and Africa, Book Two: Violence and Ethnicity*, by Bruce Berman and John Lonsdale, 315–467. Eastern African Studies. Oxford: James Currey, 1992.

Low, D. A. *Fabrication of Empire: The British and the Uganda Kingdoms, 1890–1902*. Cambridge: Cambridge University Press, 2009.

Lowe, Lisa. *The Intimacies of Four Continents*. Durham, NC: Duke University Press, 2015.

Lugard, Frederick John Dealtry. *The Dual Mandate in British Tropical Africa*. London: Forgotten Books, 2012.

Lugard, Frederick John Dealtry. *The Rise of Our East African Empire: Early Efforts in Nyasaland and Uganda*. Vol. 1. London: William Blackwood and Sons, 1893.

Luxemburg, Rosa. *The Accumulation of Capital*. Routledge Classics. London: Routledge, 2003.

MacArthur, Julie. *Cartography and the Political Imagination: Mapping Community in Colonial Kenya*. New African Histories. Athens: Ohio University Press, 2016.

Macdonald, Sir James Ronald Leslie. *Soldiering and Surveying in British East Africa, 1891–1894*. London: Edward Arnold, 1897.

Macharia, L. N. "Clientelism, Competition and Corruption: Informal Institutions and Telecommunications Reform in Kenya." PhD diss., Stanford University, 2007.

Machava, Benedito. "Reeducation Camps, Austerity, and the Carceral Regime in Socialist Mozambique (1974–79)." *Journal of African History* 60, no. 3 (2019): 429–55.

Mackenzie, Donald. *Material Markets: How Economic Agents Are Constructed*. Oxford: Oxford University Press, 2009.

Maine, Henry Sumner. *Sir Henry Maine: A Brief Memoir of His Life*. New York: Henry Holt, 1892.

Mains, Daniel. *Hope Is Cut: Youth, Unemployment, and the Future in Urban Ethiopia*. Philadelphia: Temple University Press, 2011.

Mamdani, Mahmood. *Citizen and Subject: Contemporary Africa and the Legacy of Late Colonialism*. New paperback edition. Princeton Studies in Culture/Power/History. Princeton, NJ: Princeton University Press, 2018.

Mann, Michael. "The Autonomous Power of the State: Its Origins, Mechanisms and Results." *European Journal of Sociology* 25, no. 2 (1984): 185–213.

Martin, Phyllis. *Leisure and Society in Colonial Brazzaville*. Cambridge: Cambridge University Press, 2002.

Mattei, Clara. *The Capital Order: How Economists Invented Austerity and Paved the Way to Fascism*. Chicago: University of Chicago Press, 2022.

Maurer, Bill. "Data-Mining for Development? Poverty, Payment, and Platform." In *Territories of Poverty: Rethinking North and South*, edited by Ananya Roy and Emma Shaw Crane, 126–143. Athens: University of Georgia Press, 2015.

Maurer, Bill. "Mobile Money: Communication, Consumption and Change in the Payments Space." *Journal of Development Studies* 48, no. 5 (2012): 589–604.

Maurer, Bill, Taylor C. Nelms, and Stephen C. Rea. "'Bridges to Cash': Channelling Agency in Mobile Money." *Journal of the Royal Anthropological Institute* 19, no. 1 (2013): 52–74.

Mavhunga, Clapperton Chakanetsa. *The Mobile Workshop: The Tsetse Fly and African Knowledge Production*. Cambridge, MA: MIT Press, 2018.

Mavhunga, Clapperton Chakanetsa. *Transient Workspaces: Technologies of Everyday Innovation in Zimbabwe*. Cambridge, MA: MIT Press, 2014.

Mavhunga, Clapperton Chakanetsa, ed. *What Do Science, Technology, and Innovation Mean from Africa?* Cambridge, MA: MIT Press, 2017.

Maxon, Robert M. *John Ainsworth and the Making of Kenya*. Lanham, MD: University Press of America, 1980.

Mbembe, Achille. *On the Postcolony*. Berkeley: University of California Press, 2001.

Mboya, Tom. "Kenya's Constitutional Crisis." *Africa Today* 5, no. 5 (1958): 6–10.

McDermott, P. L. *British East Africa or IBEA: A History of the Formation and Work of the Imperial British East Africa Company*. London: Chapman & Hall, 1893.

Meagher, Kate. "Cannibalizing the Informal Economy: Frugal Innovation and Economic Inclusion in Africa." *European Journal of Development Research* 30, no. 1 (2018): 17–33.

Meier, Prita. *Swahili Port Cities: The Architecture of Elsewhere*. African Expressive Cultures. Bloomington: Indiana University Press, 2016.

Meinhertzhagen, Richard. *Kenya Diary 1902-06*. London: Eland Books, 1957.

Mika, Marissa. *Africanizing Oncology: Creativity, Crisis, and Cancer in Uganda*. Athens: Ohio University Press, 2021.

Miller, Peter, and Nikolas S. Rose. *Governing the Present: Administering Economic, Social and Personal Life*. Cambridge, UK: Polity Press, 2013.

Mitchell, David T., and Sharon L. Snyder, eds. *The Body and Physical Difference: Discourses of Disability*. Ann Arbor: University of Michigan Press, 2001.

Mitchell, Timothy. *Rule of Experts: Egypt, Techno-Politics, Modernity*. Berkeley: University of California Press, 2002.

Mitchell, Timothy. "Society, Economy, and the State Effect." In *State/Culture: State-Formation after the Cultural Turn*, edited by George Steinmetz, 76-97. Ithaca, NY: Cornell University Press, 1999.

Mol, Annemarie. *The Body Multiple: Ontology in Medical Practice*. Durham, NC: Duke University Press, 2002.

Momanyi, Clara. *Ronald Ngala: Teacher with a Mission*. Nairobi: Sasa Semia Publications, 2001.

Moorman, Marissa J. *Powerful Frequencies: Radio, State Power, and the Cold War in Angola, 1931-2002*. Athens: Ohio University Press, 2019.

Moskowitz, Kara. *Seeing Like a Citizen: Decolonization, Development, and the Making of Kenya, 1945-1980*. New African Histories. Athens: Ohio University Press, 2019.

Mrázek, Rudolf. *Engineers of Happy Land: Technology and Nationalism in a Colony*. Princeton, NJ: Princeton University Press, 2018. https://doi.org/10.1515/9780691186931.

Mukerji, Chandra. *Impossible Engineering: Technology and Territoriality on the Canal du Midi*. Princeton Studies in Cultural Sociology. Princeton, NJ: Princeton University Press, 2009.

Munro, J. Forbes. *Maritime Enterprise and Empire: Sir William Mackinnon and His Business Network, 1823-93*. Suffolk, UK: Boydell Press, 2003.

Muriuki, Godfrey. *A History of the Kikuyu: 1500-1900*. Nairobi: Oxford University Press, 1974.

Murphy, Michelle. *Sick Building Syndrome and the Problem of Uncertainty: Environmental Politics, Technoscience, and Women Workers*. Durham, NC: Duke University Press, 2006.

Musson, A. E. "The Great Depression in Britain, 1873-1896: A Reappraisal." *Journal of Economic History* 19, no. 2 (1959): 199-228.

Mutongi, Kenda. *Matatu: A History of Popular Transportation in Nairobi*. Chicago: University of Chicago Press, 2017.

Mwangi, Wambui. "Of Coins and Conquest: The East African Currency Board, the Rupee Crisis, and the Problem of Colonialism in the East African Protectorate." *Comparative Studies in Society and History* 43, no. 4 (2001): 763–87.

Mwangi, Wambui. "The Order of Money: Colonialism and the East African Currency Board." PhD diss., University of Pennsylvania, 2003.

Mwatha, Regina G. "A Study of Women Small Scale Enterprise in the Urban Informal Sector: A Case of Nairobi City." Master of Arts thesis, University of Nairobi, 1990.

Nasser, Gamal Abdel. *The Philosophy of the Revolution*. Buffalo, NY: Smith, Keynes & Marshall, 1959.

Nelson, Diana. *Who Counts? The Mathematics of Death and Life after Genocide*. Durham, NC: Duke University Press, 2015.

Nyabola, Nanjala. *Digital Democracy, Analogue Politics: How the Internet Era Is Transforming Politics in Kenya*. London: Zed Books, 2018.

Offner, Amy C. *Sorting Out the Mixed Economy: The Rise and Fall of Welfare and Developmental States in the Americas*. Princeton, NJ: Princeton University Press, 2019.

Ogola, George. "The Political Economy of the Media in Kenya: From Kenyatta's Nation-Building Press to Kibaki's Local-Language FM Radio." *Africa Today* 57, no. 3 (2011): 77–95.

Oliver, Roland. "Some Factors in the British Occupation of East Africa, 1884–1894." *Uganda Journal* 15, no. 1 (1951): 49–64.

Ong, Aihwa. *Neoliberalism as Exception: Mutations in Citizenship and Sovereignty*. Durham, NC: Duke University Press, 2006.

Osborn, Emily Lynn. "From Bauxite to Cooking Pots: Aluminum, Chemistry, and West African Artisanal Production." *History of Science* 54, no. 4 (December 2016): 425–42.

Osborne, Myles. *Ethnicity and Empire in Kenya: Loyalty and Martial Race among the Kamba, c. 1800 to the Present*. New York: Cambridge University Press, 2014.

Osborne, Myles. "'The Rooting Out of Mau Mau from the Minds of the Kikuyu Is a Formidable Task': Propaganda and the Mau Mau War." *Journal of African History* 56, no. 1 (2015): 77–97.

Osseo-Asare, Abena. *Atomic Junction: Nuclear Power in Africa after Independence*. Cambridge: Cambridge University Press, 2019.

Osseo-Asare, Abena. *Bitter Roots: The Search for Healing Plants in Africa*. Chicago: University of Chicago Press, 2014.

Pallaver, Karin. "What East Africans Got for Their Ivory and Slaves: The Nature, Working and Circulation of Commodity Currencies in Nineteenth-Century East Africa." In *Currencies of the Indian Ocean World*, edited by Steven Serels and Gwyn Campbell, 71–91. Cham, CH: Palgrave Macmillan, 2019.

Park, Emma. "The Right to Sovereign Seizure? Taxation, Valuation, and the Imperial British East Africa Company." In *Imperial Inequalities: The Politics of Economic Governance across European Empires*, edited by Gurminder K. Bhambra and Julia McClure, 79–97. Manchester, UK: Manchester University Press, 2022.

Park, Emma, Derek R. Peterson, Anne Pitcher, and Keith Breckenridge. "Introduction." *Africa: Journal of the International African Institute* 91, no. 4 (2021).

Patnaik, Utsa, and Prabhat Patnaik. *Capital and Imperialism: Theory, History, and the Present*. New York: Monthly Review Press, 2021.

Peterson, Derek R. *Ethnic Patriotism and the East African Revival: A History of Dissent c. 1935–1972*. Cambridge: Cambridge University Press, 2012.

Peterson, Derek R. "Patriotism and Dissent in Western Kenya." In *Ethnic Patriotism and the East African Revival: A History of Dissent, c. 1935–1972*, by Derek R. Peterson, 127–251. Cambridge: Cambridge University Press, 2012.

Peterson, Derek R. "Wordy Women: Gender Trouble and the Oral Politics of the East African Revival in Northern Gikuyuland." *Journal of African History* 42, no. 3 (December 2001): 468–89.

Peterson, Derek R. *Creative Writing: Translation, Bookkeeping, and the Work of Imagination in Colonial Kenya*. Social History of Africa. Portsmouth, NH: Heinemann, 2004.

Phillips-Fein, Kim. *Fear City: New York's Fiscal Crisis and the Rise of Austerity Politics*. New York: Metropolitan Books, 2017.

Piketty, Thomas. *Capital in the Twenty-First Century*. Translated by Arthur Goldhammer. Cambridge, MA: Belknap Press of Harvard University Press, 2017.

Pitcher, Anne. "Entrepreneurial Governance and the Expansion of Public Investment Funds in Africa." In *Africa in World Politics: Constructing Political and Economic Order*, edited by John W. Harbeson and Donald Rothchild, 45–68. Philadelphia: Routledge, 2017.

Poggiali, Lisa. "Digital Futures and Analogue Pasts? Citizenship and Ethnicity in Techno-Utopian Kenya." *Africa: Journal of the International African Institute* 87, no. 2 (2017): 253–77.

Polanyi, Karl. *The Great Transformation: The Political and Economic Origins of Our Time*. Boston: Beacon Press, 2001.

Prahalad, C. K. *The Fortune at the Bottom of the Pyramid: Eradicating Poverty through Profits*. Upper Saddle River, NJ: Pearson FT Press, 2009.

Press, Steven. *Rogue Empires: Contracts and Conmen in Europe's Scramble for Africa*. Cambridge, MA: Harvard University Press, 2017.

Prestholdt, Jeremy. *Domesticating the World: African Consumerism and the Genealogies of Globalization*. California World History Library. Berkeley: University of California Press, 2008.

Quarshie, Nana Osei. "Psychiatry on a Shoestring: West Africa and the Global Movements of Deinstitutionalization." *Bulletin of the History of Medicine* 96, no. 2 (2022): 237–65. https://doi.org/10.1353/bhm.2022.0023.

Rajak, Dinah. *In Good Company: An Anatomy of Corporate Social Responsibility*. Palo Alto, CA: Stanford University Press, 2011.

Ranger, Terence. "The Invention of Tradition Revisited: The Case of Colonial Africa." In *Legitimacy and the State in Twentieth-Century Africa: Essays in Honour of A. H. M. Kirk-Greene*, edited by Terence Ranger and Olufemi Vaughan, 62–111. London: Palgrave Macmillan, 1993.

Reid, Richard. "Africa's Revolutionary Nineteenth Century and the Idea of the 'Scramble.'" *American Historical Review* 126, no. 4 (December 2021): 1424–47. https://doi.org/10.1093/ahr/rhab539.

Rockel, Stephen J. *Carriers of Culture: Labor on the Road in Nineteenth-Century East Africa*. Social History of Africa. Portsmouth, NH: Heinemann, 2006.

Rodney, Walter. *How Europe Underdeveloped Africa*. London: Verso, 2018.

Roitman, Janet. *Fiscal Disobedience: An Anthropology of Economic Regulation in Central Africa*. Princeton, NJ: Princeton University Press, 2018. https://doi.org/10.1515/9780691187044.

Roy, Ananya. *Poverty Capital: Microfinance and the Making of Development*. New York: Routledge, 2010.

Sandgren, David. *Mau Mau's Children: The Making of Kenya's Postcolonial Elite*. Madison: University of Wisconsin Press, 2012.

Schivelbusch, Wolfgang. *The Railway Journey: The Industrialization of Time and Space in the Nineteenth Century*. Berkeley: University of California Press, 2014.

Schram, Sanford. *The Return of Ordinary Capitalism: Neoliberalism, Precarity, Occupy*. Oxford: Oxford University Press, 2015.

Seikaly, Sherene. *Men of Capital: Scarcity and Economy in Mandate Palestine*. Stanford, CA: Stanford University Press, 2015.

Serlin, David. "Confronting African Histories of Technology: A Conversation with Keith Breckenridge and Gabrielle Hecht." *Radical History Review* 2017, no. 127 (January 2017): 87–102.

Shadle, Brett Lindsay. *"Girl Cases": Marriage and Colonialism in Gusiiland, Kenya, 1890–1970*. Social History of Africa. Portsmouth, NH: Heinemann, 2006.

Sheriff, Abdul. *Slaves, Spices, & Ivory in Zanzibar: Integration of an East African Commercial Empire into the World Economy, 1770–1873*. Eastern African Studies. London: James Currey, 1987.

Shipton, Parker MacDonald. *Credit between Cultures: Farmers, Financiers, and Misunderstanding in Africa*. Yale Agrarian Studies Series. New Haven, CT: Yale University Press, 2010.

Simone, AbdouMaliq. "People as Infrastructure: Intersecting Fragments in Johannesburg." *Public Culture* 16, no. 3 (2004): 407–29.

Slobodian, Quinn. *Globalists: The End of Empire and the Birth of Neoliberalism*. New York: Harvard University Press, 2018.

Smith, G. E. "Road-Making and Surveying in British East Africa." *Geographical Journal* 14, no. 3 (September 1899): 269–89.

Smith, James Howard. *Bewitching Development: Witchcraft and the Reinvention of Development in Neoliberal Kenya*. Chicago Studies in Practices of Meaning. Chicago: University of Chicago Press, 2008.

Smith, Marquard, and Joanne Morra, eds. *The Prosthetic Impulse: From a Posthuman Present to a Biocultural Future*. Cambridge, MA: MIT Press, 2006.

Spear, Thomas. "Neo-Traditionalism and the Limits of Invention in British Colonial Africa." *Journal of African History* 44, no. 1 (2003): 3–27.

Stern, Philip J. *The Company-State: Corporate Sovereignty and the Early Modern Foundations of the British Empire in India*. Oxford: Oxford University Press, 2011.

Stoler, Ann Laura. *Duress: Imperial Durabilities in Our Times*. Durham, NC: Duke University Press, 2016.

Sydney Dunn, Hopeton. "Telecommunications and Underdevelopment: A Policy Analysis of the Historical Role of Cable and Wireless in the Caribbean." PhD diss., City University London, 1991.

Tarus, Isaac Kipsing. "A History of the Direct Taxation of the African People of Kenya, 1895–1973." PhD diss., Rhodes University, 2004.

Taylor, Keeanga-Yamahtta. *Race for Profit: How Banks and the Real Estate Industry Undermined Black Homeownership*. Chapel Hill: UNC Press Books, 2019.

Thomas, Lynn. *The Politics of the Womb: Women, Reproduction and the State in Kenya*. Berkeley: University of California Press, 2003.

Thompson, E. P. "Time, Work-Discipline, and Industrial Capitalism." *Past & Present* 38, no. 1 (December 1967): 56–97. https://doi.org/10.1093/past/38.1.56.

Thuku, Harry. *Harry Thuku: An Autobiography*. Oxford: Oxford University Press, 1970.

Tilley, Helen. *Africa as a Living Laboratory Empire: Development, and the Problem of Scientific Knowledge, 1870-1950.* Chicago: University of Chicago Press, 2011.

Tilly, Charles. *Coercion, Capital, and European States, AD 990-1990.* Studies in Social Discontinuity. Cambridge, MA: Basil Blackwell, 1990.

Trivedi, Raj Kumar. "The Role of Imperial British East Africa Company in the Acquisition of East African Colony in the Second Half of the Nineteenth Century." *Proceedings of the Indian History Congress* 33 (1971): 616-23.

Tsing, Anna. "Sorting Out Commodities: How Capitalist Value Is Made through Gifts." *HAU: Journal of Ethnographic Theory* 3, no. 1 (2013): 21-43.

Twagira, Laura Ann. "Introduction: Africanizing the History of Technology." *Technology and Culture* 61, no. 2 (2020): S1-S19. https://doi.org/10.1353/tech.2020.0068.

Tyler, Michael, Janice Hughes, and Helena Renfrew. "Kenya: Facing the Challenges of an Open Economy." In *Telecommunications in Africa,* edited by Eli M. Noam, 79-112. Global Communications Series. New York: Oxford University Press, 1999.

Van Zwanenberg, R. M. A. *Colonial Capitalism and Labour in Kenya, 1919-1939.* Kampala: East African Literature Bureau, 1975.

Von Schnitzler, Antina. *Democracy's Infrastructure: Techno-Politics & Protest after Apartheid.* Princeton, NJ: Princeton University Press, 2016.

Wallerstein, Immanuel. *The Capitalist World-System.* Cambridge: Cambridge University Press, 1979.

Wallerstein, Immanuel. *World-Systems Analysis: An Introduction.* Durham, NC: Duke University Press, 2004.

Watkins, O. F. *Report of the Kenya Land Commission, September, 1933.* Kenya: His Majesty's Stationery Office, 1934.

Watt, Stuart, and Rachel Watt. *In the Heart of Savagedom.* London: Marshall Brothers, 1912.

Welker, Marina. *Enacting the Corporation: An American Mining Firm in Post-Authoritarian Indonesia.* Berkeley: University of California Press, 2014.

White, Luise. *The Comforts of Home: Prostitution in Colonial Nairobi.* Chicago: University of Chicago Press, 1990.

Willis, Justin. "'Men on the Spot,' Labor, and the Colonial State in British East Africa: The Mombasa Water Supply, 1911-1917." *International Journal of African Historical Studies* 28, no. 1 (1995): 25-48. https://doi.org/10.2307/221304.

Wood, Ellen Meiksins. "The Separation of the Economic and the Political in Capitalism." *New Left Review* 1, no. 127 (June 1, 1981): 66-95.

Woker, Madeline. "A Taxing Empire: Power and Taxation in the French Colonial Empire, 1850s-1950s," forthcoming.

World Bank. *World Development Report 1994: Infrastructure for Development.* New York: Oxford University Press, 1994.

Yonge, Brian. "The Uganda Road." *Kenya Past and Present* 8, no. 1 (1977): 19-26.

Young, Edward Hilton. *Report of the Commission on Closer Union of the Dependencies in Eastern and Central Africa.* London: His Majesty's Stationery Office, 1929.

Zevin, Alexander. *Liberalism at Large: The World according to the Economist.* London: Verso, 2019.

Index

Beardall, William, 52–53, 57
Berkeley, Ernest, 47, 63
Berlin Conference, 22–23
Bin Said, Barghash, 4, 13, 22, 36–37, 42, 45, 212n26, 217n115; authority of, 218n137; IBEA and, 30–31; Kirk and, 213n41; Mackinnon and, 20; public works projects of, 30, 215n91; rise to power of, 26
Breckenridge, Keith, 3, 182
Brennan, James, 123, 242n85, 244n116
British Broadcasting Corporation (BBC), 117, 125–26, 130, 240n44
British East African Association, 20, 24, 28, 30
British East African Broadcasting Co. (BEABC), 82–83, 85
British government, 20, 22, 43–44, 46, 78, 81
British India Steam Navigation Company (BI), 19, 24
broadcasters, 96, 103, 105, 126; African, 86
broadcasting, 14–15, 78–79, 81–84, 91, 93–98, 109–11, 113–20, 125, 133, 138–39, 230n57, 233n134, 239n18, 240n44; African, 112, 239n19, 239n21; Arabic, 126; autobiographies and, 234n139; colonial, 96, 110, 113, 116f, 119; development, 228n10; infrastructure, 94, 134, 136, 200, 236n169; Kiswahili-language, 15, 110, 126–27, 137, 232n127; local, 131; networks, 9, 15, 107–10, 113–14, 116f, 121, 131, 133, 202; politics of, 245n141; radio, 3, 5, 14, 79, 81–82, 100, 107, 109, 111, 113, 118–20, 123–24, 133, 137, 139, 200–201, 236n169; shortwave, 128; vernacular-language, 94, 122, 137; Wakaria and, 105–7; wired, 118; wireless, 240n42
Buxton, Thomas Fowell, 24, 220n182

Cable and Wireless Ltd. (C&W), 5, 14, 79–81, 114, 146–47, 200. See also Imperial and International Communications Ltd.
Cairo, 124, 126, 131, 138–39
capital, 7, 17, 20, 23–26, 28, 117, 146, 149–50, 157, 160–61, 187, 205, 213n48, 250n77, 253n16; accumulation, 4; British, 213n47; foreign, 148–49, 125, 155, 161, 192; global, 151; idle, 213n34; land and, 100; metropolitan, 153, 208n12; private, 4, 10, 16, 21, 25, 41, 120, 200, 206; space of, 33
capitalism, 5, 11–13, 16, 172, 209n30, 210n50, 224n84; altruistic, 191; colonial, 3; con-

temporary, 180, 205; ethical, 192, 204; historical, 9; imperial, 229n30; laissez-faire, 8; Marx's critique of, 69; neoliberalization and, 195; patriotic, 149, 168
caravan trade, 38, 56, 58. See also porters
Central Africa, 216n95, 221n27
Central Province (Kenya), 64, 69, 71–72, 84, 99, 134–35, 255n9
Champion, Arthur, 77, 86–87, 89–90, 92, 95, 234n144
Chief Native Commissioner (CNC), 74, 112, 234–35n144
Church Missionary Society (CMS), 77, 85, 93
citizenship, 143, 145, 154, 163–64, 181; all-embracing, 15, 110, 112, 238n12; common, 112, 238n12; corporate, 141; shareholder, 16, 167
civilization, 30, 65–66, 100; of Africa, 26; commerce and, 4, 13, 19–21, 24, 26, 30–31, 39, 46–49, 54, 65, 70, 72–74, 200–201; episte-mologies of, 58; Kikuyu, 102; Mau Mau and, 102–3; roads and, 48, 72–73
class, 6, 104, 166–67, 201; clerk, 119; debtor, 68; differentiation, 99; dynamics, 203; forma-tion, 100, 207n6; logic of, 166; merchant, 59; struggle, 100; warrior, 56, 202
Collymore, Bob, 173–74, 250n75
colonial austerity, 15, 79, 91, 119, 229n44
Colonial Development and Welfare Acts, 14, 113
colonialism, 15, 79, 93, 154
colonial occupation, 8, 13, 67, 108, 120; formal, 22, 43, 46, 68, 212n24; second, 234n144
Colonial Office (CO), 78, 81, 117–18
colonial officials, 73–75, 233n134, 244n116
colonial self-sufficiency, 13–14, 22, 46, 67–69, 72, 74, 79, 92; logic of, 113
colonial subjects, 10, 73, 107, 119, 244n119
colonization, 153, 202
commerce, 9, 29–30, 33, 197; civilisation/ civilization and, 4, 13, 19–21, 24, 26, 30–31, 39, 46–49, 54, 65, 70, 72–74, 200–201; legitimate, 20, 49, 211n9; roads and, 20, 40, 212n22
Communications Commission of Kenya (CCK), 148, 150
community development, 13–15, 98, 120, 126
company men, 33, 36–37, 39, 49, 51, 57, 59, 62
company rule, 5, 43, 60, 200, 215n79

entrepreneurs, 156, 178, 181

epistemology: Kikuyu, 99, 101-2, 132; Mau Mau, 105; of space and social life, 39; of wealth and power, 56

ethnicity, 6, 166-68, 194, 201, 203, 207n6; construction of, 223n58; Kikuyu, 216n108, 223n58; Thuku on, 230n57

experts, 5, 12, 15, 78-80, 85, 87-92, 98, 114, 126, 201-2, 206, 229n44, 236n169; African topographical, 11, 14, 51, 53, 55, 204; information, 125; infrastructural, 12, 201; metropolitan, 50, 89; radio, 204

exploitation, 5, 11, 72, 202, 204; waged, 205

expropriation, 5, 10-11, 17, 39-40, 72, 154, 202

extraction, 62, 68, 200, 209n47

family, the, 3, 85

famine, 33-34, 58, 64-67

films, 86-88, 91, 98, 230n69; silent, 78

finance, 22, 25, 181-82, 213n47; capital, 24; centers of, 7; circuits of, 5; costs, 171; imperial, 220n6. *See also* sound finance

Finance Against Poverty (FAP), 156-57

financialization, 8, 254n51

financial services, 184, 196; digital, 2-3, 11, 204; Safaricom's, 16, 141, 169, 181

Foreign Office (FO), 14, 43-44, 46, 48, 63, 211n12, 219-20n182

Franklin, Harry, 120, 241n62

Fraser, Nancy, 177, 204, 208n17

general strike of 1947, 109, 120, 122

Goswami, Chayya, 29, 217n131, 218n137

Goswami, Manu, 6, 207n10, 208n29, 209n39, 229n30

governance, 3-4, 6-7, 23, 26, 46, 145, 180-81; austerity, 8, 14, 199; colonial modes of, 200; fiscal, 8, 64; infrastructural, 8, 15, 212n24; market and, 198, 205

Government of Kenya, 3, 5, 82, 158-60, 162, 165, 198-99, 240n43

Grace, Joshua, 13, 22n11, 226n127, 232n93, 231n106

Great Depression, 23; imperialism and, 212n32

Great Lakes region, 26, 40, 42

Guernsey, 156, 162

guides, 50-51, 53, 202

Gusii, 93, 135

harambee, 13, 15, 146-47

hongo (tribute), 37-38, 62-63

human ATMs, 2-3, 11, 16, 179-80, 182-83, 191, 196, 201-2, 204; feminized work and, 187

Hut Tax, 68, 71

IMF. See International Monetary Fund (IMF)

immobility, 73; class, 104

Imperial British East Africa Company (IBEA), 4-5, 13-14, 19-26, 28-29, 33-34, 36-47, 56-65, 67-70, 200, 205, 213n48, 214n65; coins minted by, 60, 61f; collapse of, 146; as corporate-state, 212n24; Mackenzie and, 223n77; Mackinnon and, 219n160; roads and, 49, 51, 54, 61, 202, 211-12n22, 215n90; tolls levied by, 214n66. *See also* administrators; authority; company currency; company men;; Crown, the; sovereign authority; taxation

imperial expansion, 19, 24, 212n24; critics of, 213n34

Imperial and International Communications Ltd. (IIC), 81, 83

independence (Kenyan), 13, 15-16, 106, 135, 142, 146-49, 203-4, 255n9; ethno-regional patterns of stratification and, 6

India: in Africa, 25; British, 19; broadcasting and, 82, 124, 130-31, 139, 245n138; eastern Africa and, 224n84; laborers from, 65; state-space in, 208n29

Indian Ocean, 19-20, 22, 29

Indian rupee, 59-61, 63, 70-71

Information and Communication Technology (ICT): penetration rates, 249n50; policy, 148

Information Department, 98-99, 105

information vans, 15, 79-80, 98

information workers, 105-6, 237n191; African, 93, 98-99, 126; British, 86, 89

International Monetary Fund (IMF), 3, 148, 180

infrastructural exclusion, 6, 134

infrastructural expansion, 5, 7, 16, 62, 71, 132, 139, 161-62; private capital and, 120

infrastructural expertise, 5, 12, 48, 208n18

infrastructural networks, 14, 79, 139, 147, 201; colonial, 54; lumpy, 9, 85; nascent, 38, 71

infrastructural prosthetics, 2-3, 11, 15, 17; African topographical experts as, 55; Cable and Wireless Ltd. as, 79; IBEA as, 23; knowledge workers as, 80, 126, 201; M-Pesa agents as, 182, 195; radio as, 14; Safaricom as, 144

and, 157; IBEA and, 21, 23, 63; incapacity of, 180; infrastructures and, 6, 17, 81, 153, 173, 180-81; Kenya Post and Telecommunications Corporation (KPTC) and, 144, 148-49; knowledge-workers and, 97-98, 104; labor movement and, 123; languages and, 94, 111, 138; market and, 4, 16; Mau Mau and, 100, 104; media and, 96; mobile information unit and, 79, 88; postcolonial, 3-4, 144, 173; private capital and, 206; privatization and, 163; Safaricom and, 5, 9, 16, 143-44, 146, 148, 159, 165, 170, 173, 177, 196, 200; strategic immobility and, 73; taxation and, 67-68; technopolitics of, 134; telecommunications network of, 147; Telkom Kenya (TKL) and, 148-49; Wakaria and, 86. *See also* infrastructural state

statecraft, 3, 5, 8, 10, 17, 164; austerity as tactic of, 208n30; colonial, 208n12, 212n24; infrastructural, 200; Kenyan, 161; Moi and, 157

state formation, 4, 16-17, 200

state sovereignty, 5, 155

state space, 7, 208n29; imperial, 9

structural adjustment, 7-8, 199; Structural Adjustment Policies, 204

subsumption, 17, 196

surveyors, 50-51, 53-54; metropolitan, 48, 50

Swahili, 93-94, 109, 125-26, 131, 233n127

Swahili Coast, 30, 59, 123-24, 217n127, 245n141

Swynnerton Plan, 103, 132

Taita, 39, 93

Tanganyika, 128, 146, 239n21

Tanzania, 146-47, 222n40, 232n106

taxation, 23-25, 28-29, 49, 67-74, 220n6, 226n157, 227n159; IBEA and, 62-64, 214n72; infrastructures and, 46-47; regimes, 10, 13-14, 22-23, 48, 68-69, 71; Safaricom's fees as, 198

taxes, 47-49, 67-72, 169; IBEA and, 14, 20, 23, 26, 28, 44, 47

taxpayers, 155, 214n65; British, 20, 45; Kenyan, 149

technological experts, 12; African, 5, 90

technopolitics, 72, 211n7; delegatory, 8; infrastructural, 201; of the state, 134; of time, 131

telecommunications, 148, 181; infrastructure, 163, 198; network, 147; penetration, 248n24; sector, 153-54, 168. *See also* Cable and Wireless Ltd. (C&W); Kenya Post and Telecommunications Corporation (KPTC); Safaricom

telephony, 144, 146-47; mobile, 2, 142, 152, 180

Telkom Kenya (TKL), 16, 148-54, 157-61, 250n77. *See also* Kirui, Sammy

Thika Superhighway, 141, 204

Thuku, Harry, 84, 230n57

traders, 36, 57, 61, 75; Kamba, 31, 33; South Asian, 28, 214n72; Zanzibari, 41, 218n154

translation, 50, 80, 86, 93-95, 202, 233n128; conceptual, 95-96

translators, 85, 89, 96; African, 93, 98

transparency, 144, 149, 152, 159, 161, 168, 174

Treasury, 14, 20, 24, 68, 79, 118, 169, 199; broadcasting and, 15, 109

tropicalisation, 77, 202

Uganda, 40-43, 52, 64, 82, 128, 147, 220n182, 233n131, 239n21; communications of, 146; IBEA's occupation of, 45; Lugard and, 51; Saut el Arab and, 123; Uganda Railway, 54, 68

Uganda Road, 46, 69

Ukambani, 33-34, 53

underinvestment, 13-14, 16, 126

United Kingdom, 3, 7, 45, 89, 162, 176, 198, 203; austerity and, 209n30

valuation, 31, 37-38, 60, 70-71, 155; grammars of, 59, 75; Kamba systems of, 57; monopoly over, 47, 49, 63, 74

violence, 22, 36, 48, 61, 72-73, 164-65; election, 167; threat of, 36, 217n120

Vodafone, 3, 5, 16, 151-53, 157-62, 165, 180, 182; executives, 193. *See also* Safaricom

Vodafone Kenya Limited (VKL), 151, 160-61

Wakaria, Ambrose, 77, 85-86, 94-95, 98-99, 105-7, 201, 233n131, 255n9; independence and, 203; Matemo and, 237n191; structural adjustment policies and, 204

Wanga, 29, 215n79

West Africa, 216n95, 223n72; French, 209n47

white settler lobby, 84, 122

wiathi (self-mastery), 100–101, 104, 245n141

workers, 53, 58, 184, 195; information, 86, 93, 98–99, 105–6, 126, 237n191; infrastructural, 11, 50, 205; metropolitan, 121; Safaricom, 16, 175. *See also* knowledge-workers

World Bank, 3, 144, 148, 180, 191

Yasin, ArKan, 197–99

Zanzibar, 19, 22, 26–27, 34; British Consul General in, 24, 40; as British protectorate, 42–43; Concession, 214n72; customs house at, 29; dollar, 59; Mombasa and, 60, 121; National Anthem, 126; Radio Cairo and, 128; Stone Town, 30; Sultan of (*see* Bin Said, Barghash); Sultanate of, 20–21, 26–27, 36–37, 42, 44, 214n63; traders from, 218n154; treasury of, 22, 45, 220n184. *See also* Swahili Coast